Big Data and Artificial Intelligence in Digital
Finance

John Soldatos • Dimosthenis Kyriazis

Editors

Big Data and Artificial Intelligence in Digital Finance

Increasing Personalization and Trust in Digital Finance using Big Data and AI

 Springer

Editors
John Soldatos
INNOV-ACTS LIMITED
Nicosia, Cyprus

University of Glasgow
Glasgow, UK

Dimosthenis Kyriazis
University of Piraeus
Piraeus, Greece

ISBN 978-3-030-94592-3   ISBN 978-3-030-94590-9   (eBook)
https://doi.org/10.1007/978-3-030-94590-9

This Springer imprint is published by the registered company Springer Nature Switzerland AG
The registered company address is: Gewerbestrasse 11, 6330 Cham, Switzerland

Preface

The finance sector is among the most data-savvy and data-intensive of the global economy. The ongoing digital transformation of financial organizations, along with their interconnection as part of a global digital finance ecosystem, is producing petabytes of structured and unstructured data. The latter represent a significant opportunity for banks, financial institutions, and financial technology firms (Fin-Techs): Leveraging these data financial organizations can significantly improve both their business processes and the quality of their decisions. As a prominent example, modern banks can exploit customer data to anticipate the behaviors of their customers and to deliver personalized banking solutions to them. Likewise, data can enable new forms of intelligent algorithmic trading and personalized asset management.

To harness the benefits of big data, financial organizations need effective ways for managing and analyzing large volumes of structured, unstructured, and semi-structured data at scale. Furthermore, they need to manage both streaming data and data at rest, while at the same time providing the means for scalable processing by a variety of analytics algorithms. The latter processing may also require support for real-time analytics over the heterogeneous types of data (i.e., structured/unstructured, streaming/data-at-rest). Management of heterogeneity is therefore one of the most important concerns for big data management in the financial sector. Currently, financial organizations spend significant effort and IT resources in unifying and processing different types of data, which typically reside in different repositories such as operational databases, analytical databases, data warehouses, data lakes, as well as emerging distributed ledger infrastructures (i.e., blockchains). Moreover, several applications require also semantic interoperability across datasets that are "siloed" across different systems, while the use of big data in real-life banking applications requires also the utilization of pre-processing functions (e.g., anonymization), which boost the ever-important regulatory compliance of digital finance applications. Despite the evolution of big data technologies, there is still a need for novel solutions that can successfully confront the above-listed challenges to enable the development, deployment, and operation of big data applications for digital finance at scale.

Big data management and analytics solutions enable Artificial Intelligence (AI) solutions in finance. AI refers to the capability of machines to imitate intelligent human behaviour and act like humans. In many cases, AI systems think like humans and reason over complex contexts in order to evaluate and take optimal decisions. As such, AI systems support two main processes: (i) a learning process that allows them to produce rules about how to process and use the information they receive; and (ii) reasoning that drives their decisions and actions. Several AI systems are already deployed or planned to be integrated in applications of the financial services industry. They are usually based on one or a combination of technologies such as video processing and visual scene analysis, speech recognition, Natural Language Processing (NLP), automated translation, machine learning, deep learning, and cognitive search. Typical AI applications in digital finance include robo-advisors, AI-based personalized asset management systems, statistical credit underwriting and risk assessment applications, automated and intelligent KYC (Know Your Customer) applications, fraud detection and anti-money laundering (AML), and personalized finance applications for citizens and businesses, as well as a variety of front-office applications such as chatbots. Moreover, there are many interesting AI applications in the insurance sector such as automated insurance claims management and usage-based insurance, that is, statistical calculation of risk premiums based on data about the customers' behaviors (e.g., lifestyle or driving behavior data). Most of these use cases leverage Machine Learning (ML) and Deep Learning (DL) techniques. However, the above list of AI use cases in finance is non-exhaustive. As more data becomes available, the use of AI to improve automation and personalization and reduce costs will become more attractive. It is expected that FinTech enterprises will produce novel AI-based ideas in the years to come. Nevertheless, in several cases, AI deployments have to overcome barriers and limitations of existing big data management technologies. In other cases, integration with other emerging technologies (e.g., Robotics Process Automation (RPA) and blockchain technologies) is required. In this context, the presentation of tangible AI deployments in financial institutions is interesting in terms of the technologies employed, as well as in terms of the integration of AI in the digital transformation strategies of financial organizations.

The regulatory compliance of big data and AI solutions is also a major concern of financial institutions. Recent regulatory developments in the finance sector, such as the 2nd Payment Services Directive (PSD2) in Europe, open significant innovation opportunities for the sector, through facilitate the flow of data across financial organizations. At the same time, these regulations make compliance processes more challenging, while introducing security challenges. Other regulatory developments (e.g., MiFIDII and the 4th AML Directive) affect the way certain use cases (e.g., fraud detection) are developed. Moreover, all data-intensive use cases should comply with data protection regulation such as the General Data Privacy Regulation (GDPR) of the European Union (EU). At the same time, there are new regulations (e.g., eIDAS) that facilitate certain tasks in large-scale big data and AI use cases, such as the process of digital on-boarding and verification of customers. Overall, regulatory compliance has a two-way interaction with big data and AI applications.

On the one hand, it affects the design and deployment of big data and AI applications, while on the other big data and AI can be used to boost regulatory compliance. Indeed, AI provides many opportunities for improving regulatory compliance in the direction of greater accuracy and cost-effectiveness. For instance, the machine learning segment of AI enables the collection and processing of very large amounts of data relevant to a financial or banking workflow, including structured data from conventional banking systems and alternative data such as news and social networks. These big data can be analyzed to automate compliance against regulatory rules. This is the reason why many RegTech (Regulatory Compliance) enterprises are AI based. They leverage AI and big data analytics to audit compliance in real time (e.g., by processing streaming data in real time), while at the same boosting the accurate and richness of regulatory reporting.

Overall, the development, deployment, and operation of novel big data and AI solutions for modern digital financial organization require a holistic approach that addresses all the above-listed issues. Since October 2019, the European project INFINITECH (cofunded by the H2020 program of the European Commission) is taking such a holistic and integrated approach to designing, deploying, and demonstrating big data and AI solutions in Digital Finance. The project brings together a consortium of over forty organizations, including financial organizations, FinTechs, large vendors of AI and big data solutions, innovative high-tech Small Medium Enterprises (SMEs), as well as established research organizations with a proven track record of novel research outcomes in AI, big data, and blockchain technologies and their use in the finance sector. The present book is aimed at presenting INFINITECH's approach to big data and AI-driven innovations in Digital Finance, through a collection of scientific and technological development contributions that address most of the earlier identified challenges based on innovative solutions. Specifically, the book presents a set of novel big data, AI, and blockchain technologies for the finance sector, along with their integration in novel solutions. Furthermore, it pays emphasis on regulatory compliance issues, including technological solutions that boost compliance to some of the most important regulations of the sector.

Nicosia, Cyprus   John Soldatos
Piraeus, Greece   Dimosthenis Kyriazis

Acknowledgments

This book has received funding from the European Union's Horizon 2020 research and innovation programme under grant agreement No. 856632. Specifically, the chapters of the book present work and results achieved in the scope of the H2020 INFINITECH project. The editors acknowledge the contributions of all coauthors, as well as support from all INFINITECH project partners.

Disclaimer The contents of the book reflect only the contributors' and co-authors' view. The European Commission is not responsible for any use that may be made of the information it contains.

Contents

Editors and Contributors

About the Editors

John Soldatos (http://gr.linkedin.com/in/johnsoldatos) holds a PhD in Electrical and Computer Engineering from the National Technical University of Athens (2000) and is currently Honorary Research Fellow at the University of Glasgow, UK (2014–present). He was Associate Professor and Head of the Internet of Things (IoT) Group at the Athens Information Technology (AIT), Greece (2006–2019), and Adjunct Professor at the Carnegie Mellon University, Pittsburgh, PA (2007–2010). He has significant experience in working closely with large multinational industries (IBM Hellas, INTRACOM S.A, INTRASOFT International) as R&D consultant and delivery specialist, while being scientific advisor to high-tech startup enterprises, such as Innov-Acts Limited (Nicosia, Cyprus). Dr. Soldatos is an expert in Internet-of-Things (IoT) and Artificial Intelligence (AI) technologies and applications, including applications in smart cities, finance (Finance 4.0), and industry (Industry 4.0). He has played a leading role in the successful delivery of more than seventy (commercial-industrial, research, and business consulting) projects, for both private and public sector organizations, including complex integrated projects. He is cofounder of the open-source platform OpenIoT (https://github.com/OpenIotOrg/openiot). He has published more than 200 articles in international journals, books, and conference proceedings. He has also significant academic teaching experience, along with experience in executive education and corporate training. He is a regular contributor in various international magazines and blogs, on topics related to Artificial Intelligence, IoT, Industry 4.0, and cybersecurity. Moreover, he has received national and international recognition through appointments in standardization working groups, expert groups, and various boards. He has recently coedited and coauthored six edited volumes (books) on Artificial Intelligence, BigData, and Internet of Things related themes.

Dimosthenis Kyriazis (https://www.linkedin.com/in/dimosthenis-kyriazis-13979 19) is an Associate Professor at the University of Piraeus (Department of Digital Systems). He received his diploma from the school of Electrical and Computer

Engineering of the National Technical University of Athens (NTUA) in 2001 and his MSc degree in "Techno-economics" in 2004. Since 2007, he holds a PhD in the area of Service-Oriented Architectures with a focus on quality aspects and workflow management. His expertise lies with service-based, distributed, and heterogeneous systems, software, and service engineering. Before joining the University of Piraeus, he was a Senior Research Engineer at the Institute of Communication and Computer Systems (ICCS) of NTUA, having participated and coordinated several EU and National funded projects (e.g., BigDataStack, MATILDA, 5GTANGO, ATMOSPHERE, CrowdHEALTH, MORPHEMIC, LeanBigData, CoherentPaaS, VISION Cloud, IRMOS, etc.) focusing his research on issues related to quality of service provisioning, fault tolerance, data management and analytics, performance modelling, deployment and management of virtualized infrastructures and platforms.

Contributors

Carlos Albo WENALYZE SL, Valencia, Spain

Nuria Ituarte Aranda ATOS Spain, Madrid, Spain

Massimiliano Aschi POSTE ITALIANE, Rome, Italy

Andrea Becerra CTAG – Centro Tecnológico de Automoción de Galicia, Pontevedra, Spain

Susanna Bonura ENGINEERING, Rome, Italy

Guilherme Brito NOVA School of Science and Technology, Caparica, Portugal

Diego Burgos LeanXcale SL, Madrid, Spain

Juan Carrasco Research & Innovation, ATOS IT, Madrid, Spain

René Danzinger Privé Technologies, Vienna, Austria

Ilesh Dattani Assentian Europe Limited, Dublin, Ireland

Lambis Dionysopoulos Institute for the Future, University of Nicosia, Nicosia, Cyprus

Giovanni Di Orio NOVA School of Science and Technology, Caparica, Portugal

Nikolaos Droukas National Bank of Greece, Athens, Greece

Ignacio Elicegui Research & Innovation, ATOS IT, Madrid, Spain

Carmen Perea Escribano Research & Innovation, ATOS Spain, Madrid, Spain

Georgios Fatouros INNOV-ACTS Limited, Nicosia, Cyprus

Fabiana Fournier IBM ISRAEL - SCIENCE AND TECHNOLOGY LTD, Haifa, Israel

Jose Gato Research & Innovation, ATOS Spain, Madrid, Spain

George Giaglis Institute for the Future, University of Nicosia, Nicosia, Cyprus

Ricardo Jimenez-Peris LeanXcale SL, Madrid, Spain

Stathis Kanavos Innovation Sprint Sprl, Brussels, Belgium

Nikolaos Kapsoulis INNOV-ACTS Limited, Nicosia, Cyprus

Spyros Kasdaglis UBITECH, Chalandri, Greece

Klemen Kenda JSI, Ljubljana, Slovenia

Baran Kılıç Bogazici University, Istanbul, Turkey

Filip Koprivec JSI, FMF, Ljubljana, Slovenia

Fotis Kossiaras UBITECH, Chalandri, Greece

Dimitrios Kotios University of Piraeus, Piraeus, Greece

Pavlos Kranas LeanXcale SL, Madrid, Spain

Gregor Kržmanc JSI, Ljubljana, Slovenia

Sofoklis Kyriazakos Innovation Sprint Sprl, Brussels, Belgium

Business Development and Technology Department, Aarhus University, Herning, Denmark

Dimosthenis Kyriazis University of Piraeus, Piraeus, Greece

Antonis Litke INNOV-ACTS Limited, Nicosia, Cyprus

Craig Macdonald University of Glasgow, Glasgow, Scotland

Juan Mahíllo LeanXcale SL, Madrid, Spain

Georgios Makridis University of Piraeus, Piraeus, Greece

Pedro Maló NOVA School of Science and Technology, Caparica, Portugal

Alessandro Mamelli Hewlett Packard Enterprise, Milan, Italy

Sara El Kortbi Martinez Fundación Centro Tecnolóxico de Telecomunicacións de Galicia (GRADIANT), Vigo, Spain

Nicola Masi ENGINEERING, Rome, Italy

Richard McCreadie University of Glasgow, Glasgow, Scotland

Roland Meier Privé Technologies, Vienna, Austria

Domenco Messina ENGINEERING, Rome, Italy

Dimitris Miltiadou UBITECH, Chalandri, Greece

Georgios Misiakoulis UBITECH, Chalandri, Greece

Vittorio Monferrino GFT Technologies, Genova, Liguria, Italy

Erik Novak JSI, Ljubljana, Slovenia

Lilian Adkinson Orellana Fundación Centro Tecnolóxico de Telecomunicacións de Galicia (GRADIANT), Vigo, Spain

Ines Ortega-Fernandez Fundación Centro Tecnolóxico de Telecomunicacións de Galicia (GRADIANT), Vigo, Spain

Iadh Ounis University of Glasgow, Glasgow, Scotland

Can Özturan Bogazici University, Istanbul, Turkey

Konstantinos Perakis UBITECH, Chalandri, Greece

Eleni Perdikouri National Bank of Greece, Athens, Greece

Alexandros Perikleous Innovation Sprint Sprl, Brussels, Belgium

Stamatis Pitsios UBITECH, Chalandri, Greece

Aristodemos Pnevmatikakis Innovation Sprint Sprl, Brussels, Belgium

Andreas Politis Dynamis Insurance Company, Athens, Greece

Davide Profeta ENGINEERING, Rome, Italy

Georgia Prokopaki National Bank of Greece, Athens, Greece

Petra Ristau JRC Capital Management Consultancy & Research GmbH, Berlin, Germany

Alper Şen Bogazici University, Istanbul, Turkey

Inna Skarbovsky IBM ISRAEL - SCIENCE AND TECHNOLOGY LTD, Haifa, Israel

Maja Škrjanc JSI, Ljubljana, Slovenia

John Soldatos INNOV-ACTS Limited, Nicosia, Cyprus
University of Glasgow, Glasgow, UK

Dimitrios Spyropoulos UBITECH, Chalandri, Greece

Maanasa Srikrishna University of Glasgow, Glasgow, Scotland

Ernesto Troiano GFT Italia, Milan, Italy

Silvio Walser Bank of Cyprus, Nicosia, Cyprus

Abbreviations

ABCI	Application Blockchain Interface
ACID	Atomicity Consistency Durability
AD	Advanced Driving
ADA-M	Automatable Discovery and Access Matrix
AEs	Advanced Economies
AEMET	(Spanish) State Meteorological Agency
AI	Artificial Intelligence
AML	Anti-money Laundering
AMLD 4	4th Anti-money Laundering Directive
API	Application Programming Interface
AUC	Area Under the ROC Curve
BDA	Big Data Analytics
BDVA	Big Data Value Association
BFM	Business Finance Management
BIAN	Banking Industry Architecture Network
BIS	Bank of International Settlements
BISM	BI Sematic Metadata
BoE	Bank of England
BOS	Bank of Slovenia
BPMN	Business Process Modelling Notation
BPR	Bayesian Personalized Ranking
BSM	Basic Safety Message
BTC	Bitcoin Token
CAPEX	Capital Expenses
CAM	Cooperative Awareness Messages
CAN	Controller Area Networks
CASTLE	Continuously Anonymising STreaming Data via adaptative cLustEring
CBDCs	Central Bank Digital Currencies
CD	Continuous Delivery
CDC	Change Data Capture

CEP	Complex Event Processing
CI	Continuous Integration
CMA	Consent Management Architecture
CMS	Consent Management System
CPU	Central Processing Unit
CRISP-DM	Cross Industry Standard Process for Data Mining
CPS	Customer Profile Similarity
CSPs	Cloud Service Providers
CSS	Cascading Style Sheets
CSV	Comma Separated Values
CTF	Counter-Terrorist Financing
CTTPs	Critical ICT Third Party Providers
C-ITS	Cooperative Intelligent Transport Systems
C-VaR	Conditional Value-at-Risk
DAG	Directed Acyclic Graphs
DAIRO	Data, AI and Robotics
DBMS	Database Management Systems
DCEP	Digital Currency Electronic Payment
DeFi	Decentralised Finance
DENM	Decentralized Environmental Notification Messages
DGT	Directorate General of Traffic
DQS	Data Quality Services
DL	Deep Learning
DLT	Distributed Ledger Technology
DORA	Digital Operational Resilience Act
DOT	Polkadot Token
DPaaS	Data Pipeline as-a-Service
DPO	Data Protection Orchestrator
DSML	Data Science and Machine Learning
DUA	Data Use Ontology
DUOS	Digital User Onboarding System
EBA	European Banking Authority
EC	European Commission
ECB	European Central Bank
EEA	European Economic Area
EHR	Electronic Health Records
EMDEs	Emerging Market and Developing Economies
ERC	Ethereum Request for Comment
ES	Expected Shortfall
ESG	Environmental Social and Governance
ESMA	European Securities and Markets Authority
ETFs	Exchange-Traded Fund
EU	European Union
EUBOF	European Blockchain Observatory and Forum
FAANST	Fast Anonymising Algorithm for Numerical STreaming data

FaaS	FinTech-as-a-Service
FADS	Fast clustering-based k-Anonymisation approach for Data Streams
FAST	Fast Anonymization of Big Data Streams
FATF	Financial Action Task Force
FFT	Fast Fourier Transform
FIBO	Financial Industry Business Ontology
FIGI	Financial Industry Global Instrument Identifier
FinTech	Financial Technology
FTP	File Transfer Protocol
FX	Foreign Exchange
GAE	Graph Autoencoder
GCN	Graph Convolutional Networks
GDP	Gross Domestic Product
GDPR	General Data Protection Regulation
GPU	Graph Processing Unit
GRUs	Gated Recurrent Units
GUI	Graphical User Interface
HD	High Definition
HDFS	Hadoop Distributed File System
HDP	Hortonworks Data Platform
HIBS	Hierarchical Identity-Based Signature
HMCU	Hybrid Modular Communication Unit
HPC	High-Performance Computing
HS	Historical Simulation
HTLC	Hashed Time-Locked Smart Contracts
HTML	HyperText Markup Language
HTTP	HyperText Transfer Protocol
ICO	Initial Coin Offering
IFX	International Foreign Exchange
IIC	Industrial Internet Consortium
IIoT	Industrial Internet of Things
IIRA	Industrial Internet Reference Architecture
IoT	Internet of Things
IVI	In Vehicle Information Message
JSON	JavaScript Object Notation
KIDS	K-anonymIsation Data Stream
KPI	Key Performance Indicator
KYB	Know Your Business
KYC	Know Your Customer
LIME	Local Interpretable Model-agnostic Explanations
Lkif	Legal Knowledge Interchange Format
LTM	Local Transactional Managers
LSM	Log-Structured Merge-tree
MC	Monte Carlo
MDS	Master Data Management

MiCA	Markets in Crypto-Assets
MiFID II	Markets in Financial Instruments Directive
ML	Machine Learning
MLT	Multi-Layer Perceptron
MPI	Message Passing Interface
MPT	Modern Portfolio Theory
NACE	Nomenclature of Economic Activities
NDCG	Normalized Discounted Cumulative Gain
NFC	Near Field Communications
NFT	Non-Fungible Tokens
NGSI	Next Generation Service Interfaces
NGSI-LD	Next Generation Service Interfaces Linked Data
NIST	National Institute of Standards and Technology
NLP	Natural Language Processing
NP	Network Policy
OBD II	On-Board Diagnostic port
OBU	On-Board Unit
OD	Own Damage
ODS	Operational Data Store
OLAP	On Line Analytical Processing
OPEX	Operational Expenses
OSIN	Open Source Intelligence
PALMS	Platform for Anti-Money Laundering Supervision
PAYD	Pay As You Drive
PBoC	People's Republic Bank of China
PCC	Participant Centered Consent
PEPs	Politically Exposed Persons
PHYD	Pay How You Drive
PLM	Product Lifecycle Management
PoC	Proof of Concept
PSC	People of Significant Control
PSD2	Second Payment Services Directive
PSP	Payment Service Provider
QR	Quick Response
RA	Reference Architecture
REM	Rapid Eye Movement
RFR	Random Forest Regression
RM	Reference Model
RNN	Recurrent Neural Network
ROC	Receiver Operating Characteristic
RSUs	Road Side Units
RTGS	Real-Time Gross Settlement System
RWD	Real World Data
SARs	Suspicious Activities Reports
SAX	Symbolic Aggregate approXimation

SCA	Strong Customer Authentication
SDGs	Sustainable Development Goals
SEPA	Single Euro Payments Area
SHAP	SHapley Additive exPlanations
SKY	Stream K-anonYmity
SPAT	Signal Phase and Timing Message
SPEiDI	Service Provider eIDAS Integration
SQL	Structured Query Language
SRM	Service Request Message
SLO	Service Level Objectives
SME	Small Medium Enterprises
SPV	Simple Payment Verification
SRRI	Synthetic Risk and Reward Indicator
SSTables	String Sorted Tables
SVM	Support Vector Machine
SVR	Support Vector Regression
SWAF	Sliding Window Anonymisation Framework
TCO	Total Cost of Ownership
TGAT	Temporal Graph Attention
TF	Terrorist Financing
TPP	Third Party Provider
TRA	Transaction Risk Analysis
TTP	Trusted Third Party
UBI	Usage-Based Insurance
UI	User Interface
UMA	User Managed Access
UX	User Experience
USD	United States Dollar
UTXO	Unspent Transaction Output
VaR	Value-at-Risk
VASP	Virtual Asset Service Provider
VAT	Value-Added Tax
VGAE	Variational Graph Autoencoder
VC	Variance-Covariance
V2I	Vehicle to Infrastructure
V2V	Vehicle to Vehicle
V2X	Vehicle to Everything
XAI	eXplainable Artificial Intelligence
XML	eXtensible Markup Language
5GAA	5G Automotive Association

Part I
Big Data and AI Technologies for Digital Finance

Chapter 1
A Reference Architecture Model for Big Data Systems in the Finance Sector

John Soldatos, Ernesto Troiano, Pavlos Kranas, and Alessandro Mamelli

1 Introduction

1.1 Background

In recent years, banks and other financial institutions are accelerating their digital transformation. As part of this transformation, financial organizations produce unprecedented amounts of data about their financial and insurance processes while using advanced digital technologies (e.g., big data, artificial intelligence (AI), Internet of Things (IoT)) to collect, analyze, and fully leverage the generated data assets [1]. Furthermore, recent regulatory developments (e.g., the 2nd Payment Services Directive (PSD2) in Europe) [2] facilitate the sharing of data across financial organizations to enable the development of innovative digital finance services based on novel business models. The latter aim at lowering the barriers for new market players (e.g., payment service providers (PSPs) in the PSD2 context)

J. Soldatos (✉)
INNOV-ACTS LIMITED, Nicosia, Cyprus

University of Glasgow, Glasgow, UK
e-mail: jsoldat@innov-acts.com

E. Troiano
GFT Italia, Milan, Italy
e-mail: ernesto.troiano@gft.com

P. Kranas
LeanXcale, Madrid, Spain
e-mail: pavlos@leanxcale.com

A. Mamelli
Hewlett Packard Enterprise, Milan, Italy
e-mail: alessandro.mamelli@hpe.com

J. Soldatos, D. Kyriazis (eds.), *Big Data and Artificial Intelligence in Digital Finance*,
https://doi.org/10.1007/978-3-030-94590-9_1

to develop and roll out new services in ways that boost customers' satisfaction and create new revenue streams.

Given the proliferation of data assets in the finance sector and their sharing across financial organizations, the vast majority of digital transformation applications for the finance and insurance sectors are data-intensive. This holds for applications in different areas such as retail banking, corporate banking, payments, investment banking, capital markets, insurance services, and financial services security [3, 4]. These applications leverage very large datasets from legacy banking systems (e.g., customer accounts, customer transactions, investment portfolio data), which they combine with other data sources such as financial market data, regulatory datasets, social media data, real-time retail transactions, and more. Moreover, with the advent of Internet-of-Things (IoT) devices and applications (e.g., Fitbits, smart phones, smart home devices) [5], several FinTech and InsurTech applications take advantage of contextual data associated with finance and insurance services to offer better quality of service at a more competitive cost (e.g., personalized healthcare insurance based on medical devices and improved car insurance based on connected car sensors). Furthermore, alternative data sources (e.g., social media and online news) provide opportunities for new more automated, personalized, and accurate services [6]. Moreover, recent advances in data storage and processing technologies (including advances in AI and blockchain technologies [7]) provide new opportunities for exploiting the above-listed massive datasets and are stimulating more investments in digital finance and insurance. Overall, financial and insurance organizations take advantage of big data and IoT technologies in order to improve the accuracy and cost-effectiveness of their services, as well as the overall value that they provide to their customers. Nevertheless, despite early deployment instances, there are still many challenges that must be overcome prior to leveraging the full potential of big data and AI in the finance and insurance sectors.

1.2 Big Data Challenges in Digital Finance

1.2.1 Siloed Data and Data Fragmentation

One of the most prominent challenges faced by banks and financial organizations is the fragmentation of data across different data sources such as databases, data lakes, transactional systems (e.g., e-banking), and OLAP (online analytical processing) systems (e.g., customer data warehouses). This is the reason why financial organizations are creating big data architectures that provide the means for consolidating diverse data sources. As a prominent example, the Bank of England has recently established a "One Bank Data Architecture" based on a centralized data management platform. This platform facilitates the big data analytics tasks of the bank, as it permits analytics over significantly larger datasets [8]. The need for reducing data fragmentation has been also underlined by financial institutions following the financial crisis of 2008, where several financial organizations had no easy way to perform integrated risk assessments as different exposures (e.g., on subprime loans or ETFs (exchange-traded fund)) were siloed across different systems. Therefore, there is a need for data architectures that reduce data fragmentation and take advantage of siloed data in developing integrated big data analytics and machine learning (ML) pipelines, including deep learning (DL).

1.2.2 Real-Time Computing

Real-time computing refers to IT systems that must respond to changes according to definite time constraints, usually in the order of milliseconds or seconds. In the realm of financial and insurance sectors, real-time constraints apply where a response must be given to provide services to users or organizations and are in the order of seconds or less [9]. Examples range from banking applications to cybersecurity. Contrary to data-intensive applications in other industrial sectors (e.g., plan control in industrial automation), most real-world financial applications are not real time and are usually solved by putting more computing resources (e.g., central processing units (CPUs), graph processing units (GPUs), memory) at the problem. However, in the case of ML/DL and big data pipelines, algorithms can take a significant amount of time and become useless in practical cases (e.g., responses arrive too late to be used). In these cases, a quantitative assessment of the computing time of algorithms is needed to configure resources to provide acceptable time.

1.2.3 Mobility

The digital transformation of financial institutions includes a transition to mobile banking [10]. This refers to the interaction of customer and financial organizations through mobile channels. Therefore, there is a need for supporting mobile channels when developing big data, AI, and IoT applications for digital finance, but also when collecting and processing input from users and customers.

1.2.4 Omni-channel Banking: Multiple Channel Management

One of the main trends in banking and finance is the transition from conventional multichannel banking to omni-channel banking [11]. The latter refers to seamless and consistent interactions between customers and financial organizations across multiple channels. Omni-channel banking/finance focuses on integrated customer interactions that comprise multiple transactions, rather than individual financial transactions. Therefore, banking architecture must provide the means for supporting omni-channel interactions through creating unified customer views and managing interactions across different channels. The latter requires the production of integrated information about the customer based on the consolidation of multiple sources. Big data analytics is the cornerstone of omni-channel banking as it enables the creation of unified views of the customers and the execution of analytical functions (including ML) that track, predict, and anticipate customer behaviors.

1.2.5   Orchestration and Automation: Toward MLOps and AIOps

Data-intensive applications are realized and supported by specialized IT and data administrators. Recently, more data scientists and business analysts are involved in the development of such applications. Novel big data and AI architectures for digital finance and insurance must support data administrators and data scientists to provide the means for orchestrating data-intensive application and their management through easily creating workflows and data pipelines. In this direction, there is a need for orchestrating functionalities across different containers. Likewise, there is a need for automating the execution of data administration and data pipelining tasks, which is also a key for implementing novel data-driven development and operation paradigms like MLOps [12] and AIOps [13].

1.2.6   Transparency and Trustworthiness

During the last couple of years, financial organizations and customers of digital finance services raise the issue of transparency in the operation of data-intensive systems as a key prerequisite for the wider adoption and use of big data and AI analytics systems in finance sector use cases. This is particularly important for use cases involving the deployment and use of ML/DL systems that operate as black boxes and are hardly understandable by finance sector stakeholders. Hence, a key requirement for ML applications in the finance sector is to be able to explain their outcomes. As a prominent example, a recent paper by the Bank of England [14] illustrates the importance of providing explainable and transparent credit risk decisions. Novel architectures for big data and AI systems must support transparency in ML/DL workflows based on the use of explainable artificial intelligence (XAI) techniques (e.g., [15, 16]).

The above-listed challenges about big data systems in finance are not exhaustive. For instance, there is also a need for addressing non-technological challenges such as the need for reengineering business processes in a data-driven direction and the need to upskill and reskill digital finance workers to enable them to understand and leverage big data systems. Likewise, there are also regulatory compliance challenges, stemming from the need to comply with many and frequently changing regulations. To address the technological challenges, digital finance sectors could greatly benefit from a reference architecture (RA) [17] that would provide a blueprint solution for developing, deploying, and operating big data systems.

1.3   Merits of a Reference Architecture (RA)

RAs are designed to facilitate design and development of concrete technological architectures in the IT domain. They help reducing development and deployment risks based on the use of a standard set of components and related structuring principles for their integration. When a system is designed without an RA, organizations may accumulate technical risks and end up with a complex and nonoptimal implementation architecture. Furthermore, RAs help improving the overall communication between the various stakeholders of a big data systems. Overall, the value of RAs can be summarized in the following points:

- Reduction of development and maintenance costs of IT systems
- Facilitation of communication between important stakeholders
- Reduction of development and deployment risks

The importance of reference architectures for big data and high-performance computing (HPC) systems, has led global IT leaders (e.g., Netflix, Twitter, LinkedIn) to publicly present architectural aspects of their platforms [18]. Furthermore, over the years, various big data architectures for digital finance have been presented as well. Nevertheless, these infrastructures do not address the above-listed challenges of emerging digital finance and insurance systems. The main goal of this chapter is to introduce a novel reference architecture (RA) for big data and AI systems in digital finance and insurance, which is destined to address the presented challenges. The RA extends architectural concepts that are presented in earlier big data architectures for digital finance while adhering to the principles of the reference model of the Big Data Value Association (BDVA) [19], which has been recently transformed to the Data, AI and Robotics (DAIRO) Association. The presented RA is centered on the development and deployment of data analytics pipelines, leveraging data from various sources. It is destined to assist financial organizations and other relevant stakeholders (e.g., integrators of big data solutions) to develop novel data-intensive systems (including ML/DL systems) for the finance and insurance sectors. The merits of the architecture are illustrated by means of some sample data science pipelines that adhere to the RA. Note also that many of the systems that are illustrated in subsequent chapters of the book have been developed and deployed based on the structuring principles and the list of reference components of the presented RA. In essence, this chapter provides a high-level overview of the RA, while the following chapters provide more information on how specific big data technologies and applications have leveraged the reference architecture.

1.4 Chapter Structure

The chapter is structured as follows:

- Section 2 following this introduction reviews big data architectures for digital finance. It illustrates why and how existing architectures are mostly application-specific and not appropriate for providing broad coverage and support for big data and AI systems in digital finance.

- Section 3 introduces the RA by means of complementary viewpoints, including a logical view, as well as development and deployment considerations.
- Section 4 presents how the RA supports the development of a set of sample data pipelines (including ML/DL systems). Specifically, it illustrates that the RA can support the development of various data-intensive systems.
- Section 5 concludes the chapter and connects it to other chapters of the book. As already outlined, the RA of this chapter has served as a basis for the design and development of various big data systems and technologies that are presented in subsequent chapters.

2   Related Work: Architectures for Systems in Banking and Digital Finance

2.1   IT Vendors' Reference Architectures

Many RAs and solution blueprints for big data in digital finance have been recently proposed by prominent IT vendors. As a prominent example, IBM has introduced an RA for big data management as a layered architecture [20]. It comprises layers for data source collection, data organization and governance, as well as analysis and infusion of data. The data source layer comprises structured (e.g., relational database management systems (DBMSs), flat files), semi-structured (such as Extensible Markup Language (XML)), and unstructured (video, audio, digital, etc.) sources. The architecture specifies traditional data sources (e.g., enterprise banking systems, transactional processing systems) and new data sources (e.g., social media data and alternative data). The data collection and governance layer specifies different data management systems (e.g., data lakes, data warehouses), which support various types of data, including both data at rest and data at motion. The layer comprises real-time analytical processing functionalities, which support data transfer at a steady high-speed rate to support many zero latency ("business real time") applications. Likewise, it comprises data warehousing functionalities that provide raw and prepared data for analytics consumption. Also, a set of shared operational data components own, rationalize, manage, and share important operational data for the enterprise. Moreover, the RA specifies a set of crosscutting functionalities, which are marked as "foundational architecture principles" and are present across all the above layers. These include security and multi-cloud management functionalities.

Microsoft's RA for digital finance provides a logical banking technology architecture schema [21]. It enables high-value integration to other systems through a wide array of industry standard integration interfaces and techniques (e.g., interfaces from ISO (International Organization for Standardization), BIAN (Banking Industry Architecture Network), and IFX (International Foreign Exchange)). In this way, it reduces the costs of managing and maintaining data-intensive solutions in the banking industry. Additionally, the Microsoft RA offers an industry-leading set of robust functionalities defined and exploited in both the banks' data centers and in the cloud. This kind of functionalities extends across the overall IT stack from the crucial operations to the end-user and constitutes a valuable framework for applications like fraud detection. The Microsoft RA provides master data management (MDM), data quality services (DQS), and predefined BI semantic metadata (BISM), which overlay business intelligence (BI) capabilities delivered via pre-tuned data warehouse configurations, near real-time analytics delivered through high-performance technical computing (HPC) and complex event processing (CEP). Overall, the Microsoft RA organizes and sustains massive volumes of transactions along with robust functionalities in bank data centers.

WSO2 offers a modular platform for the implementation of connected digital finance applications [22]. The philosophy of the platform is to divide complex systems into simpler individual subsystems that can be more flexibly managed, scaled, and maintained. It emphasizes flexibility given the need to comply with a rapidly changing landscape of business and regulatory requirements. From an implementation perspective, it comprises various applications, data management systems, and toolkits. The platform architecture comprises various operational systems that feed a data warehouse to enable analytical processing and data mining. On top of the data warehouse, several enterprise applications are implemented, including accounting applications and reporting applications. The WS02 platform includes a private PaaS (platform as a service) module that supports the integration of many financial applications in the cloud. It also specifies a business process server, which orchestrates workflows of services across different business units of a financial organization, but also of services that span different banks and financial institutions. To support service-oriented architectures, WS02 prescribes an enterprise service bus (ESB).

The Hortonworks Data Platform (HDP) supports a Hadoop stack for big data applications [23]. It is a centralized, enterprise-ready platform for storage and processing of any kind of data. When combined with NoSQL databases (e.g., CouchDB), HDP creates a huge volume of business value and intelligence. The architecture boosts accuracy, through supporting superior and precise analytics insights to Hadoop. It also provides scalability and operational performance. The combination of NoSQL systems with HDP enables the implementation of many big data scenarios. For instance, it is possible to execute deep analytics when pulling data from CouchDB into Hadoop. Likewise, it is also possible to train machine learning models and then cache them in Couchbase. Big data analytics with HDP consist of three main phases, namely, data pooling and processing, business intelligence, and predictive analytics. HDP supports various finance sector use cases, including risk management, security, compliance, digital banking, fraud detection, and anti-money laundering. The combined use of HDP and NoSQL databases enables the integration of an operational data store (ODS) with analytics capabilities in the banking environment.

2.2 Reference Architecture for Standardization Organizations and Industrial Associations

In 2017, the Big Data Value Association (BDVA)[1] introduced a general-purpose reference model (RM) that specifies the structure and building blocks of big data systems and applications. The model has horizontal layers encompassing aspects of the data processing chain and vertical layers addressing crosscutting issues (e.g., cybersecurity and trust).

The BDVA reference model is structured into horizontal and vertical concerns [19]:

- Horizontal concerns cover specific aspects along the data processing chain, starting with data collection and ingestion and extending to data visualization. It should be noted that the horizontal concerns do not imply a layered architecture. As an example, data visualization may be applied directly to collected data (the data management aspect) without the need for data processing and analytics.
- Vertical concerns address crosscutting issues, which may affect all the horizontal concerns. In addition, vertical concerns may also involve nontechnical aspects.

Even though the BDVA is not an RA in the IT sense, its horizontal and vertical concerns can be mapped to layers in the context of a more specific RA. The RM can serve as a common framework to locate big data technologies on the overall IT stack. It addresses the main concerns and aspects to be considered for big data systems.

Apart from the BDVA, the National Institute of Standards and Technology (NIST) has also developed a big data RA [24] as part of its big data program. The NIST RA introduces a conceptual model that is composed of five functional components: data producer/consumer, system orchestrator, and big data application/framework provider. Data flows, algorithm/tool transfer, and service usage between the components can be used to denote different types of interactions. Furthermore, the activities and functional component views of the RA can be used for describing a big data system, where roles, sub-roles, activities, and functional components within the architecture are identified.

One more reference architecture for data-intensive systems has been defined by the Industrial Internet Consortium (IIC). Specifically, the Industrial Internet Reference Architecture (IIRA) has been introduced to provide structuring principles of industrial Internet of Things (IIoT) systems [25]. The IIRA adapts the ISO architecture specification (ISO/IEC/IEEE 42010) [26] to the needs of IIoT systems and applications. It specifies a common architecture framework for developing interoperable IoT systems for different vertical industries. It is an open, standard-based architecture, which has broad applicability. Due to its broad applicability, the IIRA is generic, abstract, and high-level. Hence, it can be used to drive the

---

[1] As of 2021, BDVA has evolved to the Data, AI and Robotics (DAIRO) initiative.

structuring principles of an IoT system for finance and insurance, without however specifying its low-level implementation details. The IIRA presents the structure of IoT systems from four viewpoints, namely, business, usage, functional, and implementation viewpoints. The functional viewpoint specifies the functionalities of an IIoT system in the form of the so-called functional domains. Functional domains can be used to decompose an IoT system to a set of important building blocks, which are applicable across different vertical domains and applications. As such, functional domains are used to conceptualize concrete functional architectures. The implementation viewpoint of the IIRA is based on a three-tier architecture, which follows the edge/cloud computing paradigm. The architecture includes an edge, a platform, and an enterprise tier, i.e., components that are applicable to the implementation of IoT-based applications for finance and insurance. Typical examples of such applications are usage-based insurance applications, which leverage data from IoT devices in order to calculate insurance premiums for applications like vehicle and healthcare insurance.

2.3 Reference Architectures of EU Projects and Research Initiatives

In recent years, various EU-funded projects and related research initiatives have also specified architectures for big data systems. As a prominent example, the BigDataStack project has specified an architecture that drives resource management decisions based on data aspects, such as the deployment and orchestration of services [27]. A relevant architecture is presented in Fig. 1.1, including the main information flows and interactions between the key components.

As presented in the figure, raw data are ingested through the gateway and unified API component to the storage engine of BigDataStack, which enables storage and data migration across different resources. The engine offers solutions both for relational and non-relational data, an object store to manage data as objects, and a CEP engine to deal with streaming data processing. The raw data are then processed by the data quality assessment component, which enhances the data schema in terms of accuracy and veracity and provides an estimation for the corresponding datasets in terms of their quality. Data stored in object store are also enhanced with relevant metadata, to track information about objects and their dataset columns.

Those metadata can be used to show that an object is not relevant to a query, and therefore does not need to be accessed from storage or sent through the network. The defined metadata are also indexed, so that during query execution objects that are irrelevant to the query can be quickly filtered out from the list of objects to be retrieved for the query processing. This functionality is achieved through the data skipping component of BigDataStack. Furthermore, the overall storage engine of BigDataStack has been enhanced to enable adaptations during runtime (i.e., self-scaling) based on the corresponding loads. Given the stored

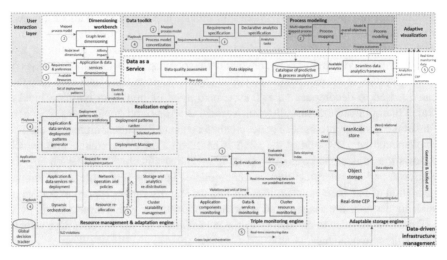

Fig. 1.1 H2020 BigDataStack architecture for big data systems

data, decision-makers can model their business workflows through the process modeling framework that incorporates two main components: the first component is process modeling, which provides an interface for business process modeling and the specification of end-to-end optimization goals for the overall process (e.g., accuracy, overall completion time, etc.). The second component refers to process mapping. Based on the analytics tasks available in the Catalogue of Predictive and Process Analytics and the specified overall goals, the mapping component identifies analytics algorithms that can realize the corresponding business processes. The outcome of the component is a model in a structural representation, e.g., a JSON file that includes the overall workflow, and the mapped business processes to specific analytics tasks by considering several (potentially concurrent) overall objectives for the business workflow. Following, through the Data Toolkit, data scientists design, develop, and ingest analytics processes/tasks to the Catalogue of Predictive and Process Analytics. This is achieved by combining a set of available or underdevelopment analytics functions into a high-level definition of the user's application. For instance, they define executables/scripts to run, as well as the execution endpoints per workflow step. Data scientists can also declare input/output data parameters, analysis configuration hyperparameters (e.g., the k in a k-means algorithm), execution substrate requirements (e.g., CPU, memory limits, etc.) as service-level objectives (SLOs), as well as potential software packages/dependencies (e.g., Apache Spark, Flink, etc.).

As another example, the H2020 BOOST 4.0 has established an RA for big data systems in the manufacturing sector [28]. The RA consists of a number of layers at its core, alongside factory dimension and a manufacturing entity dimension [29].

The core layers represent a collection of functionalities/components performing a specific role in the data processing chain and consist of integration layer, information and core big data layers, application layer, and business layer. The integration layer facilitates the access and management of external data sources such as PLM (product lifecycle management) systems, production data acquisition systems, open web APIs, and so on. The information and core big data layer is composed of components belonging in four different sublayers:

- Data management groups together with components facilitating data collection, preparation, curation, and linking
- Data processing groups together with architectures that focus on data manipulation
- Data analytics groups together with components to support descriptive, diagnostic, predictive, and prescriptive data analysis
- Data visualization groups together with algorithm components to support data visualization and user interaction

The application layer represents a group of components implementing application logic that supports specific business functionalities and exposes the functionality of lower layers through appropriate services.

The business layer forms the overall manufacturing business solution in the BOOST 4.0 five domains (networked commissioning and engineering, cognitive production planning, autonomous production automation, collaborative manufacturing networks, and full equipment and product availability) across five process life cycle stages (smart digital engineering, smart production planning and management, smart operations and digital workplace, smart connected production, smart maintenance and service). Alongside the core layers, there are a number of other crosscutting aspects that affect all layers:

- Communication aims to provide mechanisms and technologies for reliable transmission and receipt of data between the layers.
- Data-sharing platforms allow data providers to share their data as a commodity, covering specific data, for a predefined space of time, and with a guarantee of reversibility at the end of the contract.
- Development engineering and DevOps cover tool chains and frameworks that significantly increase productivity in terms of developing and deploying big data solutions.
- Standards cover the different standard organizations and technologies used by BOOST 4.0.
- Cybersecurity and trust cover topics such as device and application registration, identity and access management, data governance, data protection, and so on.

2.4   Architectures for Data Pipelining

Managing the flow of information forms an integral part of every enterprise looking to generate value from their data. This process can be complicated due to the number of sources and data volume. In these situations, pipelines can be of help as they simplify the flow of data by eliminating the manual steps and automating the process. Pipeline architectures are simple and powerful: They are inspired by the Unix technique of connecting the output of an application to the input of another via pipes on the shell. They are suitable for applications that require a series of independent computations to be performed on data [30]. Any pipeline consists of filters connected by pipes. A filter is a component that performs some operation on input data to transform them into output data. The latter is passed to other component(s) through a pipe. The pipe is a directional connector that passes a stream of data from one filter to the next. A pump is a data source and is the first element of the pipeline.

A pump could be, for example, a static file, a data lake, a data warehouse, or any device continuously creating new data. Data can be managed into two ways: batch ingestion and streaming ingestion. With batch ingestion, data are extracted following an external trigger and administered as a whole. Batch processing is mostly used for data transfer and is more suitable in cases where acquiring exhaustive insights is more important than getting faster analytics results. Instead, with the streaming ingestion, sources transfer unit data one by one. Stream processing is suitable in case real-time data is required for applications or analytics.

Finally, the sink is the last element of the pipeline. It could be another file, a database, a data warehouse, a data lake, or a screen. A pipeline is often a simple sequence of components. However, its structure can also be very complex: in fact, in principle, a filter can have any number of input and output pipes.

The pipeline architecture has various advantages: (i) It makes it easy to understand the overall behavior of a complex system, since it is a composition of behaviors of individual components. (ii) It supports the reuse of filters while easing the processes of adding, replacing, or removing filters from the pipeline. This makes a big data system easy to maintain and enhance. (iii) It supports concurrent execution that can boost performance and timeliness.

Data pipelines carry raw data from different data sources to data warehouses for data analysis and business intelligence. Developers can build data pipelines by writing code and interfacing with SaaS (software-as-a-service) platforms. In recent years, data analysts prefer using data pipeline as a service (DPaaS), which does not require coding. While using data pipelines, businesses can either build their own or use a DPaaS. Developers write, test, and maintain the code required for a data pipeline using different frameworks and toolkits, for example, management tools like Airflow and Luigi. Likewise, solutions like KNIME [31] enable the handling of pipelines without the need for coding (i.e., "codeless" pipeline development).

Apache Airflow is an open-source tool for authoring, scheduling, and monitoring workflows [32]. Airflow can be used to author workflows as directed acyclic graphs (DAGs) of tasks. Apache Airflow has an airflow scheduler that executes your tasks on an array of workers while following the specified dependencies. Its rich command line utility enables to easily perform complex surgeries on DAGs. Moreover, its rich user interface makes it easy to visualize pipelines running in production, monitor progress, and troubleshoot issues when needed.

When workflows are defined as code, they become more maintainable, versionable, testable, and collaborative. Airflow provides a simple query interface to write SQL and get results quickly, as well as a charting application letting you visualize data.

Luigi is a Python package useful for building complex pipelines of batch jobs [33]. The purpose of Luigi is to address all the plumbing typically associated with long-running batch processes. It is suitable to chain many tasks and automate them. The tasks can be anything but are typically long-running things like Hadoop jobs, dumping data to/from databases, running machine learning algorithms, and more. Luigi helps to stitch many tasks together, where each task can be a Hive query, a Hadoop job in Java, a Spark job in Scala or Python, a Python snippet, dumping a table from a database, or anything else. It makes it easy to build up long-running pipelines that comprise thousands of tasks and take days or weeks to complete. Since Luigi takes care of a lot of the workflow management, the user can focus on the tasks themselves and their dependencies. Luigi also provides a toolbox of several common task templates. It includes support for running Python MapReduce jobs in Hadoop, Hive, and Pig jobs. It also comes with file system abstractions for the Hadoop Distributed File System (HDFS), and local files, which ensures that file system operations are atomic.

The KNIME Analytics Platform is an open-source software suitable for designing data science workflows and reusable components accessible to everyone [31]. It is very intuitive, as it enables to create visual workflows with a drag-and-drop style graphical interface. The KNIME Hub offers a library of components that enable the following:

(i) *Blend data from any source*, including simple text formats (CSV, PDF, XLS, JSON, XML, etc.), unstructured data types (images, documents, networks, molecules, etc.), or time series data. It also enables to connect to a host of databases and data warehouses.

(ii) *Shape data* through deriving statistics (mean, quantiles, and standard deviation), applying statistical tests to validate a hypothesis, or make correlation analysis and more into workflows. Many components are available to clean, aggregate, sort, filter, and join data either on local machine, in database, or in distributed big data environments. In addition, features can be extracted and selected to prepare datasets for machine learning with genetic algorithms, random search, or backward and forward feature elimination.

(iii) *Apply ML/DL techniques* through building machine learning models for tasks like classification, regression, dimension reduction, or clustering, using advanced algorithms including deep learning, tree-based methods, and logistic regression.

(iv) *Optimize model performance* based on hyperparameter optimization, boosting, bagging, stacking, or building complex ensembles.

(v) *Validate models by applying performance metrics* such as the receiver operating characteristic (ROC) and the area under the ROC curve (AUC) while performing cross validation to guarantee model stability.

(vi) *Build and apply explainable AI (XAI) models* like LIME [15] and Shap/Shapley values [34].
(vii) *Discover and share insights*, based on advanced and versatile visualizations.

2.5 Discussion

Previous paragraphs presented a wide range of reference architectures and reference models for big data systems, including architectures developed for the digital finance and insurance sectors. The presented architectures have illustrated the main building blocks of a reference architecture for big data and AI in digital finance, such as interface data sources, data streaming modules, modules for handling and processing data at rest, data warehouses and analytics databases, data integration and interoperability modules, as well as data visualization components. Moreover, they outline the main crosscutting functions such as data governance and cybersecurity functions. Furthermore, they illustrate implementation architectures based on multitier systems such as edge/cloud systems comprising an edge, a platform, and a cloud tier. They also outline powerful concepts and tools, such as the data pipelining concept and frameworks that support the development and deployment of data pipelines.

Despite the presentation of this rich set of concepts, there is no single architecture that could flexibly support the development of the most representative data-intensive use cases in the target sectors. Most of the presented architectures target specific use cases of the sector (e.g., fraud detection and anti-money laundering) or come at a very abstract level that makes them applicable to multiple sectors. Moreover, most of them specify a rigorous structure for the big data applications, which limits the flexibility offered to banks, financial organizations, and integrators of big data applications in finance and insurance. Therefore, there is a need to introduce a more flexible approach based on the popular pipelining concept: Instead of providing a rigorous (but monolithic) structure of big data and AI applications in digital finance, the following sections opt to define these applications as collections of data-driven pipelines. The latter are to be built based on a set of well-defined components, spanning different areas such as data preprocessing, machine learning, data anonymizing, data filtering, data virtualization, and more. The reference architecture that is introduced in the following section provides a set of layered architectural concepts and a rich set of digital building blocks, which enable the development of virtually any big data or AI application in digital finance and insurance. This offers increased flexibility in defining data-driven applications in the sector, in ways that subsume most of the rigorous architectures outlined in earlier paragraphs.

3   The INFINITECH Reference Architecture (INFINITECH-RA)

3.1   Driving Principles: INFINITECH-RA Overview

The purpose of an RA for big data systems in digital finance and insurance is to provide a conceptual and logical schema for solutions to a very representative class of problems in this sector. The H2020 INFINITECH project develops, deploys, and validates over 15 novel data-driven use cases in digital finance and insurance, which constitute a representative set of use cases. The development of the RA that is presented in the following paragraphs (i.e., the INFINITECH-RA) is driven by the requirements of these applications. Likewise, it has been validated and used to support the actual development of these use cases, some of which are detailed in later chapters of the book.

The INFINITECH-RA is the result of the analysis of a considerable number of use cases, including their requirements (i.e., users' stories) and constraints (e.g., regulatory, technological, organizational). Furthermore, the INFINITECH-RA considers state-of-the-art technologies and similar architectures in order to provide best practices and blueprints that enable relevant stakeholders (e.g., end users in financial organizations, business owners, designers, data scientists, developers) to develop, deploy, and operate data-driven applications. The INFINITECH-RA is largely based on the concept of data pipelines. Therefore, it can be used to understand how data can be collected and how models and technologies should be developed, distributed, and deployed.

The INFINITECH-RA is inspired by state-of-the-art solutions and technologies. It is based on the BDVA RM to provide an abstraction that solves a general class of use cases, including the ones of the INFINITECH project. It exploits microservice technologies and DevOps methodologies. Specifically, data components in the INFINITECH-RA are encapsulated in microservices in line with a loosely coupled approach. The latter is preferred over tightly coupled intertwined applications. In the context of the INFINITECH-RA, data can be pipelined into different microservices to perform different types of processing, while data can be streamed or stored in data stores. In line with big data management best practices, data movements are limited as much as possible, in which cross cutting services are specified to value-added functionalities across all different layers of the architecture.

To understand the development and use of the INFINITECH-RA, the following concepts are essential:

- *Nodes*: In the INFINITECH-RA, a node is a unit of data processing. Every node exhibits interfaces (APIs) for data management in particular for consuming and producing data (i.e., IN and OUT). From an implementation perspective, nodes are microservices that expose REST APIs.
- *BDVA RM Compliance*: The INFINITECH-RA layers can be mapped to layers of the BDVA RM. Each node belongs to some layer of the RA.

- *Pipeline Concept*: Nodes with IN/OUT interfaces can be stacked up to form data pipelines, i.e., a pipelining concept is supported. Moreover, nodes are loosely coupled in the RA, i.e., they are not connected until a pipeline is created. Data can flow in all directions (e.g., an ML node can push back data into a data layer).
- *Vertical Layers of Nodes*: Every node stack with other compatible nodes, i.e., whether nodes can be connected or not, depends on their IN-OUT interfaces. Therefore, the RA can be segmented into vertical layers (called bars) to group compatible nodes. A node can belong to one or more vertical bars.

In line with the following concepts, the INFINITECH-RA is:

- *Layered*: Layers are a way of grouping nodes in the same way as the BDVA RM has "concerns."
- *Loosely Coupled*: There are no rules to connect nodes in a predefined way or in a rigid stack.
- *Distributed*: Computing and nodes can be physically deployed and distributed anywhere, e.g., on premise, on a cloud infrastructure, or across multiple clouds.
- *Scalable*: Nodes provide the means for distributing computing at edges, at HPC nodes (GPUs), or centrally.
- *Multi-workflow*: The INFINITECH-RA allows for simple pipelines and/or complex data flows.
- *Interoperable*: Nodes can be connected to other nodes with compatible interfaces in a way that boosts interoperability.
- *Orchestrable*: Nodes can be composed in different ways allowing creation of virtually infinite combinations.

3.2 The INFINITECH-RA

The INFINITECH reference architecture has been specified based on the "4+1" architectural view model [35]. The "4+1" architectural view model is a methodology for designing software architectures based on five concurrent "views." These views represent different stakeholders who could deal with the platform and the architecture, from the management, development, and user perspectives. In this context, the *logical viewpoint* of the RA illustrates the range of functionalities or services that the system provides to the end users. The following paragraphs illustrate the logical views of the architecture while providing some development and deployment considerations as well.

3.2.1 Logical View of the INFINITECH-RA

The logical views of the RA are presented in Fig. 1.2. Rather than identifying and specifying functional blocks and how they are interconnected, the INFINITECH-RA is defined by a set of "principles" to build pipelines or workflows. These

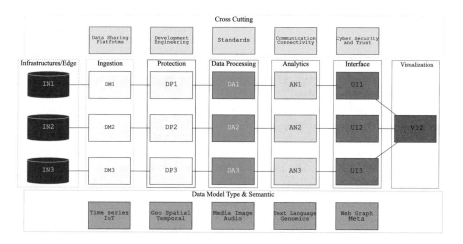

Fig. 1.2 Logical view of the INFINITECH-RA and mapping to BDVA RM

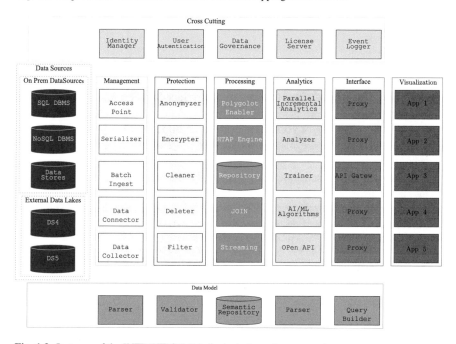

Fig. 1.3 Instance of the INFINITECH-RA: logical view of components

principles constitute the guidelines that drive specific implementation of digital finance and insurance. At a high level, the RA can be seen as a pipelined mapping of nodes referring to the BDVA reference model and crosscutting layers.

Figure 1.3 provides a high-level logical view of a specific solution instance that is structured according to the INFINITECH-RA. The generic nodes that are depicted in the figure are only examples and do not refer to any specific tool or application. They represent generic services that belong to a general class of application performing the functionality of the corresponding layer in the BDVA RM. Hence, the RA is generic and makes provision for flexibly specifying and implementing nodes to support the wide array of requirements of data-intensive applications in digital finance.

In this logical view of the INFINITECH-RA, the various components are grouped into the following layers:

- *Data Sources*: This layer comprises the various data sources (e.g., database management systems, data lakes holding nonstructural data, etc.) of a big data application.
- *Ingestion*: This is a data management layer that is associated with data import, semantic annotation, and filtering data from the data sources.
- *Security*: This layer manages data clearance before any further storing or elaboration. It provides functions for security, anonymization, cleaning of data, and more.
- *Management*: This layer is responsible for data management, including persistent data storage in the central repository and data processing enabling advanced functionalities such as hybrid transactional and analytical processing (HTAP).
- *Analytics*: This layer comprises the ML, DL, and AI components of the big data system.
- *Interface*: This layer defines the data to be produced and provided to the various user interfaces.
- *Crosscutting*: This layer comprises service components that provide function-alities orthogonal to the data flows such as authentication, authorization, and accounting.
- *Data Model*: This is a crosscutting layer for modeling and semantics of data in the data flow.
- *Presentation and Visualization*: This layer is associated with the presentation of the results of an application in some form like a dashboard.

The RA does not impose any pipelined or sequential composition of nodes. However, each different layer and its components can be used to solve specific problems of the use case.

3.2.2 Development Considerations

From a development viewpoint, the RA complies with a microservice architec-ture implementation, including services that interact through REST APIs. All microservices run in containers (e.g., Docker) managed by an orchestrator platform (e.g., Kubernetes). Development and testing activities are based on a DevOps methodology, which emphasizes a continuous integration (CI) approach. Every time a developer pushes changes to the source code repository, a CI server triggers automatically a new build of the component and deploys the updated container to an integration environment on Kubernetes. This enables continuous testing of the various components against an up-to-date environment, which speeds up development and avoids painful integration problems at the end of the cycle. Continuous delivery (CD) is an extension of that process, which automates the release process so that new code can be deployed to target environments, typically to test environments, in a repeatable and automated fashion.

This process is enhanced based on DevSecOps approach, which integrates security in the software development life cycle since the beginning of the development process. With DevSecOps, security is considered throughout the process and not just as an afterthought at the end of it, which produces safer and more solid design. Moreover, this approach avoids costly release delays and rework due to non-compliance, as issues are detected early in the process. DevSecOps introduces several security tools in the CI/CD pipeline, so that different kinds of security checks are executed continuously and automatically, giving to developers quick feedback whether their latest changes introduce a vulnerability that must be corrected.

To facilitate communication and collaboration between teams and to improve model tracking, versioning, monitoring, and management, there is also a need for standardizing and automating the machine learning process. In practice, this is often very complicated, because ML works in heterogeneous environments: For example, ML models are often developed on a data scientist's notebook. ML training is done in the cloud to take advantage of available resources, while execution of the software in production takes place on premise. The first step to establishing an MLOps approach for INFINITECH-RA compliant applications requires the standardization of the various development environments. In this direction, the Kubernetes platform along with the Docker containers provides the abstraction, scalability, portability, and reproducibility required to run the same piece of software across different environments. As a second step, it is necessary to standardize the workflow used for constructing and building of ML models. In the case of the development of INFINITECH-RA compliant applications, software like Kubeflow enables the development of models and boosts their portability. ML workflows are defined as Kubeflow pipelines, which consist of data preparation, training, testing, and execution step. Each step is implemented in a separate container, and the output of each step is the input to the following step. Once compiled, this pipeline is portable across environments.

3.2.3   Deployment Considerations

From a deployment perspective, the INFINITECH-RA leverages a microservice architecture. Microservices are atomic components, which can be developed and updated individually, yet being part of a wider end-to-end solution. This eases the development of individual components, yet it can complicate the management of a complete solution that comprises multiple microservices. In this context, INFINITECH-RA provides a tool for managing the life cycle of microservices, including their deployment, scaling, and management. This is based on Kubernetes and uses two main concepts:

- *Namespaces*, i.e., a logical grouping of a set of Kubernetes objects to whom it's possible to apply some policies, for example, to set limits on how many hardware resources can be consumed by all objects or to constraints on whether the namespace can be accessed by or can access other namespaces.
- *POD* which is the simplest unit in the Kubernetes object. A POD encapsulates one container yet in cases of complex applications can encapsulate more than one container. Each POD has its own storage resources, i.e., a unique network IP, an access port, and options related to how the container/s should run.

The Kubernetes namespace enables the logical isolation of objects (mainly PODs) from other namespaces.

An environment that enables testing of services/components as a separate namespace in the same Kubernetes cluster is provided to support CI/CD processes. This development environment is fully integrated with the CI/CD tools. Moreover, automated replication functionalities are provided, which facilitate the creation of a similar environment using appropriate scripts. In this way, an "infrastructure as code" paradigm is supported.

4   Sample Pipelines Based on the INFINITECH-RA

The following subsections present some sample reference solutions for common use case scenarios.

4.1   Simple Machine Learning Pipeline

The reference architecture enables the design and execution of classical machine learning pipelines. Typical machine learning applications are implemented in line with popular data mining methods like the CRISP-DM (cross-industry standard process for data mining) [36]. CRISP-DM includes several phases, including the phases of data preparation, modeling, and evaluation. In typical machine learning pipeline, data are acquired from various sources and prepared in line with the needs of the target machine learning model. The preparation of the data includes their segmentation into training data (i.e., data used to train the model) and test data (i.e., data used to test the model). The model is usually evaluated against the requirements of the target business application. Figure 1.4 illustrates how a typical machine learning pipeline can be implemented/mapped to the layers and functionalities of the INFINITECH-RA. Specifically, the following layers/parts are envisaged:

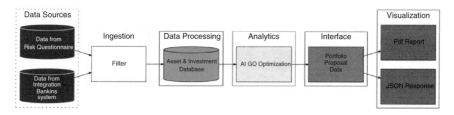

Fig. 1.4 Simple machine learning pipeline based on the INFINITECH-RA

- *Data Sources*: At this level, reside the data sources of the big data/AI application. These may be of different types including databases, data stores, data lakes, files (e.g., spreadsheets), and more.
- *Ingestion*: At this level of the pipeline, data are accessed based on appropriate connectors. Depending on the type of the data sources, other INFINITECH components for data ingestion can be used such as data collectors and serializers. Moreover, conversions between the data formats of the different sources can take place.
- *Data Processing*: At this level, data are managed. Filtering functions may be applied, and data sources can be joined toward forming integrated datasets. Likewise, other preprocessing functionalities may be applied, such as partitioning of datasets into training and test segments. Furthermore, if needed, this layer provides the means for persisting data at scale, but also for accessing it through user-friendly logical query interfaces. The latter functionalities are not however depicted in the figure.
- *Analytics*: This is the layer where the machine learning functions are placed. A typical ML application entails the training of machine learning models based on the training datasets, as well as the execution of the learned model-based test dataset. It may also include the scoring of the model based on the test data.
- *Interface and Visualization*: These are the layers where the model and its results are visualized in line with the needs of the target application. For example, in Fig. 1.4, a portfolio construction proposal is constructed and presented to the end user.

4.2   Blockchain Data-Sharing and Analytics

The INFINITECH-RA can also facilitate the specification, implementation, and execution of blockchain empowered scenarios for the finance sector, where, for example, several discrete entities (including banks, insurance companies, clients, etc.) may be engaged in data-sharing activities. The latter can empower financial organizations to update customer information as needed while being able to access an up-to-date picture of the customer's profile. This is extremely significant for

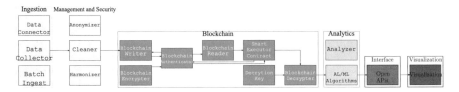

Fig. 1.5 Blockchain data-sharing and analytics pipeline

a plethora of FinTech applications, including yet not limited to Know Your Customer/Business, Credit Risk Scoring, Financial Products Personalization, Insurance Claims Management, and more. Let's consider a use case in which Financial Organization A (e.g., a bank or an insurance company) wishes to be granted access to a customer of another financial organization (e.g., another bank within the same group or a different entity altogether). Supposing that legal constraints are properly handled (e.g., consents for the sharing of the data have been granted, and all corresponding national and/or European regulations such as GDPR are properly respected and abided by) and that all data connectors (required for the retrieval of the data from the raw information sources) and data curation services (e.g., the cleaner, required for cleaning the raw data sources; the anonymizer, required for anonymizing the same information sources; and the harmonizer, required for mapping the cleaned and properly anonymized information sources to the Common INFINITECH Information Model) are in place, enabling the collection and preprocessing of the raw information sources, the following flow is envisaged (Fig. 1.5):

- The blockchain encryptor encrypts the properly preprocessed data so that they can be saved in the private collection of Organization A.
- The blockchain authenticator authenticates (the user of) Organization A, so that access to update the ledger is granted.
- Once authenticated, the blockchain writer inserts the encrypted data to the private collection of Organization A, and the hash of the data is submitted to the ledger.
- Organization B requests access to the data of a customer from Organization A.
- The blockchain authenticator authenticates (the user of) Organization B that initiates the request so as to grant or decline the request to access private-sensitive data.
- Once (the user of) Organization B is authenticated, the smart contract executor translates the query submitted and queries the private collection of Organization A.
- The blockchain authenticator authenticates (the user of) Organization B, so that access to read the ledger is granted.
- The blockchain reader retrieves the queried information from Organization A's private collection.

- The smart contract executor generates a new transaction and triggers the blockchain authenticator so as to authenticate (the user of) Organization A, in order to grant access to update the ledger.
- Once authenticated, the blockchain writer submits the encrypted data to the ledger, and a new record on the same ledger is created, containing the metadata of the contract (organizations involved, data created, metadata of the encrypted data transmitted, validity of the contract) and the actual encrypted data.
- Organization A sends out of band the decryption key to Organization B, and the blockchain decryptor decrypts the queried data.

4.3 Using the INFINITECH-RA for Pipeline Development and Specification

The INFINITECH-RA provides a blueprint for constructing big data solutions according to standards-based flows and components. The development of any use case based on INFINITECH-RA can be carried out based on the following considerations:

- *Use Case Design*: A use case is considered as a transformation from some data sources to some data destinations. This is called the "stretching phase" where a pipeline or workflow is defined from the different sources to the resulting data. The INFINITECH-RA is the total transformation of the sources into destinations once the appropriate workflow is defined and implemented.
- *Data Access and Ingestion*: Each data source is ingested via a first layer of data management components (e.g., serializers and database connectors). Data sources are accessible through different types of data technologies (e.g., SQL, NoSQL, IoT streams, blockchains' data), but also as raw data such as text, images, videos, etc. A connector must be provided to support all the variety of the supported target data store management systems.
- *Data Preprocessing*: Data must not be stored in the INFINITECH platform data store unless they are properly preprocessed and "cleared." Clearing can involve filtering, deletion, anonymization, encrypting of raw data, and more. The "clearing" can be managed via the security layer of the RA model or the crosscutting services.
- *Data Storage*: Most use cases ask for storing data in some data store. In the case of big data applications, data stores that handle big data must be used. Moreover, these data stores must be able to scale out to support the diversity of data sources and workloads.
- *Analytics*: Analytics work on top of the data (stored or streamed), consume, and might produce data, which are stored in some data repository as well.

Nevertheless, real-world use cases can be much more complicated than the methodology proposed above. They are intertwined with existing infrastructures and data processing components and cannot be easily stretched into a "pipelined" workflow as suggested. In these cases, the boundaries of "computational nodes" that provide basic interfaces must be identified. A complex infrastructure can be more than one microservice with many endpoints distributed among the different microservices so as to keep the functionalities homogeneous. Furthermore, a huge infrastructure can be encapsulated (wrapped) into a microservice interface to exhibit basic functions.

5   Conclusions

A reference architecture for big data systems in digital finance can greatly facilitate stakeholders in structuring, designing, developing, deploying, and operating big data, AI, and IoT solutions. It serves as a stakeholders' communication device while at the same time providing a range of best practices that can accelerate the development and deployment of effective big data systems. This chapter has introduced such a reference architecture (i.e., the INFINITECH-RA), which adopts the concept and principles of data pipelines. The INFINITECH-RA is in line with the principles of the BDVA RM. In practice, it extends and customizes the BDVA RA with constructs that permit its use for digital finance use cases.

The chapter has illustrated how INFINITECH-RA can support the development of common big data/AI pipelines for digital finance applications. In this direction, the deliverable has provided some sample pipelines. The presented RA served as a basis for the implementation of the applications and use cases that are described in later chapters of the book, including various digital finance and insurance use cases. The INFINITECH-RA is therefore referenced in several of the subsequent chapters of the book, which is the reason why the INFINITECH-RA has been introduced as part of the first chapter.

Acknowledgments This work has been carried out in the H2020 INFINITECH project, which has received funding from the European Union's Horizon 2020 Research and Innovation Programme under Grant Agreement No. 856632.

References

1. Hasan, M. M., Popp, J., & Oláh, J. (2020). Current landscape and influence of big data on finance. *Journal of Big Data, 7*, 21. https://doi.org/10.1186/s40537-020-00291-z
2. Botta, A., Digiacomo, N., Höll, R., & Oakes, L. (2018). *PSD2: Taking advantage of open-banking disruption*. Mc Kinsey. https://www.mckinsey.com/industries/financial-services/our-insights/psd2-taking-advantage-of-open-banking-disruption. Accessed 20 July 2021.
3. Troiano, E., et al. (2020). Security challenges for the critical infrastructures of the financial sector. In J. Soldatos, J. Philpot, & G. Giunta (Eds.), *Cyber-physical threat intelligence for critical infrastructures security: A guide to integrated cyber-physical protection of modern critical infrastructures* (pp. 2–12). Now Publishers. https://doi.org/10.1561/9781680836875.ch1

4. Troiano, E., Soldatos, J., Polyviou, A., et al. (2019). Big data platform for integrated cyber and physical security of critical infrastructures for the financial sector: Critical infrastructures as cyber-physical systems. *MEDES*, 262–269.
5. Soldatos, J. (2020, December). *A 360-degree view of IoT technologies*. Artech House. ISBN: 9781630817527.
6. Dierckx, T., Davis, J., & Schoutens, W. (2020). *Using machine learning and alternative data to predict movements in market risk*. arXiv: Computational Finance.
7. Polyviou, A., Velanas, P., & Soldatos, J. (2019). Blockchain technology: Financial sector applications beyond cryptocurrencies. *Proceedings, 28*(1), 7. https://doi.org/10.3390/proceedings2019028007
8. Cloudera. *Bank of England: Using data analytics to build a stable future*. https://www.cloudera.com/content/dam/www/marketing/resources/case-studies/bank-of-england-customer-success-story.pdf.landing.html. Accessed 20 July 2021.
9. Hussain, K., & Prieto, E. (2016). Big data in the finance and insurance sectors. In J. Cavanillas, E. Curry, & W. Wahlster (Eds.), *New horizons for a data-driven economy*. Springer. https://doi.org/10.1007/978-3-319-21569-3_12
10. Bons, R. W. H., Alt, R., Lee, H. G., et al. (2012). Banking in the Internet and mobile era. *Electronic Markets, 22*, 197–202. https://doi.org/10.1007/s12525-012-0110-6
11. Komulainen, H., & Makkonen, H. (2018). Customer experience in omni-channel banking services. *Journal of Financial Services Marketing, 23*, 190–199. https://doi.org/10.1057/s41264-018-0057-6
12. Tamburri, D. A. (2020). *Sustainable MLOps: Trends and challenges*. In 2020 22nd international symposium on symbolic and numeric algorithms for scientific computing (SYNASC), pp. 17–23. https://doi.org/10.1109/SYNASC51798.2020.00015
13. Dang, Y., Lin, Q., & Huang, P. (2019). *AIOps: Real-world challenges and research innovations*. In 2019 IEEE/ACM 41st international conference on software engineering: Companion proceedings (ICSE-Companion), pp. 4–5. https://doi.org/10.1109/ICSE-Companion.2019.00023
14. Bracke, P., Datta, A., Jung, C., & Sen, S. (2019, August). *Machine learning explainability in finance: An application to default risk analysis* (Staff working paper No. 816). Bank of England.
15. Ribeiro, M., Singh, S., & Guestrin, C. (2016). "Why should I trust you?" Explaining the predictions of any classifier, pp. 1135–1144. https://doi.org/10.1145/2939672.2939778
16. Adadi, A., & Berrada, M. (2018). Peeking inside the black-box: A survey on Explainable Artificial Intelligence (XAI). *IEEE Access, 6*, 52138–52160. https://doi.org/10.1109/ACCESS.2018.2870052
17. Angelov, S., Grefen, P., & Greefhorst, D. (2009). *A classification of software reference architectures: Analyzing their success and effectiveness*. In 2009 Joint working IEEE/IFIP conference on software architecture & european conference on software architecture, pp. 141–150. https://doi.org/10.1109/WICSA.2009.5290800
18. Manciola, R. M., Ré, R., & Schwerz, A. L. (2018). *An analysis of frameworks for microservices*. In 2018 XLIV Latin American Computer Conference (CLEI), pp. 542–551. https://doi.org/10.1109/CLEI.2018.00071
19. European Big Data Value Strategic Research and Innovation Agenda, Version 4.0, October 2017. https://bdva.eu/sites/default/files/BDVA_SRIA_v4_Ed1.1.pdf. Accessed 20 July 2021.
20. IBM Data Reference Architecture. https://www.ibm.com/cloud/architecture/architectures/dataArchitecture/reference-architecture/. Accessed 20 July 2021.
21. Microsoft Industry Reference Architecture for Banking (MIRA-B), Microsoft Corporation Whitepaper, May 2012.
22. Abeysinghe, A. (2015, September). *Connected finance reference architecture*. WS02 Whitepaper. https://wso2.com/whitepapers/connected-finance-reference-architecture/. Accessed 20 July 2021.
23. Cloudera, Hortonworks Data Platform (HDP). https://www.cloudera.com/downloads/hdp.html. Accessed 20 July 2021.

24. Boid, D., & Chang, W. (2018). *NIST big data interoperability framework: Volume 6, RA Version 2*. NIST Big Data Program. https://bigdatawg.nist.gov/_uploadfiles/NIST.SP.1500-6r1.pdf. Accessed 20 July 2021.
25. The Industrial Internet Consortium Reference Architecture, IIRA, v1.9. https://www.iiconsortium.org/IIRA.htm. Accessed 20 July 2021.
26. ISO/IEC/IEEE 42010:2011, Systems and software engineering – Architecture description. https://www.iso.org/standard/50508.html. Accessed 20 July 2021.
27. Kyriazis, D., et al. (2018). *BigDataStack: A holistic data-driven stack for big data applications and operations*. In BigData Congress, pp. 237–241.
28. Guerreiro, G., Costa, R., Figueiras, P., Graça, D., & Jardim-Gonçalves, R. (2019). A self-adapted swarm architecture to handle big data for "factories of the future". *IFAC-Papers OnLine, 52*(13), 916–921, ISSN 2405-8963. https://doi.org/10.1016/j.ifacol.2019.11.356
29. BOOST4.0, D2.5 – BOOST 4.0 Reference Architecture Specification v1. Available at: https://cordis.europa.eu/project/id/780732/results. Accessed 20 July 2021.
30. Raj, A., Bosch, J., Olsson, H. H, & Wang, T. J. (2020). *Modelling data pipelines*. In 2020 46th Euromicro conference on software engineering and advanced applications (SEAA), pp. 13–20. https://doi.org/10.1109/SEAA51224.2020.00014
31. Berthold, M. R., et al. (2008). KNIME: The Konstanz information miner. In C. Preisach, H. Burkhardt, L. Schmidt-Thieme, & R. Decker (Eds.), *Data analysis, machine learning and applications. Studies in classification, data analysis, and knowledge organization*. Springer. https://doi.org/10.1007/978-3-540-78246-9_38
32. Apache Airflow. https://airflow.apache.org/. Accessed 20 July 2021.
33. Luigi. https://github.com/spotify/luigi. Accessed 20 July 2021.
34. Lundberg, S. M., & Lee, S.-I. (2017). *A unified approach to interpreting model predictions*. In Proceedings of the 31st international conference on neural information processing systems (NIPS'17), pp. 4768–4777. Curran Associates Inc.
35. Kruchten, P. (1995). Architectural blueprints – The "4+1" view model of software architecture. *IEEE Software, 12*(6), 42–50.
36. Shearer, C. (2000). The CRISP-DM model: The new blueprint for data mining. *Journal of Data Warehousing, 5*, 13–22.

Chapter 2
Simplifying and Accelerating Data Pipelines in Digital Finance and Insurance Applications

Pavlos Kranas, Diego Burgos, Ricardo Jimenez-Peris, and Juan Mahíllo

1 Introduction

The traditional finance and insurance companies' back ends rely on complex processes to execute their daily activities. Some examples are risk, debt, or credit scoring computation. As companies are more and more data-driven and customer relationships are not just off-line, but omnichannel, other use cases are popping up, such as highly personalized marketing or driving style control based on IoT sensors. The current trend is that organizations make increasing use of all the available information.

Since data volumes become large, data analysis takes longer than the business ideal. To partially speed up processing time, complex data platforms combine different data technologies to compute these operations. Organizations have many different data sources and typically keep many different operational databases, of varying natures, to store their data. This includes both SQL and NoSQL databases. These operational databases only keep recent data, for instance, the data from the last 12 months. On the other hand, they use slower analytical databases such as data warehouses, data marts, and data lakes to perform analytical tasks over this data, such as analysis, reporting, and machine learning. In some cases, they also use data streaming technologies to process data on the fly and compute continuous queries over streaming data. These analytical databases are designed to house large volumes of historical data, typically, data older than 12 months.

Data pipelines combine current and historical data while, at the same time, enriching and transforming this data. These transformations often involve aggregating data to compute KPIs and summarize information for reporting purposes.

P. Kranas (✉) · D. Burgos · R. Jimenez-Peris · J. Mahíllo
LeanXcale SL, Madrid, Spain
e-mail: pavlos@leanxcale.com; diego.burgos@leanxcale.com; rjimenez@leanxcale.com; juan@leanxcale.com

© The Author(s) 2022
J. Soldatos, D. Kyriazis (eds.), *Big Data and Artificial Intelligence in Digital Finance*,
https://doi.org/10.1007/978-3-030-94590-9_2

For these transformation processes, it is a common practice to create intermediate staging databases to enrich and transform the data before generating the targeted final reports or other output.

As a result of this complexity, platform development and maintenance cost increases, which may hurt the process agility needed for the business requirements.

2 Challenges in Data Pipelines in Digital Finance and Insurance

McKinsey has noted that "Yesterday's data architecture can't meet today's need for speed, flexibility and innovation" [1]. The key [...] is the agility. The right architecture provides:

- IT cost savings
- Productivity improvements
- Reduced regulatory and operational risks
- Delivery of new capabilities and services

Furthermore, McKinsey highlights the issues with a complex data architecture, showing that a midsize organization with $5 billion in operating costs ends up spending $250 million on its IT architecture [2].

In what follows, we will briefly review these challenges.

2.1 IT Cost Savings

The cost of operation and support are proportional to processing time. The shorter the processing, the smaller the technical teams' expenses are. This cost structure, where the expenses are proportional to processing time, is exemplified by cloud platforms. Thus, increasing performance implies IT cost reduction.

2.2 Productivity Improvements

The time spent by data analysts and data scientists is split between the time when they manually do work with the data and when computer algorithms do data processing. Intuitively, the productivity of data analysts and data scientists increases as processing time is reduced.

2.3 *Reduced Regulatory and Operational Risks*

Digital finance and insurance companies typically handle the information of millions to hundreds of millions of customers, which results in massive data. In these companies, it is very common to have daily, weekly, and monthly batch processes in order to process such massive data. These processes raise many challenges as they involve ingesting huge amounts of data, while they are limited by a time window. Daily processes normally have to be completed overnight to prevent disturbing the operational databases when extracting the daily updates and before the next business day. Any problem in the night process will make the data unavailable the next day, thereby causing business problems. For this reason, the process should be short enough so it can be repeated in case something goes wrong. For the weekly processes, the time window is the weekend, so the processes take more than one day, and if they fail, they cannot be completed over the weekend. The monthly processes, because they have to deal with one and two orders of magnitude more data than the daily processes, have 1–2 weeks of processing, thus potentially impacting business very seriously.

2.4 *Delivery of New Capabilities and Services*

Competition forces companies to release new services and products that are better and less costly for the customers. Data-driven finance and insurance companies leverage their capability to process more information in less time, thus creating a competitive advantage. Current state-of-the-art data platforms do not offer effective ways for managing all the different data points and datasets, which are typically managed through different types of databases and data stores. Thus, the cost of an opportunity to develop a new product can jeopardize the market share of a company.

After seeing the importance of speeding up these data pipelines, what can be done to accelerate these processes? In order to understand how these processes can be accelerated, we need to understand what takes time in each of their steps. This will be discussed in the following section.

3 Regular Data Pipeline Steps in Digital Finance and Insurance

Banking and insurance data pipelines typically consist of three steps: data intaking, data transformation, and generating the required output.

3.1   Data Intaking

The first step is the ingestion of data from the relevant data sources, typically stored in operational databases. This data ingestion is massive and takes many hours, a significant amount of time. The challenge here is to be able to accelerate data ingestion without sacrificing the querying speed that becomes crucial for the next steps. A NoSQL solution, in particular, a key-value data store, can accelerate data ingestion between one and two orders of magnitude. But why such acceleration? To answer that question, we need to dig into how traditional SQL and NoSQL solutions ingest data.

Traditional SQL solutions are based on a persistent data structure, called B+ tree, to store database tables. A B+ tree is a search tree that enables data to be found with logarithmic complexity. Figure 2.1 illustrates how this tree works. The data that can be reached from the child nodes is split across subtrees. The keys in the node tell how the data is split across subtrees that provide the split points. There are three children. The first one keeps all the rows with primary keys lower than k1, the last one those higher than or equal to k2, and the middle one the rest of the rows. Since the B+ tree is a balanced tree, it is guaranteed that the row with the search primary key will be accessed in a logarithmic number of accessed nodes, as illustrated in Fig. 2.1.

B+ trees store data in the leaf nodes. They also implement a read-write cache of tree blocks, an LRU (least recently used) cache, which is illustrated in Fig. 2.2. This cache typically keeps the upper part of the tree.

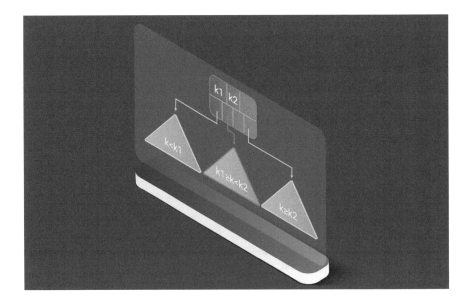

Fig. 2.1 Logarithmic search in a B+ tree

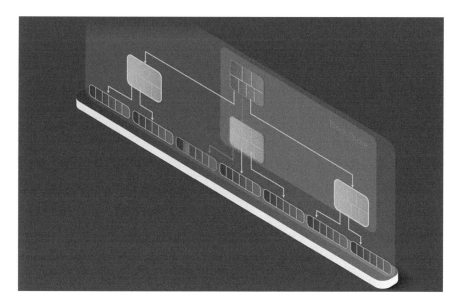

Fig. 2.2 B+ tree and associated block cache

When inserting rows in a scattered manner, the leaf node needs to be read from disk, which means that a block has to be evicted from the cache and written to disk (with high ingestion, all blocks are dirty) and then the leaf node is read from disk. In large tables, inserting a row will typically mean reading multiple blocks from disk and evicting multiple blocks from the cache, since several levels will be missing. Therefore, the cost of inserting one row will be that of reading and writing one or more blocks, i.e., two or more IOs per row, which is very expensive.

The fastest NoSQL databases to ingest data are key-value data stores and wide column databases. Hash-based solutions do not support even simple range queries so are unusable for the vast majority of data pipeline use cases. Other solutions are based on the approach of string sorted tables (SSTables), mimicking what is done in Google Big Table. This approach relies on storing inserts and updates into a cache (see Fig. 2.3). This cache is typically based on a skip list, which is a kind of probabilistic search tree. When this cache reaches a threshold, it is serialized into a buffer sorted by the primary key (hence the name string sorted table, since the key is considered just a string, actually an array of bytes), and the buffer is written into a file. This process is repeated. As can be seen in Fig. 2.3, every block that is written carries as many rows as they fit in the block, so it is impossible to be more efficient in writing. However, there is an important trade-off. Reads now require scanning all the files related to the range that is being queried. The aforementioned process typically results in tens of files, which makes reading one to two orders of magnitude more expensive. To avoid further degradation of reads, there is also a compaction process that merges the files related to the same range. This compaction process is typically very intrusive and is often a stop-the-world process.

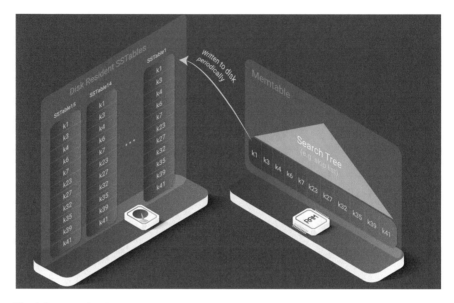

Fig. 2.3 Key-value data stores based on SSTables

In traditional SQL databases, queries are quite efficient since the read of the first row is logarithmic, thanks to the search tree nature of the B+ tree. Then, from there, the next read rows are just consecutive reads. Thus, range queries in SQL databases are substantially more efficient than in NoSQL ones.

Dealing with both current and historical data is very challenging because table size increases dramatically, which degrades the performance of ingestion in traditional SQL operational databases. The reason is that as table size grows, ingesting data becomes more and more expensive because there are more levels in the B+ tree and each row costs more and more blocks to be read and evicted from the cache. Finally, rows become more and more scattered, further degrading the efficiency of data ingestion. The behavior is depicted in Fig. 2.4.

3.2   Data Transformation

The second step is the process of analyzing and enriching the data, which typically involves aggregating information to attain KPIs, using aggregation SQL queries. Since they are dealing with large amounts of data, these queries may take a lot of time to be computed since they have to traverse millions to billions of rows. Data warehouses are good at answering these kinds of queries. However, they do not support processes that enrich and transform the data since they do not support updates, only read-only queries.

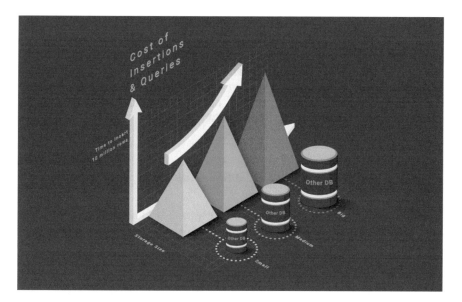

Fig. 2.4 Speed of ingestion with increasing DB sizes

NoSQL databases provide no, or very limited, aggregation support. In hash-based solutions, full scans of the table are necessary since there is no ability of performing range queries. The NoSQL solutions able to perform range queries are mostly based on SSTables, resulting in them being more than one order of magnitude more expensive and, thus, not competitive. Additionally, NoSQL solutions fail to work in the case of concurrent updates of the data due to the lack of ACID properties, so many enrichment processes that require updating data are simply not supported. Doing aggregation in real time with NoSQL does not work since concurrent updates result in the lost update problem. For instance, two concurrent operations trying to add 10 and 20 to an aggregation row with a value of 0 would result in both concurrent operations reading 0, each of them writing 10 and 20, respectively. One would be executed first and the other, second, for instance, the first 10 is written and then 20 is written. The final result would be 20 instead of the expected 30. This lost update problem is due to the lack of ACID properties.

3.3   Generate the Required Output

The third step is to generate reports or output files that are required to perform large simple queries to extract the data of interest. Again, these kinds of queries are not the most appropriate for traditional operational SQL databases since they are optimized for many short queries instead of large analytical queries. NoSQL databases are simply bad for this process since they typically do not support the expressiveness necessary to query and extract the data.

Finally, all these three steps need to be performed at different scales: from small to very large scale. NoSQL databases can scale well the part they manage well, that is, data ingestion, but they do not support other steps. SQL databases either do not scale out (if they are centralized) or scale out logarithmically (the case for shared storage or cluster replication, the two available approaches).

4   How LeanXcale Simplifies and Accelerates Data Pipelines

LeanXcale is a SQL database that can be integrated with all the applications that have been developed over the last 40 years using relational databases, making it simple to transform and use data. However, beyond the regular capabilities of an SQL database, LeanXcale has a set of unique features that makes it optimal for simplifying and accelerating data pipelines. These features have two effects: the speeding up of data insertion, thanks to an efficient key-value store engine and bidimensional partitioning, and the capability to parallelize the process in several execution threads while avoiding database locking due to online aggregation and linear horizontal scalability.

LeanXcale overcomes the challenges of data pipelines based on the features that are detailed in the following paragraphs.

4.1   High Insertion Rates

The first feature addresses the first of the above-listed challenges, which lies in how to be able to efficiently ingest data. The goal is to ingest large amounts of data in times that are at least an order of magnitude lower than those of traditional SQL operational databases. LeanXcale is a NewSQL database, and it adopts an approach similar to LSM trees. To maximize the locality of accesses, LeanXcale also has a cache of updates/inserts like we saw in the SSTable approach adopted by some NoSQL databases. However, instead of writing the updates into SSTable files that highly degrades read performance, LeanXcale propagates the updates into the B+ tree. Since there are many updates/inserts, many of them go to the same leaf node, thus amortizing disk block accesses across many rows instead of a single row, as is the case in traditional SQL databases. It must be noted that the cache of read blocks is kept as in the case of traditional SQL databases. This approach is depicted in Fig. 2.5.

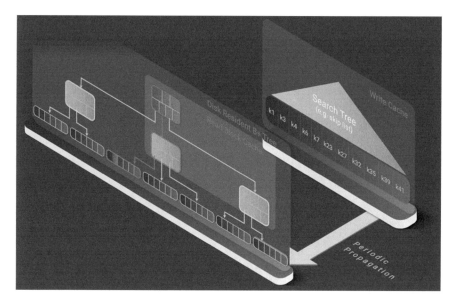

Fig. 2.5 LeanXcale B+ tree and write and read caches

4.2   Bidimensional Partitioning

LeanXcale also deals efficiently with long historical data. For this purpose, it uses a novel mechanism called bidimensional partitioning. The rationale behind this mechanism is that insertions often have some temporal locality around a timestamp column or auto-increment column. This mechanism leverages this observation to split data fragments automatically, so they do not get too big. The fragments contain the rows within a primary key range that is how data within a table is partitioned across nodes as shown in Fig. 2.6.

The auto partitioning is done on a different dimension, namely, time based on the timestamp column as depicted in Fig. 2.7.

With this mechanism, the achieved effect is that the newly data ingested will happen on the new data fragment and the previous data fragment will cool down, get evicted from the cache and written to disk. Thus, only the hot data fragments are kept in memory, and a load that was inherently IO-bound becomes CPU-bound and reaches high efficiency.

4.3   Online Aggregates

As mentioned above, one of the typical tasks in data pipelines is computing aggregates over the ingested data. These aggregates can be very expensive since they

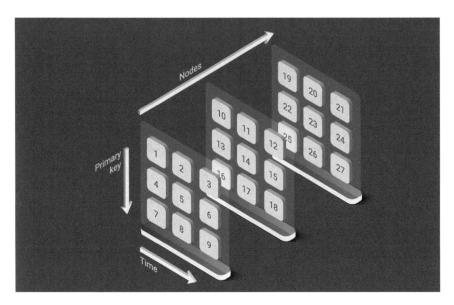

Fig. 2.6 Bidimensional partitioning in a distributed setting

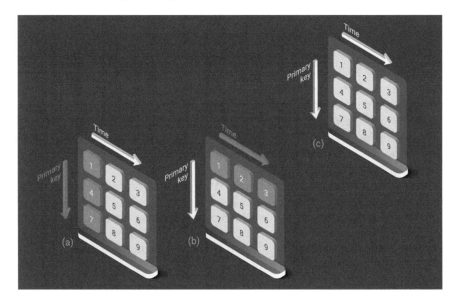

Fig. 2.7 Bidimensional partitioning

are performed over millions to billions of rows. In LeanXcale, we invented a novel mechanism, online aggregates, to alleviate this problem. Online aggregates leverage a new concurrency control method (patent filed by LeanXcale) called semantic multiversion concurrency control. This invention makes it possible to aggregate over a row in real time, with high levels of concurrency (e.g., many thousands per second) and without any contention, while providing full ACID consistency. SQL databases do not do aggregations in real time since they result in high levels of contention due to traditional concurrency control, based on locking, that forces each update of the aggregate to happen one by one. When one update is performed over the aggregation row, until that transaction does not commit, the next update cannot be applied and will be blocked on a lock held over the row.

In LeanXcale, all aggregations over a row do not conflict because we leverage semantic multiversion concurrency control, which prevents any contention. Thus, the aggregations can be computed progressively as data is ingested. The idea is that the aggregate tables are created for the targeted aggregations. When a row is inserted, the aggregations in which it participates are updated by updating the corresponding rows in the relevant aggregate tables. When the data ingestion ends, the aggregations are already computed. Thus, a heavy and long aggregation analytical query in a traditional SQL database is transformed by LeanXcale into a super light query that reads just one value, or a list of them, rendering these queries almost costless.

4.4 Scalability

LeanXcale provides horizontal scalability by scaling the three layers of the database, namely, storage engine, transactional manager, and query engine (see Fig. 2.8). The storage layer is a proprietary technology called KiVi, which is a relational key-value data store. It has the same principles as key-value data stores/wide column data stores and splits data into data fragments according to primary key ranges. However, unlike NoSQL data stores, it implements all relational algebra operators, except joins. This allows be pushed down in the query plan any operator below a join to KiVi. In this way, it is possible to obtain a high level of efficiency in processing SQL, despite the distribution. For instance, let's assume we wanted to compute the sum of a column over a large table with one billion rows that is held across ten servers. In an SQL database, the billion rows would be sent to the query engine where they would be summed. In LeanXcale, each of the ten servers would perform the sum of 1/10th of the rows locally and would send a single message per server to the query engine with the locally aggregated value. In this way, LeanXcale avoids sending billion rows over the network.

The query engine scales out by distributing the database clients across different query engine instances. Each query engine instance takes care of a fraction of the clients and executes the queries sent by them.

The transactional manager is one of the main inventions of LeanXcale. Scaling out transactional management has been unsolved for three decades. LeanXcale solved the problem by decomposing the ACID properties and scaling out each of them in a composable manner. The ACID properties are:

Fig. 2.8 Architecture of
LeanXcale platform

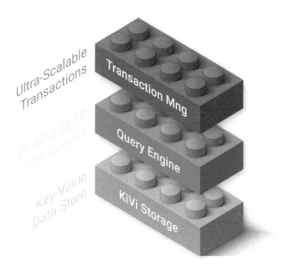

- Atomicity: In the advent of failure, a transaction is all or nothing, that is, all the updates take effect, or it is as if the transaction was not executed.
- Consistency: Transaction code should be correct. If it takes a correct database state, after the transaction the database state should remain correct.
- Isolation: The effect of concurrent execution of transactions should fulfill a correctness criterion, i.e., serialization, which means that it should be equivalent to a serial execution of the transactions.
- Durability: After committing a transaction, its updates cannot be lost, even in the advent of failures.

Traditional SQL operational databases rely on a centralized transactional manager (see Fig. 2.9), which becomes one of the main bottlenecks that prevents scaling out.

In LeanXcale, these ACID properties get decomposed as illustrated in Fig. 2.10.

Consistency is enforced by the application. The database simply helps by providing the ability to set integrity constraints that are automatically checked. Thus, we focus on the other three ACID properties, and we split isolation into the isolation of reads and writes that are attained by means of different mechanisms. Atomicity is scaled out by means of using different transaction manager instances called local transaction managers (LTMs). Each of them handles the lifecycle of a subset of the transactions, and atomicity is enforced as part of this lifecycle. Isolation of writes has to handle write-write conflicts. They are distributed by means of hashing, and each conflict manager handles the conflict management of a set of buckets of primary keys. Durability is scaled out by means of loggers. Each logger handles a subset of the log records, for instance, the log records from a set of LTMs. Isolation of reads is a little bit more involved. We actually use two components, commit sequencer and snapshot server, that do not need to be scaled out due to the tiny

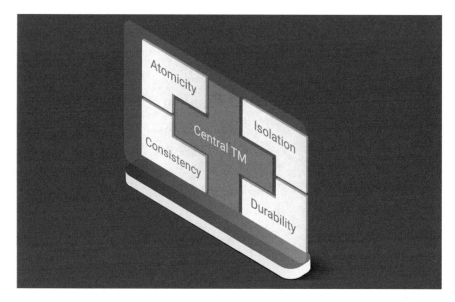

Fig. 2.9 Central transactional manager providing all the ACID properties

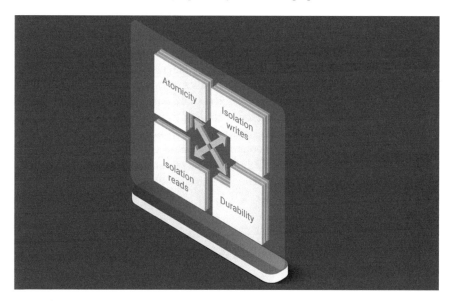

Fig. 2.10 LeanXcale decomposition of the ACID properties and scale out of them

amount of work performed on a per update transaction basis that enables them to handle hundreds of millions of update transactions per second.

LeanXcale avoids the contention during commit management by avoiding committing the transactions sequentially (Fig. 2.11). LeanXcale basically commits all

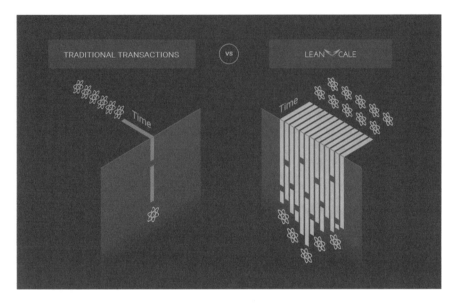

Fig. 2.11 Traditional transactional databases vs. LeanXcale

transactions in parallel without any coordination and thus deals with commitment as an embarrassingly parallel problem. However, there is one aspect from transaction commit that is decoupled from it that lies in the visibility of committed state. Basically, transactions are committed in parallel, but their updates are not made visible right away. Transactions are assigned a commit timestamp (by the commit sequencer). Transactions are serialized according to this order. As far as there are not gaps in the commit order, the state is made visible by advancing the current snapshot. All transactions with a commit timestamp equal to, or lower than, this snapshot are visible. In this way, LeanXcale can commit, in parallel, an arbitrarily high number of commits, which enable it to scale out linearly up to hundreds of nodes.

5   Exploring New Use Cases: The INFINITECH Approach to Data Pipelines

As described above, LeanXcale's features speed up data pipeline time processing, which results in IT cost savings, productivity improvements, and reduced operational risks. Furthermore, LeanXcale may tackle new use cases. LeanXcale participates in the H2020 INFINITECH, as technical leader. Below we discuss the work done in INFINITECH to uncover new ways to satisfy financial and insurance customer and internal needs.

INFINITECH aims to break through the current technology barriers that keep modern AI architectures and solutions in their infancy, as there is no adequate support for real-time business intelligence in the finance and insurance sector. Data tends to be located in isolated sources, and data pipelines are not efficient for real-time analytical processing over datasets that span across different data stores (hot and historical data combined with data streams). This limits the creation of novel business models and their validation via scenarios that could exploit real-time risk assessment algorithms, the detection of fraudulent finance transactions on the fly, or the identification of potential business opportunities.

Toward this aim, INFINITECH is developing a unified data management layer via the use of intelligent data pipelines in order to fulfill the needs of the finance and insurance sector efficiently. This means providing a unified access over fragmented data stored in different silos and allowing for integrated query processing that combines static (or persistently stored) data, with data *in-flight* that comes from streaming sources. This will allow for a cost-effective execution of advanced data mining algorithms and real-time identification of events of interest. Moreover, by combining a multitude of advanced analytic techniques, via the use of its intelligent data pipelines, it facilitates the access and data processing to analytical functions, providing low-latency results that enable the algorithms to provide real-time analytics.

Using the INFINITECH data pipelines, analytic results do not suffer from having to access only an outdated snapshot of the dataset that resides in a data warehouse after the execution of a periodic batch process. This barrier forbids real-time business intelligence. Additionally, INFINITECH aims to provide a streaming processing framework that allows the combination of query executions over both data *at rest* and *in-flight*. As analytical queries are cost-expensive, their typical response times are not suitable for streaming operators, which makes the databases as the bottleneck for such architectures. Moreover, the ingestion of streaming data into persistent data stores cannot be supported by traditional SQL databases, while the use of NoSQL stores would sacrifice transactional semantics. This is not an option when having applications in the insurance and finance domains. The use of the INFINITECH data pipelines removes these existing barriers of current architectures, while it automates the parallelization of data streaming operators. This allows those operators to be dynamically and horizontally scaled at run time, relying on the fact that its data management layer can support the increase of the input workload that might also force it to scale out.

6 Conclusion

Data is the pivot around which financial and insurance companies revolve, in order for them to develop their competitive advantages. To process their ever-increasingly massive volumes of data, these organizations have built data pipelines. Typically, these platforms are not agile enough, even when organizations develop complex and expensive architectures that blend different kinds of data stores. Beyond the benefits that a regular SQL database provides, LeanXcale can reduce data access time independently of data size and allows efficient process parallelization. This combination of capabilities helps to reduce the data pipeline complexity and the total cost of ownership. However, more importantly, it uncovers new ways of generating value with new use cases that were previously not possible.

Acknowledgments The research performed to lead the above results has received funding from the European Union's Horizon 2020 Research and Innovation Programme under INFINITECH project, Grant Agreement No. 825632, and Comunidad de Madrid Industrial Doctorates Grants No. IND2017_TIC-7829 and IND2018_TIC-9846.

References

1. McKinsey. (2020a). *How to build a data architecture*. Retrieved 6 2021, from https://www.mckinsey.com/business-functions/mckinsey-digital/our-insights/how-to-build-a-data-architecture-to-drive-innovation-today-and-tomorrow
2. McKinsey. (2020b). *Reducing data costs*. Retrieved 6 2021, from https://www.mckinsey.com/business-functions/mckinsey-digital/our-insights/reducing-data-costs-without-jeopardizing-growth

Chapter 3
Architectural Patterns for Data Pipelines in Digital Finance and Insurance Applications

Diego Burgos, Pavlos Kranas, Ricardo Jimenez-Peris, and Juan Mahíllo

1 Introduction

1.1 Motivation

Data is the new oil that moves businesses. Banking, financial services, and insurance organization's value proposition are dependent on the information they can process. Companies require to process more data in less time in a more personalized way. However, current database technology has some limitations to provide in a single database engine the required intaking speed, the capacity to transform it in a usable way in the volume a corporation needs.

To overcome these constraints, companies use approaches based on complex platforms that blend several technologies. This complexity means an increment of the total cost of ownership, a longer time to market, and creates long-run friction to adapt to the new business opportunities. According to McKinsey, a midsize organization (between \$5B and \$10B on operating expenses) may spend around \$90M–\$120M to create and maintain these architectures, mainly because of the architecture complexity and data fragmentation. The advice from McKinsey lies in simplifying the manner that financial and insurance organizations use information that will greatly impact the way that a company does business.

McKinsey also notes that a company can reduce in up to 30% of the expenses by simplifying the data architecture, in combination with other activities, such as data infrastructure off-loading, engineer's productivity improvement, and pausing expensive projects.

D. Burgos · P. Kranas (✉) · R. Jimenez-Peris · J. Mahíllo
LeanXcale SL, Madrid, Spain
e-mail: diego.burgos@leanxcale.com; pavlos@leanxcale.com; rjimenez@leanxcale.com; juan@leanxcale.com

J. Soldatos, D. Kyriazis (eds.), *Big Data and Artificial Intelligence in Digital Finance*,
https://doi.org/10.1007/978-3-030-94590-9_3

1.2 Data Pipelining Architectural Pattern Catalogue and How LeanXcale Simplifies All of Them

In the current data management landscape, there are many different families of databases since they have different capabilities. Because of the use of different databases, data needs to be moved from one database into another. This practice is called data pipelining. In the next section, we first do a brief taxonomy of the different kinds of databases and the functions for which they are used, and after, we describe the most common data pipelines.

In the next sections, we discuss groups of data pipelining architectural patterns targeting specific tasks.

At LeanXcale, we are looking at how to simplify these data pipelines and adopt a uniform simple approach for them. Data pipelines get complicated mainly due to the mismatch of capabilities across the different kinds of systems. Data pipelines may get very complex because of real-time requirements. There are many architectural patterns commonly used for solving different data pipeline constraints.

LeanXcale envisions a holistic solution to the issue of data pipelining that ingests data as fast as needed, works with current and historic data, handles efficiently aggregates, and can handle them at any scale. This holistic solution aims at minimizing the TCO (total cost of ownership) of the number of storage systems needed to develop a data pipeline and minimize the duration of the data pipelining or even perform it in real time. In this section, we will address all of the above identified architectural patterns for data pipelining and see how by means of LeanXcale we can highly simplify them.

Finally, we discuss instantiations of these patterns in the context of the INFINITECH European project where a large number of pilots are being run. We also discuss how LeanXcale has been leveraged to simplify these data pipelines in those pilots.

2   A Taxonomy of Databases for Data Pipelining

This section surveys the building blocks of data pipelines, databases, giving a taxonomy at different levels of abstraction, from high-level capabilities operational vs informational to the different flavors of NoSQL databases.

2.1 Database Taxonomy

2.1.1 Operational Databases

Operational databases store data in persistent media (disk). They allow to update the data while the data is being read. The consistency guarantees that are given with concurrent reads and writes vary. Because they can be used for critical mission, operational databases might provide capabilities for attaining high availability that tolerates node failures, and, in some cases, they can even tolerate data center disasters that would otherwise lead to the whole loss or lack of availability of a whole data center. The source of these disasters can be from a natural event like a flood or a fire, the loss of electric power, the loss of Internet connectivity, a denial-of-service attack resulting in the loss of CPU power and/or network bandwidth, the saturation of some critical resource like DNS (domain name service), and more.

2.1.2 Data Warehouses

Data warehouses are informational databases, typically with much bigger persistent storage than operational databases. They are designed only to query data after ingesting it. They do not allow modifications, simply load the data, and then after query the data. They focus on speeding up queries by means of OLAP (online analytical processing) capabilities, attained by introducing intra-query parallelism typically using parallel operators (intra-operator parallelism). They often specialize the storage model to accelerate the analytical queries, e.g., using a columnar model, or an in-memory architecture.

2.1.3 Data Lakes

Data lakes are used as scalable cheap storage where to keep historical data at affordable prices. The motivation of keeping this historical data might be legal requirements of data retention, but more recently the motivation is from the business side to have enough data to be able to train machine learning models in a more effective way by reaching a critical mass of data in terms of time and detail. Some organizations use data lakes as cheap data warehouses when the queries are not especially demanding. A data lake might require more than an order of magnitude higher resources for an analytical query with a target response time higher than a data warehouse, while the price follows an inverse relationship.

2.2 Operational Database Taxonomy

Operational databases can be further divided into three broad categories.

2.2.1   Traditional SQL Databases

Traditional SQL (structured query language) databases are characterized by two main features. First, they provide SQL as query language. Second, they provide the so-called ACID guarantees over the data. ACID (atomicity, consistency, durability) properties will be discussed in detail in the following paragraphs. The main limitation of traditional SQL databases is their scalability: they either do not scale out or only logarithmically, meaning that their cost grows exponentially with the scale of the workload to be processed. They typically provide mechanisms for high availability that guarantee the ACID properties, which is technically known as one-copy consistency guarantees. The second limitation is that they ingest data very inefficiently, so they are not able to insert or update data at high rates.

2.2.2   NoSQL Databases

NoSQL databases have been introduced to overcome traditional SQL database limitations. There are four main kinds of NoSQL databases as explained above. Basically, they address the lack of flexibility of the relational schema that is too rigid and forces to know in advance all the fields of each row in the database, and they are very disruptive when this schema needs to be changed, typically resulting in having the database or at least the involved tables not available during the schema change. NoSQL databases fail to provide ACID consistency guarantees. On the other hand, most of them can scale out, although not all kinds have this ability. Some of them can scale out but either not linearly or not to large numbers of nodes.

2.2.3   NewSQL Databases

NewSQL databases appear as a new approach to address the requirements of SQL databases, trying to remove part or all their limitations. NewSQL databases are designed to bring new capabilities to old traditional SQL databases by leveraging approaches from NoSQL and/or new data warehouse technologies. Some try to improve the scalability of storage. That is normally achieved by relying on some NoSQL technology or adopting an approach similar to some NoSQL databases. Scaling queries was an already solved problem. However, scaling inserts and updates had two problems. The first one is the inefficiency of ingesting data. The second one is the inability to scale out to large scale the ACID properties, that is, transactional management. Others have tried to overcome the lack of scalability of data ingestion, while others address the lack of scalability of transactional management.

2.3 NoSQL Database Taxonomy

NoSQL databases are usually distributed (execute on a shared-nothing cluster with multiple nodes), have different flavors, and are typically divided into four categories that are explained in the following subsections.

2.3.1 Key-Value Data Stores

Key-value data stores are schemaless and allow any value associated with a key. In most cases, they attain linear scalability. Basically, each instance of the key-value data store processes a fraction of the load. Since operations are based on an individual key-value pair, the scalability does not pose any challenge and most of the times is achieved. The schemaless approach provides a lot of flexibility. As a matter of fact, each row can potentially have a totally different schema. Obviously, that is not how key-value data stores are used. But they allow to evolve the schema without any major disruption. Of course, the queries have to do the extra work of being able to understand rows with different schema versions, but since normally, the schemas are additive, they add new columns or new variants, it is easy to handle. Key-value data stores excel at ingesting data very efficiently. Since they are schemaless, they can just store the data as is. This is very inefficient for querying, and this is why they provide very little capabilities for querying such as getting the value associated with a key. In most cases, they are based on hashing, so they are unable to perform basic range scans and only provide full scans that are very expensive since they traverse all the table rows. Example of key-value data stores are Cassandra and DynamoDB.

2.3.2 Document-Oriented Databases

Document-oriented databases are technologies that support semi-structured data written in a popular language such as JSON or XML. Their main capability is being able to store data in one of these languages efficiently and perform queries for these data in an effective way. Representing these data in a relational database is just a nightmare and doing queries of this relational schema even a worse nightmare. This is why they have succeeded. Some of them scale out in a limited way and not linearly, while some others do better and scale out linearly. However, they do not support the ACID properties and are inefficient at querying data that is structured in nature. Structured data can be queried one to two orders of magnitude more efficiently with SQL databases. Examples in this category are MongoDB and Couchbase.

2.3.3 Graph Databases

Graph databases are specialized on storing and querying data modeled as a graph. Graph data represented in a relational format becomes very expensive to query. The reason is that to traverse a path from a given vertex in the graph, one has to perform many queries, one per edge stemming from the vertex and as many times as the longest path sought in the graph. This results in too many invocations to the database. If the graph does not fit in memory, then it is even a bigger problem since disk accesses will be involved for most of the queries. Also, the queries cannot be programmed in SQL and have to be performed programmatically. Graph databases, on the other hand, have a query language in which with a single invocation solves the problem. Data is stored to maximize locality of a vertex with contiguous vertexes. However, when *they do not fit in a single node, graph databases start suffering from the same problem* when they become distributed by losing their efficiency and any performance gain as the system grows. At some point, a relational schema solution becomes more efficient than the graph solution for a large number of nodes. A widely used graph database is Neo4J.

2.3.4 Wide-Column Data Stores

Wide-column data stores provide more capabilities than key-value data stores. They typically perform range partitioning, thus supporting range queries. In fact, they might support some basic filtering. They are still schemaless. They also support vertical partitioning that can be convenient when the number of columns is very high. They have some notion of schema, yet quite flexible. Examples of this kind of data stores are Bigtable and HBase.

3 Architectural Patterns Dealing with Current and Historical Data

In this section, an overview of two common patterns used to deal with current and historical data is given.

3.1 Lambda Architecture

The lambda architecture combines techniques from batch processing with data streaming to be able to process data in a real time. The lambda architecture is motivated by the lack of scalability of operational SQL databases. The architecture consists of three layers as illustrated in Fig.3.1.

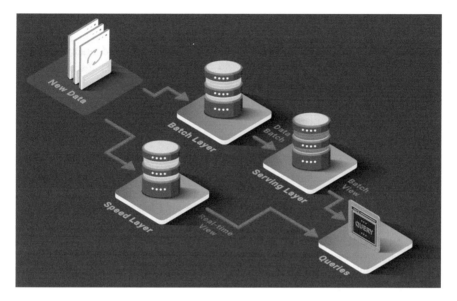

Fig. 3.1 Lambda architecture

1. Batch Layer

This layer is based on append-only storage, typically a data lake, such as the ones based on HDFS. Then, it relies on MapReduce for processing new batches of data in the forms of files. This batch layer provides a view in a read-only database. Depending on the problem being solved, the output might need to fully recompute all the data to be accurate. After each iteration, a new view of the current data is provided. This approach is quite inefficient but solves a scalability problem that used to have no solution, the processing of tweets in Twitter.

2. Speed Layer

This layer is based on data streaming. In the original system at Twitter, it was accomplished by the Storm data streaming engine. It basically processes new data to complement the batch view with the most recent data. This layer does not aim accuracy, but to provide more recent data to the global view achieved with the architecture.

3. Serving Layer

This layer processes the queries over the views provided by both the batch and speed layers. Batch views are indexed to be able to answer queries with low response times and combine them with the real-time view to provide the answer to the query, combining both real-time data and historical data. This layer typically uses some key-value data store to implement the indexes over the batch views.

The main shortcoming of the lambda architecture is its complexity and the need to have totally different code bases for each layer that have to be coordinated to be fully synchronized. Maintenance of the platform is very hard since debugging implies understanding the different layers that typically rely on totally different natures, technologies, and approaches.

3.2   Beyond Lambda Architecture

By means of LeanXcale, the lambda architecture is totally trivialized by substituting the three data management technologies and three different code bases with ad hoc code for each of the queries with a single database manager with declarative queries in SQL. In other words, the lambda architecture is simply substituted by the LeanXcale database that provides all the capabilities of the lambda architecture without any of its complexities and development and maintenance cost. LeanXcale scales out linearly operational storage, thus solving one of the key shortcomings of operational databases that motivate the lambda architecture.

The second obstacle from operational databases was its inefficiency in ingesting data that makes them too expensive even for data ingestions they can manage. As the database grows, the cache is rendered ineffective, and each insert requires to read a leaf node from the B+ tree that requires first to evict a node from the cache and write it to disk. This means that every write requires at least two IOs. LeanXcale solves this issue by providing the efficiency of key-value data stores in ingesting data, thanks to the blending of SQL and NoSQL capabilities using a new variant of LSM trees. With this approach, updates and inserts are cached in an in-memory search tree and periodically propagated all together to the persisted B+ tree. With this approach, the locality of updates and inserts on each leaf of the B+ tree is greatly increased amortizing the cost of each IO among many rows.

The third issue solved by LeanXcale that is not solved by the lambda architecture is the ease to query. The lambda architecture requires to develop programmatically each query with three different code bases for each of the three layers. In LeanXcale, queries are simply written in SQL. SQL queries are automatically optimized unlike the programmatic queries in the lambda architecture that require manual optimization across three different code bases for each of the layers. The fourth issue that is solved is the cost of recurrent aggregation queries. In the lambda architecture, this issue is typically solved in the speed layer using data streaming. In LeanXcale, the online aggregates enable real-time aggregation without the problems of operational databases and provide a low-cost solution with low response time.

3.3 Current Historical Data Splitting

Other more traditional architectures are based on combining an operational database with a data warehouse as shown in Fig. 3.2. The operational database deals with more recent data, while the data warehouse deals with historical data. In this architecture, queries can only see either the recent data or historical data, but not a combination of both as shown in the lambda architecture. In this architecture, there is a periodic process that copies data from the operational database into the data warehouse. This periodic process has to be performed very carefully since it can hamper the quality of service of the operational database. This periodic process is most of the time achieved by ETL tools. This process is typically performed over the weekends in businesses where their main workload comes during weekdays.

Another problem with this architecture is that the data warehouse typically cannot be queried while it is being loaded, at least the tables that are being loaded. This forces to split the time of the data warehouse into loading and processing. When the loading process is daily, finally the day is split into loading and processing. The processing time consumes a fraction of hours of the day that depends on the analytical queries that have to be answered daily. It leaves a window of time for loading data that is the remaining hours of the day. At some point, data warehouses cannot ingest more data because the loading window is exhausted. We call this architectural pattern current historical data splitting.

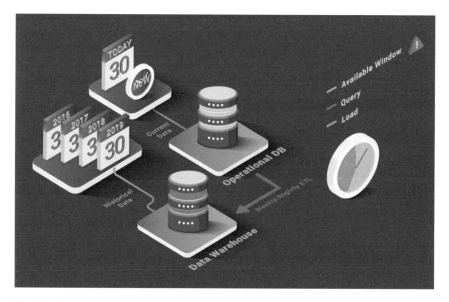

Fig. 3.2 Current historical data splitting architectural pattern

3.4 *From Current Historical Data Splitting to Real-Time Data Warehousing*

In this pattern, data is split between an operational database and a data warehouse or a data lake. The current data is kept on the operational database and historic data in the data warehouse or data lake. However, queries across all the data are not supported with this architectural pattern. With LeanXcale, a new pattern, called *real-time data warehousing*, will be used to solve this problem. This pattern will be solved by an innovation that will be introduced in LeanXcale, namely, the ability to split analytical queries over LeanXcale and an external data warehouse. Basically, it will copy older fragments of data into the data warehouse periodically. LeanXcale will keep the recent data and some of the more recent historical data. The data warehouse will keep only historical data. Queries over recent data will be solved by LeanXcale, and queries over historical data will be solved by the data warehouse. Queries across both kinds of data will be solved using a federated query approach leveraging LeanXcale capabilities to query across different databases and innovative techniques for join optimization. In this way, the bulk of the historical data query is performed by the data warehouse, while the rest of the query is performed by LeanXcale. This approach enables to deliver real-time queries over both recent and historical data giving a 360° view of the data.

4 Architectural Patterns for Off-Loading Critical Databases

4.1 *Data Warehouse Off-Loading*

Since the saturation of the data warehouse is a common problem, another architectural pattern has been devised to deal with the so-called data warehouse off-loading (Fig. 3.3). This pattern relies in creating small views of the data contained by the data warehouse and stores them on independent databases, typically called data marts. Depending on the size of the data and the complexity of the queries, data marts can be handled by operational SQL databases, or they might need a data manager with OLAP capabilities that might be another data warehouse or a data lake plus an OLAP engine that works over data lakes.

4.2 *Simplifying Data Warehouse Off-Loading*

Data warehouse off-loading is typically motivated either to cost reasons or due to the saturation of the data warehouse. For data warehouse off-loading, data marts are used using other database managers and making a more complex architecture that requires multiple ETLs and copies of the data. With LeanXcale, this issue can

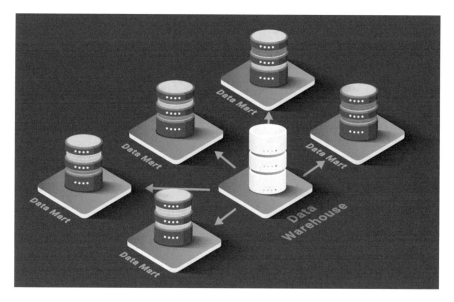

Fig. 3.3 Data warehouse off-loading architectural pattern

be solved into two ways. One way is to use operational database off-loading to
LeanXcale with the dataset of the data mart. The advantage of this approach with
respect to data warehouse off-loading is that the data mart contains data that is real
time, instead of obsolete data copied via a periodic ETL. The second way is to use
database snapshotting by taking advantage of the fast speed and high efficiency of
loading of LeanXcale. Thus, periodically a snapshot of the data would be stored
in LeanXcale with the same or higher freshness than a data mart would have. The
advantage is that the copy would come directly from the operational database instead
of coming from the data warehouse, thus resulting in fresher data.

4.3   Operational Database Off-Loading

As deeply explained, there are cases with real-time or quasi-real-time requirements,
where the database snapshotting pattern does not solve the problem. In this case, a
CDC (Change Data Capture) system is used that captures changes in the operational
data and inject them into another operational database. The CDC is only applied
over the fraction of the data that will be processed by the other operational database.
The workload is not performed over the operational database due to technical or
financial reasons. The technical reason is that the operational database cannot handle
the full workload and some processes need to be off-loaded to another database. The
financial reason is that the operational database can handle the workload but the

Fig. 3.4 Operational database off-loading architectural pattern

price, typically with a mainframe, is very high. This architectural pattern is called operational database off-loading and is illustrated in Fig. 3.4.

4.4   Operational Database Off-Loading at Any Scale

One of the main limitations of operational database off-loading is that only a fraction of data can be off-loaded to a single database due to the lack of scalability of traditional SQL operational databases. Typically, this approach is adopted to reduce the financial cost of processing by mainframes that can process very high workloads. However, traditional SQL database with much more limited capacity and little scalability limit what can be done by means of this pattern. LeanXcale can even support the full set of changes performed over the mainframe, thanks to its scalability, so it does not set any limitation on the data size and rate of data updates/inserts.

4.5   Database Snapshotting

In some cases, the problem is that the operational database cannot handle the whole workload due to its lack of scalability and part of this workload can be performed without being real time. In these cases, a copy of the database or the relevant part of

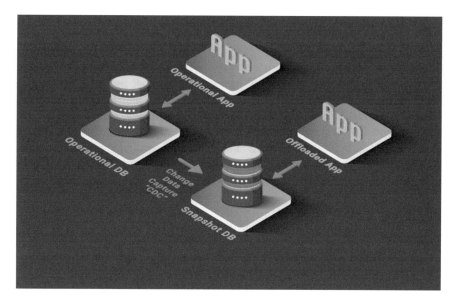

Fig. 3.5 Database snapshotting architectural pattern

the data of the database is copied into another operational database during the time that the operational is not being used, normally, weekends or nights, depending on how long the copy of the database takes. If the copy process takes less than one night, then it is performed daily. If the copy process takes more than one day, then it is performed during the weekend. Finally, if the copy process takes more than weekend, it cannot be done with this architectural pattern. This architectural pattern is called database snapshotting and is depicted in Fig. 3.5.

4.6   Accelerating Database Snapshotting

The LeanXcale can avoid the database snapshotting if it is used as operational database. This can be done thanks to the linear scalability of LeanXcale that does not require off-loading part of the workload to other databases. However, in many cases, organizations are not ready to migrate their operational database because of the large amount of code relying on specific features of the underlying database. This is the case with mainframes with large COBOL programs and batch programs in JCL (job control language). In this case, one can rely on LeanXcale to provide a more effective snapshotting or even substitute snapshotting by operational database off-loading, thus with real-time data. In the case of snapshotting, thanks to the efficiency and speed of data ingestion of LeanXcale, snapshotting can be performed daily instead of weekly since load processes that used to take days are reduced to minutes. The main benefit is that data freshness changes from weekly to real time. This speed in ingestion is achieved thanks to LeanXcale capability of ingesting and querying data with the same efficiency independent of the dataset size. This is achieved by means of bidimensional partitioning. The bidimensional partitioning exploits the timestamp in the key of historical data to partition tables on a second dimension. Tables in LeanXcale are partitioned horizontally through the primary key. But then, they are automatically split on the time dimension (or an auto-increment key, whatever is available) to guarantee that the table partition fits in memory and thus the load becomes CPU-bound, which is very fast. Traditional SQL databases get slower as data grows since the B+ tree used to store data becomes bigger in both number of levels and number of nodes. Thanks to bidimensional partitioning, LeanXcale keeps the time to ingest data constant. Queries are also speeded up thanks to the parallelization of all algebraic operators (intra-operator parallelism) below joins.

5   Architectural Patterns Dealing with Aggregations

5.1   In-Memory Application Aggregation

Other systems tackle the previous problem of recurrent aggregate queries by computing the aggregates on the application side in memory. These in-memory aggregates are computed and maintained as time progresses. The recurrent aggregation queries are solved by reading the in-memory aggregations, while access to the detailed data is solved by reading from the operational database, typically using sharding. This pattern is depicted in Fig. 3.6 and is called in-memory application aggregation.

5.2   From In-Memory Application Aggregation to Online Aggregation

LeanXcale does not need in-memory application aggregations while allowing to remove all the problems around like the loss of data in the advent of failures and more importantly all the development and maintenance cost of the code required to perform the in-memory aggregations. In-memory aggregations work as far they can be computed in a single node. However, when multiple nodes are required, they become extremely complex and often out of reach of technical teams. Leveraging the online aggregates from LeanXcale will be leveraged to compute the aggregations for recurrent aggregation queries. LeanXcale keeps internally the relationship between tables (called parent tables) and aggregate tables built from the inserts in these tables (called child aggregate tables). When aggregation queries are issued, the query optimizer uses new rules to automatically detect which aggregations on the parent table can be accelerated by using the aggregations in the child aggregate

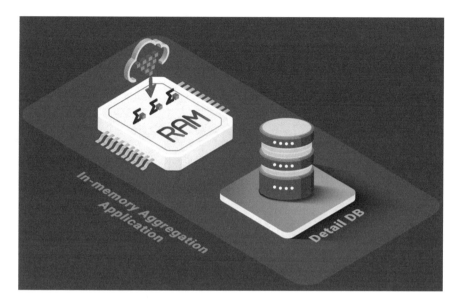

Fig. 3.6 In-memory application aggregation

table. This results in transparent improvement of all aggregations in the parent table by simply declaring a child aggregate table (obviously of the ones that can exploit the child table aggregates).

5.3 Detail-Aggregate View Splitting

A typical and important workload is to ingest high volumes of detailed data and compute recurrent aggregate analytical queries over this data. This workload has been addressed with more specific architectures. The architectural pattern uses sharding to store fractions of the detailed data and then a federator at the application level that basically queries the individual sharded database managers to get the result sets of the individual aggregate queries and then aggregate them manually to compute the aggregate query over the logical database. These aggregated views are generated periodically by means of an ETL (extract, transform, load) process that traverses the data from the previous period in the detailed operational database, computes the aggregations, and stores them in the aggregate operational database. The recurrent queries are processed over the aggregation database. Since the database contains already the pre-computed aggregates, the queries are light enough to be computed at an operational database. This architectural pattern is called detail-aggregate view splitting and is depicted in Fig. 3.7. One of the main shortcomings of this architectural patterns is that the aggregate queries have an obsolete view of

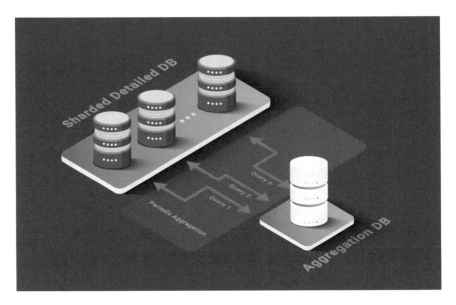

Fig. 3.7 Detail-aggregate view splitting architectural pattern

the data since they miss the data from the last period. Typical period lengths go from 15 min to hours or a full day.

5.4 Avoiding Detail-Aggregate View Splitting

LeanXcale does not need detail-aggregate view pattern. As a matter of fact, by taking advantage of LeanXcale's online aggregates, aggregate tables are built incrementally as data is inserted. This implies to increase the cost of ingestion, but since LeanXcale is more than one order of magnitude more efficient than the market operational SQL database leader, it can still ingest the data more efficiently despite the online aggregation. Then, recurrent aggregation analytical queries become costless since they only have to read a single row or a bunch of rows to provide the answer since each aggregation has been already computed incrementally.

6 Architectural Patterns Dealing with Scalability

6.1 Database Sharding

Databases typically used for the above architecture are SQL operational databases, and since they do not scale, they require to use an additional architectural pattern

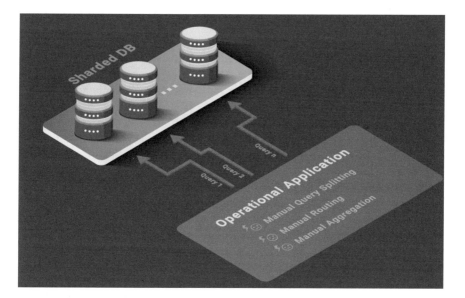

Fig. 3.8 Database sharding

which is called database sharding (Fig. 3.8). Sharding overcomes the lack of scalability or linear scalability of an operational database by storing fractions of the data on different database-independent servers. Thus, each database server handles a workload small enough, and by aggregating the power of many different database manager instances, the system can scale. The main shortcoming of this architecture is that now queries cannot be performed over the logical database, since each database manager instance only knows about the fraction of data is storing and cannot query any other data. Another major shortcoming is that there are no transactions across database instances; they do not have consistency guarantees neither in the advent of concurrent reads and writes nor in the advent of failures.

6.2 Removing Database Sharding

LeanXcale does not need the database sharding architectural pattern (Fig. 3.8) thanks to its linear scalability. Thus, what used to require programmatically splitting the data ingestion and queries across independent database instances is not needed anymore. LeanXcale can scale out linearly to hundreds of nodes.

7   Data Pipelining in INFINITECH

Modern applications being currently used by data-driven organizations, such as those belonging to the finance and insurance sector, require processing data streams along with data persistently stored in a database. A key requirement for such organizations is that the processing must take place in real time providing real-time results, alerts, or notifications in order, for instance, to detect fraud finance transactions the moment they are being occurred, detect possible indications for money laundering, or provide real-time risk assessment among other needs. Toward this direction, streaming processing frameworks have been used during the last decade in order to process streaming data coming from various sources, in combination with data persistently stored in a database that can be considered as data at rest. However, processing data at rest introduces an inherit significant latency, as data access involves expensive I/O operations, which are not suitable for streaming processing. Due to this, various architectural designs have been proposed and are used in the modern landscape that deals with such problems. They tend to formulate data pipelines, moving data from different sources to other data management systems, in order to allow for an efficient processing in real time. However, they are far from being considered as intelligent, and each of the proposed approaches comes with their own barriers and drawbacks, as it has been widely explained in the previous section.

A second key requirement for data-driven organizations in finance and insurance sector is to be able to cope with diverse workloads and continue to provide results in real time even when there is a burst of incoming data load from a stream. This might happen in case of having a stream consuming data feeds from social media in order to perform a sentiment analysis and an important event or incident takes place which will make the social community to response by posting an increased number of tweets or articles. Another example is the unexpected currency devaluation that will most likely trigger numerous of finance transactions with people and organizations change their currencies. The problem with the current landscape is that modern streaming processing frameworks allow for static deployments of data streams that consist of several operators, in order to serve an expected input workload. To make things worse, data management solutions used in such scenarios are difficult to scale out to support an increased load. In order to cope with this, architectural solutions as the ones described previously are being adopted, with all the inherit drawbacks and technology barriers.

In order to cope with those two requirements and overcome the current barriers of the modern landscape, we envision the INFINITECH approach for intelligent data pipelines that can be further exploited by its parallelized data stream processing framework. In INFINITECH, we provide a holistic approach for data pipelines that relies on the key innovations and technologies provided by LeanXcale. The INFINITECH intelligent data pipelines break through the current technological barriers when having to deal with the different types of storage, use of different types of databases for persistently store data and allowing for efficient query processing, handling aggregates, and dealing with snapshots of data. By having LeanXcale as the base for the INFINITECH intelligent data pipelines, we can solve the problem of data ingestion in very high rates, removing the need for database off-loading. Moreover, the online aggregates remove all issues when having to pre-calculate the results of complex analytical queries, which lead to inconsistent and obsolete results. The integration of LeanXcale with Apache Flink streaming processing framework and with tools for Change Data Capture (CDC) enables the deployment of such intelligent data pipelines. Finally, our solution allows for parallelized data stream processing, via the ability for the deployed operators to save and restore their state, thus allowing for online reconfiguration of the streaming clusters, which enables elastic scalability by programmatically scaling those clusters.

In what follows, we describe the use cases of data pipelining in INFINITECH, which data pipeline patterns they use, and how they have been improved by means of LeanXcale.

The current historical data splitting is an architectural pattern that in INFINITECH is being used by a use case handling real-world data for novel health insurance products, where data is split between an operational database and a data warehouse. The main issue of this approach is that analytic algorithms can only rely on historical data moved and cannot provide real-time business intelligence, as analysis is always performed on obsolete datasets. With LeanXcale now, the AI tools of this insurance pilot can perform the analysis on real-time data that combines both the current and historical data.

The data warehouse off-loading architectural pattern is being used by INFINITECH a use case from Bank of Cyprus related to business financial management that delivers smart business advice. The data stored in its main data warehouse are being periodically moved to other data marts, and the analytical queries used to provide smart business advice are targeting the data marts which contain small views of the overall schema. The main drawback with this approach is having to maintain several data stores at the same time. In this use case, a single database with all the information from data marts is being held by LeanXcale, thus avoiding the ETL processes between data warehouse and data marts and avoiding storing some information multiple times. At the same time, the analytical processing now will access more recent data since it can be updated as frequently as needed, even in real time.

There are several use cases in INFINITECH that rely on the database snapshotting architectural pattern. One of them is from the National Bank of Greece that provides personalized investment portfolio management for their retailed customers, and it is periodically copying parts from their entire operational database into different databases. Their analytical algorithms make use of a snapshot of this operational database. Since the snapshot is not updated very frequently, it prevents them from performing real-time business intelligence. By introducing LeanXcale, the personalized investment portfolio management provided by the National Bank of Greece can have a database snapshot that is updated as frequently as needed and even in real time using the CDC architectural pattern. That way, the analysis can be as close to real time as they need or even in real time.

The operational DB off-loading architectural pattern is being used in an INFINITECH pilot related to personalized insurance products based on IoT data from connected vehicles. Data ingested from the IoT sensors cannot be processed in the operational database. After raw data has been ingested, they are preprocessed and further stored into an additional database that is accessed by the AI algorithms to prevent interference from the AI processes with the operational database. This pilot is now using LeanXcale as its only database, removing the need to maintain different data management technologies and all data accesses, either for data ingestion or for analytical processing can target a single instance.

Regarding database sharding, there are several use cases in the use case organizations participating in INFINITECH, but since the focus is on analytical pipelines, none of the targeted use cases is actually relying on this pattern.

INFINITECH has scenarios used for real-time identification of financial crime and anti-money laundering supervision from CaixaBank in Spain and Central Bank of Slovenia that require to compute aggregated data over streaming data. The drawback of this architectural approach is that the result of the aggregated data is stale as it has been calculated in point of time previously than the time of the current value. LeanXcale is being used for its online aggregations to solve this issue and have real-time aggregations that are fully persisted.

In INFINITECH, the detail-aggregate view splitting architectural pattern was initially used for real-time risk assessment in investing banking, implemented by the JRC Capital Management Consultancy. The real-time risk assessment solution provided by the JRC Capital Management Consultancy has been evolved using LeanXcale. Now the incoming data stream is being ingested into LeanXcale, exploiting its ability to support data ingestion in very high rates, while on the other hand, its online aggregates are being used by its AI algorithms to retrieve aggregated results with a latency of milliseconds. Also, the real-time identification of financial crime and the anti-money laundering supervision use cases are benefiting from using LeanXcale and its online aggregates. Both scenarios implement detection over real-time data that is also updated in real time.

8   Conclusions

Financial services and insurance companies are in a race trying to be more efficient in processing all the available information. Despite the popularity of a wide set of highly specialized data engines for specific challenges, none of them solved more of the frequent use cases independently. A group of complex architectural patterns blending different kinds of databases have emerged to solve the most common situations. Nevertheless, they highly increase the total cost of ownership, mainly due to their complexity. They also reduce the business value since most of them result in exploiting stale data from batch processes performed weekly or daily.

LeanXcale, thanks to its disruptive technology, leads to simpler architectures with more data freshness or even real-time data. These architectures speed up the development process due to their simplicity, reducing the time between requirement collection and idea inception to production-ready software. They are more afford-able to maintain since fewer specialists, different servers, and database licenses are required. Additionally, their processing capabilities provide a more agile way to speed up business processes while reducing operational risk. Finally, these architectures can really support the creation of new revenue streams by building new ways to satisfy customer needs, by using the oil of nowadays: the data.

Acknowledgments The research performed to lead the above results has received funding from the European Union's Horizon 2020 Research and Innovation Programme under INFINITECH project, Grant Agreement No. 825632, and Comunidad de Madrid Industrial Doctorates Grants No. IND2017_TIC-7829 and IND2018_TIC-9846.

Chapter 4
Semantic Interoperability Framework for Digital Finance Applications

Giovanni Di Orio, Guilherme Brito, and Pedro Maló

1 Introduction

Data are today the most valuable and precious asset within organizations; however, they also represent an ongoing challenge for businesses since data is continually growing in variety, complexity, as well as fragmentation. As a matter of fact, most of the data collected and possessed by financial organizations reside in a wide array of "siloed" (i.e., fragmented) and heterogeneous systems and databases. Online Transaction Processing (OLTP) databases, Online Analytical Processing (OLAP) databases, data warehouses, and data lakes are only few examples within the data landscape. Furthermore, intensive and heavy data consumption tasks are usually performed over OLAP systems, which lead financial organizations in transferring data from OLTP, data lakes, and other systems to OLAP systems based on intrusive and expensive extract-transform-load (ETL) processes. In several cases, ETLs consume 75–80% of the budget allocated to data analytics while being a setback to seamless interoperability across different data systems using up-to-date data. Beyond the lack of integrated OLTP and OLAP processes, financial and insurance organizations have no unified way of accessing and querying vast amounts of structured, unstructured, and semi-structured data, which increases the effort and cost that are associated with the development of big data analytics and artificial intelligence (AI) systems. Except for data fragmentation, there is also a lack of interoperability across diverse datasets that refer to the same data entities with similar semantics. This is a main obstacle to datasets sharing across different stakeholders and to enabling more connected applications and services that span multiple systems across the financial supply chain. The impacts these aspects

G. Di Orio (✉) · G. Brito · P. Maló
NOVA School of Science and Technology, Caparica, Portugal
e-mail: gido@uninova.pt; guilherme.brito@uninova.pt; pmm@uninova.pt

67
J. Soldatos, D. Kyriazis (eds.), *Big Data and Artificial Intelligence in Digital Finance*,
https://doi.org/10.1007/978-3-030-94590-9_4

have on data management are huge and are forcing both financial and insurance organizations to research, rethink, and apply new strategies and approaches for data management.

Data management can be defined as [1] "the development, execution, and supervision of plans, policies, programs, and practices that deliver, control, protect, and enhance the value of data and information assets throughout their life cycles." The inclusion of these disciplines triggers a transformation process that allows organizations to become data-driven [2], i.e., to ensure the right usage of data at the right time to support operational business intelligence (BI). The activities related to data management span multiple areas from data architecture and data modeling and design to data governance and security passing through data interoperability and data persistence operations. This chapter focuses on the data integration, data interoperability, and data modeling aspects of data management.

2   Background: Relevant Concepts and Definitions for the INFINITECH Semantic Interoperability Framework

2.1   Interoperability

There is no unique definition of interoperability in the literature since the concept has different meanings depending on the context. As a matter of fact, according to ISO/IEC 2382-01 [3], interoperability is defined as "The capability to communicate, execute program, or transfer data among various functional units in a manner that requires the user to have little or no knowledge of the unique characteristics of those units." According to ETSI's technical committee TISPAN [4], interoperability is "the ability of equipment from different manufacturers (or different systems) to communicate together on the same infrastructure (same system), or on another." Moreover, EICTA defines interoperability as [5] "the ability of two or more networks, systems, devices, applications or components to exchange information between them and to use the information so exchanged." Based on these definitions, interoperability is always about making sure that systems are capable of sharing data between each other and of understanding the exchanged data [6]. In this context, the word "understand" includes the content, the format, as well as the semantics of the exchanged data [7]. Interoperability ranges over four different levels [8], namely:

1. Physical/technical interoperability: It is concerned with the physical connection of hardware and software platforms.
2. Syntactical interoperability: It is concerned with data format, i.e., it relates on how the data are structured.
3. Semantic interoperability: It is concerned with the meaningful interaction between systems, devices, components, and/or applications.
4. Organizational interoperability: It is concerned with the way organizations share data and information (Fig. 4.1).

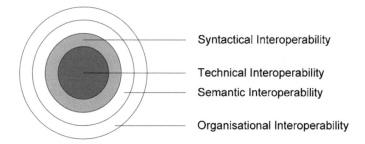

Syntactical Interoperability

Technical Interoperability

Semantic Interoperability

Organisational Interoperability

Fig. 4.1 Different interoperability levels according to [8]

2.1.1   Semantic Interoperability

Semantics plays a main role in interoperability for ensuring that exchanged information between counterparts are provided with sense. For computer systems, this notion of semantic interoperability translates in the ability of two or more systems to exchange data between them, by means of precise unambiguous and shared meaning, which enables readily access and reuse of the exchanged data. The concept of Semantic Web [9] was introduced by World Wide Web (WWW) founder Tim Berners-Lee in the 1990s and has been widely used in research and industry contexts. It has also given rise to the development of the concepts of Semantic Web services and more recently of Semantic Internet of Things (IoT) concepts [10–12]. These concepts aim to facilitate collaboration across semantically heterogeneous environments toward contributing to a connected world of consuming and provisioning devices. This connected world can potentially exchange and combine data to offer new or augmented services. However, the accomplishment of this vision is associated with several challenges due to the existence of a variety of standards, legacy systems constraints, and tools. The semantic interoperability process can, therefore, focus on different viewpoints of semantic aspects, such as the exchanged data description or the systems interaction terms. As a prominent example, in IoT systems, the interoperability specifications can be used to define the meaning of a given sensor or IoT device, but also to provide information on the units of its value and on the protocols that can be used to connect and extract the value from the device.

2.1.2   Semantic Models

The provision of semantic information modeling can be implemented in various ways, including key-value, markup scheme, graphics, object-role, logic-based, and ontology-based models [13]. From this set, the key-value type offers the simplest data structure but lacks expressivity and inference. On the other hand, the ontology-based model provides the best way to express complex concepts and interrelations, being therefore the most popular model for elaborating semantic models.

2.1.3 Ontologies

The inherent semantic interoperability features of the Semantic Web have been mostly grounded on the use of ontologies for knowledge representation. In most cases, there exists a top-level ontology (or domain ontology) and multiple sub-domain ontologies, each one representative of a more specific domain. With the use of ontologies, entities can be described in very comprehensive ways [14].

2.1.4 Semantic Annotations

Semantic annotation is the process of attaching additional information to any element of data comprised of some sort of document. Ontologies on their own are not sufficient to fulfill the semantic interoperability requirements needed to enable data readability by machines. This is because of the differences and inconsistencies across different data sources and their ontologies. Semantic annotation has been widely used to fill this gap by creating links between the disparate ontologies and the original sources [15].

2.2 Methodologies for Ontology Engineering

2.2.1 METHONTOLOGY

METHONTOLOGY has been developed by the Ontology Engineering Group at the Universidade Tecnica de Madrid. It is a structured method to build ontologies initially developed in the domain of chemicals [16]. The methodology guides the ontology development process throughout the whole ontology life cycle. It consists of the following main development activities:

- *Specification*: It is concerned with the definition of the objectives of the ontology and of the end users, i.e., it frames the domain.
- *Conceptualization*: It is concerned with developing an initial conceptual representation/model of a perceived view of the application domain. A set of intermediate representations are here used to organize the concepts to be easily understood by both ontology and domain experts.
- *Formalization*: It is concerned with the implementation of a semi-computable model from the conceptual model generated in the previous activity.
- *Integration*: It is concerned with knowledge reuse, i.e., extracting and integrating definitions and concepts from already built ontologies.
- *Implementation*: It is concerned with the implementation of fully computational models using various ontology languages.
- *Maintenance*: It is concerned with any update to the ontology.

Furthermore, as part of the methodology, several orthogonal supporting activities are also identified to manage and support the development ones. These activities span knowledge acquisition, documentation, and evaluation.

2.2.2 SAMOD

The Simplified Agile Methodology for Ontology Development (SAMOD) [17] focuses on designing and developing well-developed and well-documented models from significant domain data and/or descriptions. It consists of three simple and small steps that are part of an iterative process aimed to produce preliminary and incremental results. The three steps can be labeled as:

1. *Test case definition*: This is about writing down a motivating scenario, being as close as possible to the language commonly used for talking about the domain.
2. *Merging current model with modelet*: This merges the modelet included in the defined test case with the current model.
3. *Refactoring current model*: This refactors the current model shared among all the defined test cases.

2.2.3 DILIGENT

The methodology for distributed, loosely controlled, and evolving engineering of ontologies (DILIGENT) [18] is a methodological approach intended to support domain experts in a distributed setting to engineer and evolve ontologies. It is based on rhetorical structure theory, viz., the DILIGENT model of ontology engineering by argumentation. The process comprises five main activities, namely:

1. *Build*: It concerns with the development of ontologies by having different stakeholders, with different needs and purposes that are typically distributed.
2. *Local Adaptation*: It concerns with the usage and adaptation of the developed ontology. By using the ontology, many updates can be necessary due, for example, to new business requirements and/or new arising needs.
3. *Analysis*: It concerns with the analysis of any local request for update. As a matter of fact, local ontologies can be updated, but the shared ontology will be updated only after the analysis of the update request.
4. *Revision*: It concerns with the constant revision of the shared ontology to guarantee the alignment with the local ones.
5. *Local Update*: It concerns with the update of the local ontologies after a new shared ontology is available.

2.2.4 UPON Lite

The lightweight unified process for ontology building (UPON Lite) methodology [19] is a simple, agile ontology engineering approach and/or method that is intended to place the end users and domain experts at the center of the overall ontology building process while avoiding the presence of ontology engineers. Therefore, the main pillars of the process are (i) the adoption of a fully user-centered approach, (ii) the adoption of a social collaborative approach to collect domain expert knowledge to achieve all the steps in the method, and (iii) an ontology building method based on six main activities. The six activities and/or steps of the UPON Lite method are the following (named and/or labeled according to the produced outcome):

1. *Domain terminology*: It is concerned with producing the list of all the fundamental domain terms that characterize the observed domain.
2. *Domain glossary*: It provides the definition and possible synonyms of the domain terms.
3. *Taxonomy*: It concerns with the organization of the domain terms according to an "ISA" hierarchy.
4. *Predication*: It concerns with the identification of those terms that represent properties and/or relations between other terms and/or concepts.
5. *Parthood*: It concerns with the analysis of the structure of the identified concepts and/or entities in order to elicit their (de-)composition hierarchies.
6. *Ontology*: It concerns with the production of the formally encoded ontology.

3   INFINITECH Semantic Interoperability Framework

The INFINITECH Semantic Interoperability Framework is a commonly agreed approach to enable semantic interoperability between applications and services within the INFINITECH platform while defining basic interoperability guidelines in the form of common principles, models, and recommendations. Furthermore, as part of the framework, ontology mapping processes are also considered to establish a common platform to deal with multiple ontologies. The proposed framework has been designed by combining a top-down and bottom-up approach (hybrid approach) as shown in Fig. 4.2. The latter – also called *pilot characterization* – is aimed to describe the specific application domain for each one of the test beds and pilot within the project. The main objective here is the identification, the definition, and the clear description of the context of application in terms of domain terminologies, glossaries, and taxonomies. The former – also called *state-of-the-art (SotA) analysis* – is aimed to identify reference ontologies for considered domain (finance and insurance); these ontologies are not linked to a specific application domain. The main objective here is the identification of a common and above all generic set of core concepts and relationships between them that can be used as top ontology, i.e., the glue between diverse specific domain

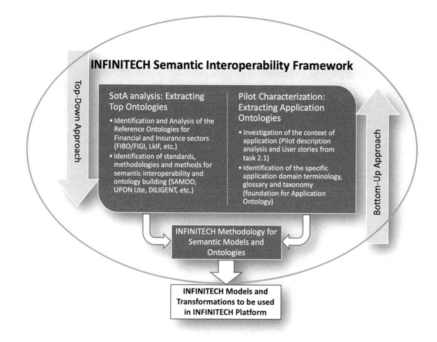

Fig. 4.2 Hybrid approach for interoperability in INFINITECH

ontologies for the same context of application. In both cases, the combination of the results of the *pilot characterization* and *SotA analysis* is used as inputs of the INFINITECH methodology for semantic models and ontologies and used for generating INFINITECH models, as well as baseline for the development of transformers that needs to be used to exploit all the features and full potentiality of the INFINITECH platform. Therefore, the hybrid approach aims firstly to design a high-level normalized and domain-specific enterprise model and secondly to link this model to the business-specific data.

The INFINITECH Semantic Interoperability Framework focuses primarily on the planning/designing and development/implementation of the objectives, infrastructure, and necessary deliverables to ensure interoperability in digital finance applications. The successful execution of the planning/development and development/implementation phases – of the framework – strictly depends on the presence of a data governance layer. As pointed out in [20], data governance is a business-driven process where specific business data representations are aligned to data domain. To do that, several data governance procedures have been carried out, namely (see Fig. 4.3):

- Identifying potential data sources and data owners
- Assigning roles and responsibilities to data management processes

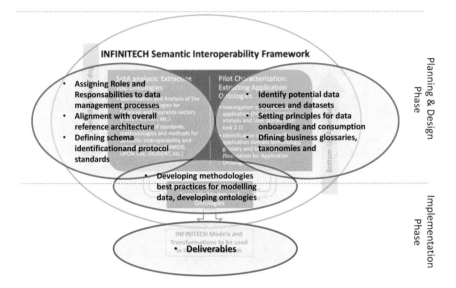

Fig. 4.3 Mapping data governance activities with proposed hybrid approach

- Defining the granularity of the data according to the type of applications needed to deliver it
- Alignment with the overall reference architecture
- Defining schema identification requirements and protocol standards for common data formats
- Developing methodologies and best practices for modeling data, developing ontologies, defining business glossaries, etc.
- Setting principles for data onboarding and consumption

3.1 Methodology for Semantic Models, Ontology Engineering, and Prototyping

Ontologies are the baseline for developing semantic interoperability applications. Ontologies are conceptual models – constituted by interlinked concepts related to a specific domain – of an observed reality. Since ontologies play a fundamental role in INFINITECH while providing the necessary mechanisms for describing test beds and pilot application domain, then a systematic engineering approach is needed to facilitate the design and development of high-quality and, above all, pilot-aligned ontologies to reference top-level ontologies for the domain.

As shown in Fig. 4.4, the INFINITECH methodology for ontology engineering shares terminology, definitions, and activities and/or steps with the SAMOD

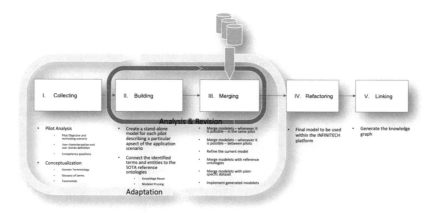

Fig. 4.4 Methodology for semantic models and ontology engineering

methodology. It is an iterative process that is aimed at building semantic models and ontologies. It is organized as a sequence of four sequential steps, namely:

1. *Collecting*. This step collects all the information about the application domain. It involves the following tasks and/or activities:

   - *Pilot Analysis*: Write down the motivating scenario and identify user expectation by writing down *user stories* and clarifying everything by using a set of competency questions (user characterization).
   - *Conceptualization*: Write down domain terminology, glossary of terms, and taxonomies of concepts.

2. *Building*. This step builds a new interoperability test case (*aka modelet*). The *modelet* is a stand-alone model describing the application domain for the considered pilot and/or test bed. The step involves the following tasks and/or activities:

   - Creation of a stand-alone model for the pilot or test bed describing the relevant aspects of the application domain.
   - Connection with the top reference ontology(ies). This activity is aimed to reuse as much as possible already defined concepts, relations, and properties while pruning all the elements that are superfluous.

3. *Merging*. This step refines the generated *modelet* with concepts and relations extracted from reference ontologies for the domain to determine more generic domain ontologies. The step involves the following tasks and/or activities:

   - Merge *modelets* in the same pilot/test bed.
   - Merge *modelets* between different pilots/test beds within the same application domain.

- Refinement of the current modelet.
- Merge *modelets* with reference ontologies.
- Implement generated *modelets*.

4. *Refactoring*. This step provides the final ontology and semantic model as conceptual schema to be used within INFINITECH. This model delivers the complete description and characterization of the application domain aligned with reference ontologies while enabling any user of the INFINITECH application to seamlessly access diverse ontologies and thus concrete data.
5. *Linking*: This step links the refactored models to real data while generating the so-called linked knowledge graph.

Two iteration cycles (analysis and revision and adaptation) are part of the methodology. The analysis and revision iteration (executed essentially during the *building* step) aims at analyzing and reviewing the building process to guarantee the alignment with the domain expert's expectations and requirements. The result of this step and its related iterations is a preliminary model also called *modelet*. The adaptation iteration includes the steps *collecting*, *defining*, and *merging* and aims at refining the generated *modelets* to cope with new knowledge and/or any change in user characterization, user needs, application domain, or, more in general, any change that could directly have an impact on the way domain experts describe their own business and – thus – application domains.

Generated modelets are very specific and targeted conceptual models that need to be filled and populated with dynamic data from typically heterogeneous and distributed resources. Here is where the semantic graphs and/or knowledge graphs play a fundamental role.

3.1.1 Modeling Method

The main result of applying the methodology for semantic models and ontology engineering is an evolving conceptual schema (e.g., ontology) and linked knowledge graph that empowers the INFINITECH platform to access, query, use and process/analyze data and/or information from heterogeneous and distributed sources.

The conceptual schema is determined by using an evolving prototyping (i.e., a foundation of agile software methodologies like DevOps) approach, where it grows up by layers while continuously delivering software prototypes. In particular, the conceptual model is the combination of three layers, according to [2]:

- Top-level ontology: Describes at very high-level concepts of interest for the domain
- Domain ontology: Describes specific concepts typically related to sub-domains of the top-level model
- Application ontology: Describes very specific concepts related to the particular application and scenario

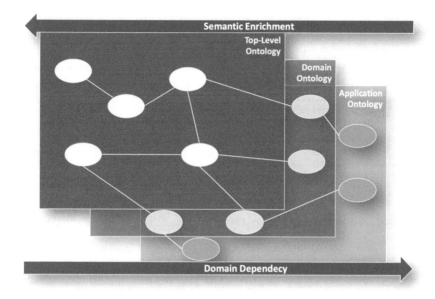

Fig. 4.5 Semantic and ontology modeling method

The layered model allows easy adaptation and extension while enabling for knowledge reuse, i.e., to reuse as much as possible already available ontologies and models. As a matter of fact, this model facilitates the adaptation to various applications as well as new domains (Fig. 4.5).

3.1.2 Envisioned Roles and Functions in Semantic Models, Ontology Engineering, and Prototyping

Several actors are typically involved in the process of defining, specifying, and developing semantic models and ontologies. The ontology engineering process is a collaborative process among several stakeholders. Since the main objective of the INFINITECH methodology for semantic models and ontology engineering is to provide a stakeholder-centric approach, it is necessary to identify the main roles and functions of the distinct actors of the process (see Fig. 4.6). The engineering process starts by having a small group composed by the following stakeholders: domain experts, end users, knowledge, and ontology engineers. The actors assume a distinct role considering the specific step within the methodology. In particular, during the *collecting* step, they assume the role of *data user* since they are essentially individuals who intend to use data for a specific purpose and are accountable for setting requirements. During the *building*, *merging*, and *refactoring*, ontology engineers play the role of *data trustee* since they are responsible for managing the data and

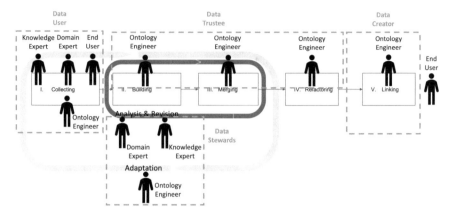

Fig. 4.6 Data governance: actors, roles, and functions

related metadata, data classifications, and conceptual schemas. Furthermore, during the *analysis and revision* process, the actors play the role of *data stewards* since they are responsible in ensuring the data policies and data standards are adhered to. Finally, during the *linking* step, the ontology engineer plays the role of *data creator* by physically creating and linking data to models as defined by the *data trustee*. This data is then ready to be consumed by end users.

4 Applying the Methodology: Connecting the Dots

This section illustrates the INFNITECH methodology for semantic models, ontology engineering, and prototyping, using exemplary data from considered pilots and selected supporting technologies. INFINITECH pilots have typically their own very specific data with different formats, data structure and differently organized. To establish the foundation for interoperability between those pilots, in the same application domain, ontologies are needed. However, most of them have not a well-defined and well-established conceptual model of their own application domain (the so-called application ontology). Furthermore, the usage of reference ontologies (such as FIBO, LKIF, FIGI, etc.) alone becomes practically impossible due to the lack of a connection with the application ontology. Therefore, it is peremptory to provide firstly pilots with application ontologies.

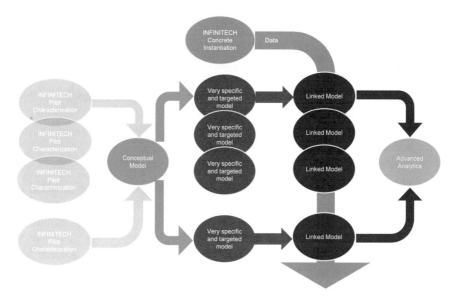

Fig. 4.7 Workflow and main deliverables and their connection

4.1 Workflow and Technological Tools for Validation of the Methodology

The proposed methodology is an iterative process that aims at providing very specific and targeted models to be used by advanced analytic applications (typically outside to the INIFITECH platform). Figure 4.7 shows the main output of the methodology starting from the pilot characterization and how it is connected to the INIFINITECH platform.

The *collecting* and *building* steps are initially used to define and build the conceptual model (*modelet*). The *modelet* needs to be merged, from one side to reference ontologies and from the other side to specific pilot conceptual model, and refactored to provide a very specific and highly targeted model. Finally, the model needs to be linked to real data from the pilot as the result of the *linking* step.

4.2 Collecting

The *collecting* step is the first step of the INFINITECH methodology and is aimed to characterize the application domain by providing three fundamental deliverables, namely (see Fig. 4.8):

- Domain terminology: The complete list of terms that are relevant for the application domain.

Fig. 4.8 *Collecting* step inputs and outputs

- Glossary of terms: The domain terminology enriched with the description of the term as well as possible synonyms. Furthermore, the object process actor modeling language (OPAL) semantic is also used at this stage that provides a first high-level classification of concepts.
- Taxonomy of identified concepts: The list of terms represented/organized into hierarchies according to the "ISA" relationship.

The output of this activity is then used – together with the description of available data – to create a conceptual model around the application domain during the next step: *building*.

4.3   Building and Merging

During the *building* step, a semantic approach is used where collected data are analyzed to understand the meaning of the data, as well as to identify the main concepts and relationships between them. The main result is a knowledge graph (see Fig. 4.9) that, in turn, represents the first step toward harmonization and standardization of data assets.

However, the knowledge graph needs to be further analyzed and refined to be connected to reference ontologies. This is done during the *merging* step where FIBO reference ontology is considered considering the domain of application.

At this stage, both pilot-specific and FIBO models have been analyzed to identify:

- Common concepts
- Connections and relations between the two models

The final knowledge graph is shown in Fig. 4.10.

The following FIBO concepts have been used for modeling the final knowledge graph of the INFINITECH real-time risk assessment pilot system:

- *Currency*: Medium of exchange value, defined by reference to the geographical location of the monetary authorities responsible for it.

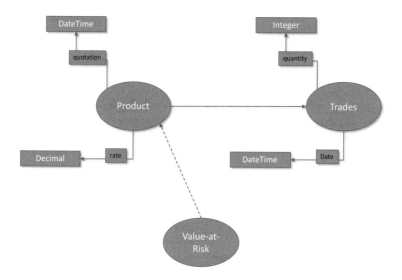

Fig. 4.9 Pilot-specific knowledge graph

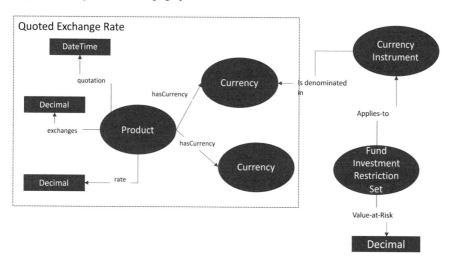

Fig. 4.10 Pilot-specific knowledge graph aligned with FIBO

- *Quoted Exchange Rate*: An exchange rate quoted at a specific point in time, for a given block amount of currency as quoted against another (base) currency. An exchange rate of R represents a rate of R units of the quoted currency to 1 unit of the base currency.
- *Value at Risk*: Measures and quantifies the level of financial risk within a firm, portfolio, or position over a specific time frame.

- *Fund Investment Restriction Set*: Limitations that apply to the fund as a whole, such as risk factors. These are used to determine whether the fund is appropriate for a given type of investor to invest in.
- *Currency Instrument*: Financial instrument used for the purposes of currency trading.

The result of the *merging* step is a connected graph – aligned with FIBO ontology – capable of spanning organizational concepts that are relevant for the selected application scenario and use cases.

4.4 Refactoring and Linking

The *refactoring* and *linking* stages aim at concretely developing and implementing knowledge graph produced after the *merging* stage while also linking it to the pilot-specific real data. Therefore, it is mainly focused on the selection of the model serialization format (RDF, JSON-LD, etc.) and concrete supporting technology for creating linked knowledge graphs. During these steps, selected technology is applied to support the overall process of data harmonization (according to FIBO) and data-sharing and provisioning to any external application that needs to use them. The main result of these steps is a linked knowledge graph that can be queried.

At this point, two technologies have been selected to show the repeatability of the process, regardless of the specific environment deployed within the pilot, namely, (1) dataworld.com[1], a cloud-based collaborative data catalog aimed to supply data integration and analysis, and (2) *GraphDB*[2], an RDF database for knowledge graphs with extended features for providing integration and transformation of data. With both solutions, the execution of the validation stages of the workflow was attempted, with support of other technologies used in INFINITECH, such as *LeanXcale* database for the origin data, *Node-RED* for client application, or *Docker* for deployment. As a result, a simple architecture was designed. The architecture specifies the necessary elements of the system, along with their internal interfaces and communications (Fig. 4.11).

4.4.1 Data Ingestion

The first necessary step for validation of the modeling methodology is to pull the real data from the source into the selected technology. Currently, it is already possible to import data from a wide range of sources (local files, cloud locations, different

---

[1] https://data.world

[2] https://graphdb.ontotext.com

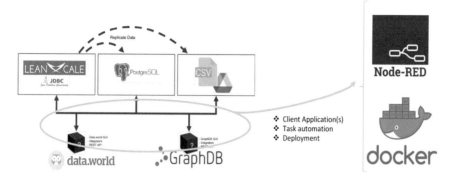

Fig. 4.11 Communication architecture of the validation setup

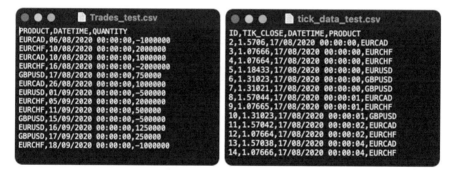

Fig. 4.12 Snippets of the origin datasets (trades and ticks), in CSV format

databases such as PostgreSQL or MySQL using JDBC connectors, etc.) and formats (CSV, JSON Lines, etc.). In Fig. 4.12, a snippet of the datasets used can be seen:

After importing the datasets, the data is then able to be accessed, combined (as seen in Fig. 4.13, where the imported data was preprocessed into a single dataset), and used on the next steps.

4.4.2 Semantic Alignment: Building and Merging

This is the stage where the semantic alignment model obtained by using the modeling methodology is executed. In the example given in Fig. 4.14, the mapping is constructed by using the mapping tool, by associating the items of the imported datasets with the modeled semantics.

On the other hand, the developer of the semantic mappings can, for example, replicate the semantic alignment by using SPARQL queries (Fig. 4.15):

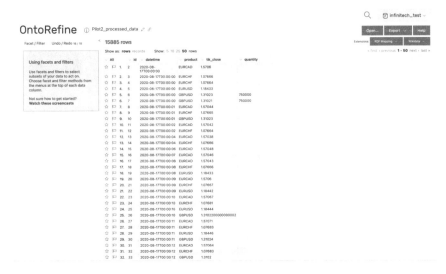

Fig. 4.13 Imported datasets from the test pilot

Fig. 4.14 Semantic mapping performed with GraphDB mapping tool

4.4.3 Semantic Transformation: Generating a Queryable Knowledge Graphs

Once the mapping is specified, it can be used against the original dataset, to enable the transformation and, thus, to produce the semantic data. This implies that the resulting knowledge graph follows the semantic alignment that is provided by the merging stage. In the specific cases of the adopted frameworks, the user can invoke

Fig. 4.15 Semantic mapping with SPARQL query

```
1   @base <http://example.com/base/> .
2   @prefix fibo-fnd-acc-cur: <https://spec.edmcouncil.org/fibo/ontology/FND/Accounting/CurrencyAmount/> .
3   @prefix fibo-fnd-qt-qtu: <https://spec.edmcouncil.org/fibo/ontology/FND/Quantities/QuantitiesAndUnits/> .
4   @prefix fibo-ind-ind-ind: <https://spec.edmcouncil.org/fibo/ontology/IND/Indicators/Indicators/> .
5   @prefix fibo-ind-fx-fx: <https://spec.edmcouncil.org/fibo/ontology/IND/ForeignExchange/ForeignExchange/> .
6   @prefix inf-spec-ont-pilot-2: <https://spec.infinitech.org/ontology/application/pilot_2/> .
7   @prefix xsd: <http://www.w3.org/2001/XMLSchema#> .
8
9   inf-spec-ont-pilot-2:2 a fibo-ind-fx-fx:QuotedExchangeRate;
10     inf-spec-ont-pilot-2:CurrencyPairTag "EURCAD";
11     fibo-ind-ind-ind:hasQuotationDateTime "2020-08-17T00:00:00"^^xsd:dateTime;
12     fibo-fnd-acc-cur:hasRateValue "1.5706"^^xsd:long .
13
14  inf-spec-ont-pilot-2:3 a fibo-ind-fx-fx:QuotedExchangeRate;
15     inf-spec-ont-pilot-2:CurrencyPairTag "EURCHF";
16     fibo-ind-ind-ind:hasQuotationDateTime "2020-08-17T00:00:00"^^xsd:dateTime;
17     fibo-fnd-acc-cur:hasRateValue "1.07666"^^xsd:long .
18
19  inf-spec-ont-pilot-2:4 a fibo-ind-fx-fx:QuotedExchangeRate;
20     inf-spec-ont-pilot-2:CurrencyPairTag "EURCHF";
21     fibo-ind-ind-ind:hasQuotationDateTime "2020-08-17T00:00:00"^^xsd:dateTime;
22     fibo-fnd-acc-cur:hasRateValue "1.07664"^^xsd:long .
23
24  inf-spec-ont-pilot-2:5 a fibo-ind-fx-fx:QuotedExchangeRate;
25     inf-spec-ont-pilot-2:CurrencyPairTag "EURUSD";
```

Fig. 4.16 RDF results resulting from the semantic alignment applied to the origin dataset

them to, for example, execute the relevant SPARQL queries, by using the REST
API endpoints or any other solution that may be developed for the effect. In the
current implementation, either running SPARQL queries or streaming data through
HTTP request can fulfill the data transformation, which results on the generation of
an RDF format dataset, such as the code snippet shown in Fig. 4.16.

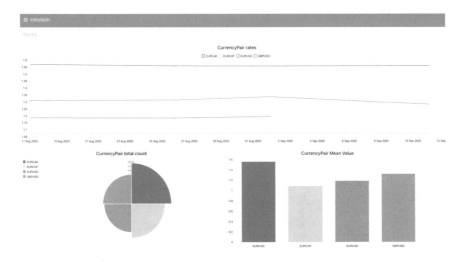

Fig. 4.17 Analytics over RDF data generated from the origin datasets

4.4.4 Data-Sharing/Provisioning

Finally, it is important that end users possess the necessary means to access and further use the generated data. As such, the set of solutions in use allow to directly download the RDF results from transformation process. Moreover, they enable the storage of the results as knowledge graphs which can later be retrieved using the REST APIs or other access points created.

Moreover, RDF compliant client applications can be developed to make use of such capability using the data to execute the desired analytics over the data.

In the scope of this validation process, a simple node.js client application was developed for demonstration, where a few analytic algorithms were applied to the RDF data. The latter data were consumed through one of the REST APIs, as presented in Fig. 4.17.

5 Conclusions

This chapter presented a methodology for generating models that enable the transformation of data to semantically annotated data, in alignment with reference ontologies of financial and insurance sectors.

The outcomes (*modelets*) of the proposed methodology provide a simpler and more trustable way to generate and transform the original data into semantically meaningful data. In this way, the proposed methodology enhances the interoperability of different sources.

To validate the approach, state-of-the-art technological tools for the processing and management of tabular and RDF data were used. Leveraging these tools, it is demonstrated that the mapping process can be extremely simplified and, also, that it is applicable to different types of available solutions. As a matter of fact, the same results were obtained when applying the methodology into different sources of data or when using other types of mapping technologies.

Our future work involves the application and validation of the methodology at a larger scale, starting from its deployment and use in the digital finance systems that have been developed, deployed, and validated in the scope of the INFINITECH project.

Acknowledgments The research leading to the results presented in this chapter has received funding from the European Union's funded Project INFINITECH under Grant Agreement No. 856632.

References

1. Cupoli, P., Earley, S., & Henderson, D. (2012). *DAMA-DMBOK2 framework* (p. 27).
2. Strengholt, P. (2020). *Data management at scale: Best practices for enterprise architecture*. O'Reilly Media Inc.
3. C. Iso and others. (1993). *ISO/IEC 2382-1: 1993 information technology-vocabulary-part 1: Fundamental terms*.
4. ETSI. (2011). *ETSI TS 186 020 V3.1.1 – Telecommunications and Internet converged Services and Protocols for Advanced Networking (TISPAN); IMS-based IPTV interoperability test specification*. Technical.
5. EICTA. (2004). *EICTA INTEROPERABILITY WHITE PAPER*. [Online]. Available: http://www.agoria.be/www.wsc/webextra/prg/nwAttach?vWebSessionID=8284&vUserID=99 9999&appl=Agoriav2&enewsdetid=73029& attach=DescFile10.720108001095774198.pdf
6. Maló, P. M. N. (2013). *Hub-and-spoke interoperability: An out of the skies approach for large-scale data interoperability*. [Online]. Available: http://run.unl.pt/handle/10362/11397. Accessed 18 Jan 2016.
7. Martidis, A., Tomasic, I., & Funk, P. (2014). Deliverable 5.3: CREATE interoperability. August 25, 2014.
8. van der Veer, H., & Wiles, A. (2008). *Achieving technical interoperability*. European Telecommunications Standards Institute.
9. Berners-Lee, T., Hendler, J., & Lassila, O. (2001). The semantic web. *Scientific American, 284*(5), 34–43.
10. Palavalli, A., Karri, D., & Pasupuleti, S. (2016). Semantic internet of things. In *2016 IEEE tenth international Conference on Semantic Computing (ICSC)* (pp. 91–95).
11. Elkhodr, M., Shahrestani, S., & Cheung, H. (2013). The Internet of Things: Vision & challenges. In *IEEE 2013 Tencon-Spring* (pp. 218–222).
12. McIlraith, S. A., Son, T. C., & Zeng, H. (2001). Semantic web services. *IEEE Intelligent Systems, 16*(2), 46–53.
13. Slimani, T. (2013). Semantic annotation: The mainstay of semantic web. *arXiv preprint arXiv:1312.4794*.
14. Andrews, P., Zaihrayeu, I., & Pane, J. (2012). A classification of semantic annotation systems. *Semantic Web, 3*(3), 223–248.

15. Liao, Y., Lezoche, M., Panetto H., & Boudjlida, N. (2011). Semantic annotation model definition for systems interoperability. In *OTM confederated international conferences "On the Move to Meaningful Internet Systems"* (pp. 61–70).
16. Fernández-López, M., Gómez-Pérez, A., & Juristo, N. (1997). *Methontology: from ontological art towards ontological engineering.* Stanford University.
17. Peroni, S. (2016). A simplified agile methodology for ontology development. In *OWL: Experiences and directions–reasoner evaluation* (pp. 55–69). Springer.
18. Pinto, H. S., Staab, S., & Tempich, C. (2004). DILIGENT: Towards a fine-grained methodology for DIstributed, Loosely-controlled and evolvInG Engineering of oNTologies. In *Proceedings of the 16th European conference on artificial intelligence* (pp. 393–397).
19. De Nicola, A., & Missikoff, M. (2016). A lightweight methodology for rapid ontology engineering. *Communications of the ACM, 59*(3), 79–86.
20. Allen, M., & Cervo, D. (2015). *Multi-domain master data management: Advanced MDM and data governance in practice.* Morgan Kaufmann.

Part II
Blockchain Technologies and Digital Currencies for Digital Finance

Chapter 5
Towards Optimal Technological Solutions for Central Bank Digital Currencies

Lambis Dionysopoulos and George Giaglis

1   Understanding CBDCs

1.1   A Brief History of Definitions

The European Blockchain Observatory and Forum (EUBOF), in their recent report titled "Central Bank Digital Currencies & a Euro for the Future", defines CBDCs as "a form of digital money that is issued by a central bank". This lean and abstractive definition is the most recent in line with exploratory work on CBDCs. The Bank for International Settlements (BIS) was among the first to highlight the features of, and concerns surrounding, electronic money. Many of the concepts presented in the paper remain surprisingly topical when applied to cryptocurrencies and even inform today's dominant narrative that views CBDCs as a secure alternative to private and decentralised money [4].

BIS' papers have remained a notable facilitator of the discussion on new forms of central bank money. More recently, a paper from BIS sparked the discussion around CBDCs. In the 2017 report titled "Central Bank Cryptocurrencies", the authors lean heavily on the concepts introduced by cryptocurrencies and present the concept of a central bank cryptocurrency (CBCC). CBCCs are defined as "an electronic form of central bank money that can be exchanged in a decentralised manner, [...] meaning that transactions occur directly between the payer and the payee without the need for a central intermediary". The paper also distinguishes CBCC from other prevalent forms of money cash, deposits and e-money [5]. The report can be read as a partial response to concerns raised in a 2015 paper by the Committee on Payments and Market Infrastructure regarding the implications of cryptocurrencies for central

L. Dionysopoulos (✉) · G. Giaglis
Institute for the Future, University of Nicosia, Nicosia, Cyprus
e-mail: dionysopoulos.c@unic.ac.cy; giaglis.g@unic.ac.cy

© The Author(s) 2022
J. Soldatos, D. Kyriazis (eds.), *Big Data and Artificial Intelligence in Digital Finance*,
https://doi.org/10.1007/978-3-030-94590-9_5

banks, the financial market infrastructure, its intermediaries and the wider monetary policy [8].

Most recently, in 2020, the central banks of Canada, Europe, Japan, Sweden, Switzerland, England and the Federal Reserve and the BIS collaborated on a report that addressed the foundational principles and core features of CBDCs [14]. Therein, they are defined as "a digital form of central bank money that is different from balances in traditional reserve or settlement accounts" and "a digital payment instrument, denominated in the national unit of account, that is a direct liability of the central bank". The report also highlights the influence of the private sector on CBDC design, predominantly in the form of increased payment diversity. Individually, the BoE [20] has defined CBDCs as "an electronic form of central bank money that could be used by households and businesses to make payments and store value", again noting the importance of the ever-changing payment landscape and impact of new forms of money such as stablecoins. The Central Bank of Sweden [34], leaning heavily on prior work of the BIS, describes their national version of a CBDC, known as an e-krona, as "money denominated in the national value unit, the Swedish krona, a claim on the central bank, electronically available at all times, and accessible by the general public". The European Central Bank (ECB) [23] defines their version of a CBDC, also known as a digital euro, as "a central bank liability offered in digital form for use by citizens and businesses for their retail payments [that would] complement the current offering of cash and wholesale central bank deposits" and also underlines how competition in payments and money necessitate a new euro to address modern and future needs.

The common denominator of virtually every report on digital versions of sovereign money accessible by the private sector is that they were motivated by advancements and challenges emerging from the private sector itself. By examining this nonexhaustive collection of definitions, we can denote that CBDCs are digital forms of money denominated in the national currency of the issuing country/authority, accessible by either parts of or the wider private sector. Ultimately, CBDCs are to provide an alternative to the offerings of the private sector, through broad access to a central bank money and its accompanying protection and guarantees, while serving the needs of end consumers and the wider economy.

1.2   How CBDCs Differ from Other Forms of Money

However, the above definitions do not make immediately apparent how CBDCs differ from other forms of money prevalent in today's world. Intuitively, CBDCs are novel neither because of their digital nature nor due to their issuance by a central bank. In fact, electronic forms of money have existed for the better part of the century, and money is issued by a central authority, such as a bank for millenniums. The novelty of CBDCs relies on two primary factors [22]:

1. The extent to which this digital claim with the central bank is extended to the wider private sector – and the accompanying implications of this shift
2. The technology behind CBDCs – and the new options that it might enable

By studying the nature of claims of modern money, we can observe that one relational connection is notably absent, as there is no way for the private sector to maintain a direct claim with the central bank electronically. Cash, meaning physical notes and coinage, constitute a liability of the central bank and asset of the wider private sector, including commercial banks, households and individuals. Deposits are a liability of commercial banks and asset for the private sector, while central bank reserves a liability of the central bank and asset for commercial banks. E-money, meaning currency stored in software or hardware and backed by a fiat currency, does not necessarily involve bank accounts; hence, it serves as a liability of payment service providers (PSPs) from the private sector and an asset for businesses, households and individuals. Finally, cryptocurrencies while an asset for the private, and more recently the public sector [26], are a liability of neither. CBDCs occupy a new space, one that offers the option for central bank money with modern features to be widely accessible by the private sector.

1.3   Wholesale and Retail CBDCs

Depending on how this electronic claim of the central bank is made available to the private sector, CBDCs are separated into a "wholesale" and "retail" variant. Wholesale CBDCs are devised for use between designated entities and the central bank. Those entities can include commercial banks, payment service providers and other financial institutions. Thus, wholesale CBDCs can be thought as an iteration on the existing reserve system. As nonconsumer facing money, their scope would be limited to streamlining intrabank operations such as improving efficiency of payments, minimising counterparty risk of the settlement and interbank lending process and reducing overhead costs. As reserves with the central bank already exist, the wholesale variant of CBDCs is not an entirely novel concept. Conversely, retail CBDCs would fulfil the functions of money, serving as a unit of account, store of value and medium of exchange, backed by the protection and guarantees of the central bank.

The premise of an electronic form of money to maintain, or even further, the relationship of the private sector with the central bank is seemingly superfluous. However, multifaceted factors that relate financial stability and inclusion, payment efficiency and safety and implementation of monetary policy among other satellite considerations contribute to the appeal of CBDCs.

1.4   Motivations of CBDCs

While at different stages in development and research, approximately 90% of central banks globally are exploring the issuance of CBDCs. Across the different national and regional motives, a broad agreement on the factors that necessitated the issuance of CBDCs and informed their design and, ultimately, their utility can be observed [7]. We broadly categorise factors to those that relate to financial stability and monetary policy and those that relate to increased competition in payments and threats to financial sovereignty.

1.4.1   Financial Stability and Monetary Policy

The 2008 recession officially ended in 2010, with annual global GDP rates demonstrating steady growth coupled with low volatility throughout the decade. The American economy experienced the longest expansion in its history and 50-year low unemployment rates [29]. The S&P 500 tripled in value, Apple, Amazon, Alphabet and Microsoft reached a market capitalisation of $1 trillion, while start-up growth was unprecedented. Yet, echoing the lasting effects of the great recession, global GDP growth levels have retraced from their 2000–2008 pace [35]. Wages, labour productivity [36] and the quality of jobs deteriorated [27], the overall number of IPOs fell [33], and the start-up miracle gave its place to some of the biggest scandals of the decade [24]. The COVID-19 pandemic spurred the largest plunge [32] for the Dow Jones in its history, not once, but thrice within less than a week. Surging unemployment and plummeting asset prices followed, igniting fears of a looming global economic recession.

Over the past decades, government, central banks and policy-makers adopted increasingly unorthodox approaches to keep economies and societies afloat, employing tools ranging from forward guidance to quantitative easing and, more recently, even handouts or "helicopter money", that overall proved less effective than initial estimates [31]. These gave room rise to CBDCs as mediums for enhancing financial stability through streamlining existing monetary policy tools and creating new ones. While an in-depth analysis of the role of CBDCs for monetary policy is beyond the scope of the present, it is generally accepted that their digital nature could provide better insight into the inner workings of economies, contribute towards elimination of the lower zero-bound interest rate, offer granularity and control over social benefits and streamline the monetary policy transmission mechanism [14]. Additionally, CBDCs can also offer new options for monetary policy, such as a new obligation clearing system enabling a collaborative economy [22]. As demonstrated in the latest BIS report/survey, the above are well realised and of increasing importance especially for central banks in advanced economies (AEs).

1.4.2 Increased Competition in Payments and Threats to Financial Sovereignty

Besides the need for solutions that relate to monetary policy, the promises of new technological advancements are, in many cases, too substantial to ignore. On January 3, 2009, Bitcoin emerged as the birth child of converging technological advancements and rising dismay for the then unfolding economic events. Through its novel technological backbone, decentralised structure and reliance on social consensus, it showcased the advantages of money birthed by, and for the needs of, the internet. Disregarding the accompanying anti-systemic narrative, its promise for fast, cheap, reliable, inclusive and borderless value transfer is still appealing and, in many cases, unattainable by existing systems. Over the years, countless imitators and iterators materialised, and while most faded to obscurity, others such as Ethereum expanded upon Bitcoin's functionality. Ethereum showcased the full range of applications that can be achieved with programmable DLT-enabled forms of electronic value and systems built upon these. Programmable money, new innovative financial services by the name of decentralised finance (DeFi) and the enabling of novel forms of commerce, such as machine-to-machine economy, digital identities and decentralised governance schemes, are only some of the concepts largely nurtured in the decentralised space, which could benefit the economies through CBDCs. Decentralised stablecoins showcased how the advances of cryptocurrencies can be brought to digital money denominated in sovereign currencies, free from the volatility of other digital assets. Over the years, stablecoins emerged as a blockchain native unit of account while maintained most of the desirable characteristics of their non-stable counterparts, including fast borderless payments, programmability and reliability. Their popularity and explosive growth sparked initial discussions around CBDCs, as evident by early statements from central banks globally. An indicative example is the joint press release by the European Council and the European Commission in 2019, in which the significance of stablecoins as fast, cost-effective and efficient payment mediums to address consumer's needs is outlined [18]. In the same press release, the possibility of CBDCs as a sovereign alternative to stablecoins is also discussed. Similar reports from central banks further solidify how stablecoins influenced the design and accelerated the discussion around CBDCs [1, 6, 20]. Yet, before central banks could offer an alternative to stablecoins, free of the ambiguity and risks of their decentralised issuance and nature, the private sector was quick to react.

An alternative to the decentralised issuance-induced ambiguity and lack of synergy with existing infrastructure came with Facebook's announcement of Libra (later rebranded to Diem) and its network of partners. Despite a significant scale-back, Libra/Diem aims to build a blockchain-backed financial network to enable open, instant and low-cost movement of money and further financial inclusion by enabling universal access to related services [15].

In parallel to Facebook's announcement, China's pilots for the country's version of a CBDC, the Digital Currency Electronic Payment (DCEP), commenced through the People's Bank of China (PBoC), also involving four state-owned banks and the public. A few months later, Fan Yifei, deputy governor of the PBoC in his speech at the Sibos 2020 conference, confirmed that more than 110,000 consumer wallets and 8000 corporate digital wallets were opened for citizens of Shenzhen, Suzhou and Xiong'an. With over €150 million worth of DCEP processed, China became the first major economy to successfully employ CBDCs in that scale [25].

In light of privately issued global currencies and escalating interest in CBDCs from other nations, the mobilisation of regulators was unprecedented. Within days of Libra's announcement, top financial authorities globally issued statements foreshadowing the regulatory scrutiny that Libra/Diem would face in the following months and years. Perhaps most notably, the G7, consisting of financial leaders from the world's seven biggest economies, openly opposed the launch of Libra/Diem in the absence of proper supervision and regulatory framework [28]. Lastly in the EU in particular, top representatives from France, the UK and Germany were quick to underline that Facebook's Libra/Diem will be held at the highest regulatory standards. Moreover, the Markets in Crypto-Assets (MiCA) regulation developed in response aims at regulating Libra/Diem and similar deployments.

The driving motives behind cryptocurrencies, decentralised and global stable-coins, fintech and neobanks solidified the need for enhances in payment efficiency and innovation, as well as the furthering of financial inclusion [16]. As demonstrated by BIS recent report/survey, the above constitute fundamental design principles of CBDCs. Specifically, while financial stability and inclusion have over the years attracted the interest of emerging market and developing economies (EMDEs), both AEs and EMDEs are equally motivated by the implications of retail CBDCs for payment efficiency and robustness. Finally, maintaining access to central bank money in light of the declining use of cash is another important factor behind the issuance of CBDCs.

With regard to wholesale CBDCs, as noted in the previous section, those are not an entirely novel concept, and thus, many of their potential benefits are already enjoyed by commercial banks. As such, they are not a priority for most central banks, with this insouciance reflected in the report/survey of BIS. The motives behind wholesale CBDCs generally follow that of their retail counterparts, save for the notion of financial inclusion.

2   From Motivations to Design Options

The motives presented above have informed a collection of foundational principles and specific characteristics for CBDCs. Indicatively, the European central bank presented a comprehensive collection of core principles and scenario-specific and general requirements for a potential digital euro, Europe's name for a CBDC [19]. Across those three-category issues that relate to monetary policy, financial stability and sovereignty, efficiency, security, cost-effectiveness and innovation are highlighted. The Bank of England noted 15 design principles across reliable and resilient, fast and efficient, innovative and open to competition [20]. The Central

Bank of Sweden, in their second project report on the country's CBDC, the e-krona, highlighted its desirable functions. The report cites comprehensive range of services, wide availability and ease of use, transaction capacity and performance, robustness, security, reliability and integrity as well as trackability [34]. Similar considerations have been highlighted in reports produced by the Bank of Thailand [10] and others.

The convergence of central banks pertains to not only the definition and motives behind CBDCs but also the foundational principles that should characterise their design. This is intuitive to the extent that the overarching mandate of central banks is universal and relates to the monetary and financial stability, as well as the provision of trusted money. The reports presented above fall in line with BIS' research on the desirable characteristics of CBDC [14]. As a collaborative effort involving the central banks of Canada, Europe, Japan, Sweden, Switzerland, England and the Federal Reserve, the factors presented have largely informed the reports on a national and supranational level and vice versa. As such, we feel that they can be treated as a middle ground of universally shared and generally accepted principles of CBDC design, save for the caveats stemming from national and local considerations.

The report identifies a total of 14 desirable characteristics that relate to *instrument features*, meaning the characteristics of the CBDC itself; *system features*, which relate to the technology enabling the CBDC; and *institutional features*, meaning the policy underpinnings of a CBDC. Of those, 5 are promoted as *core features*. The features fall in line with the three common objectives of all central banks as identified in the report. Those are:

(i) Do no harm, meaning that new money should not interfere with the central banks mandate for financial and monetary stability.
(ii) Coexistence, as CBDC should complement and not compete with private solutions and other forms of money, as long as those serve a purpose for the public.
(iii) Innovation and efficiency, in a safe and accessible manner.

The desirable core CBDC features include the following:

Instrument features	System features	Institutional features
Convertible	Secure	Robust legal framework
Convenient	Instant	Standards
Accepted and available	Resilient	
Low cost	Available	
	(High) Throughput	
	Scalable	
	Flexible and adaptable	

2.1 The Design Space of CBDCs

As demonstrated, a lot of ink has been spilled on the factors that necessitate the issuance of CBDCs and their desirable features. Yet little effort has been put towards identifying universally preferable technology options for CBDC to satisfy their desirable characteristics.

As with the definition for CBDC, the EUBOF has introduced a lean framework for mapping out the design space and options for a digital Euro, Europe's version of a CBDC. The framework considers previous work on the area and predominately draws for a study by BIS titled "The technology of retail central bank digital currency" [3]. The framework highlights three primary choices for CBDCs, namely, the evidence of ownership (E-Own), the ledger infrastructure and the management scheme, which collectively make up the design space of a digital euro. For the E-Own, they identify the alternatives of account-based and token-based systems as potential mechanisms for "verifying ownership of value and authorisation of transactions". In terms of the technology that underpins the digital euro, EUBOF proposes the alternatives of Europe's existing real-time gross settlement system (RTGS) and TIPS and that of a novel distributed ledger technology (DLT) or blockchain infrastructure. Finally, regarding the management scheme, the authors coin the umbrella terms *centralist* and *federalist* to respectively describe a CBDC managed by the central bank alone or by the central bank in collaboration with commercial banks and other entities. The entire design space defined by EUBOF is presented in the Fig. 5.1.

While the framework was initially devised for assessing the design space and options for a digital euro, there is nothing to suggest that the same principles do not directly apply to any other potential CBDCs. Moreover, while wholesale CBDC options are not considered, as discussed above, those are largely overshadowed as redundant especially when compared to their retail counterparts. As such, we opt to evaluate the desirable characteristics for a CBDC introduced by the relevant BIS report, against the derivative options stemming from EUBOF's framework. To facilitate for this analysis, we assign unique names to every derivative option. We utilise the same naming scheme proposed by EUBOF and substitute the term digital Euro for CBDC. As an example, a CBDC utilising an account-based E-own, RTGS infrastructure and centralist management scheme would be abbreviated as "ARC CBDC". While not the most elegant option, this naming scheme is intuitive in its simplicity.

2.2 Assessing Design Space Against Desirable Characteristics

In accessing the options, we consider the instrument features and system features, against the design space presented above, as they are directly relevant to those technical options, unlike the institutional features. Additionally, we will opt to

Fig. 5.1 CBDC design space. (Source: EU Blockchain Observatory & Forum)

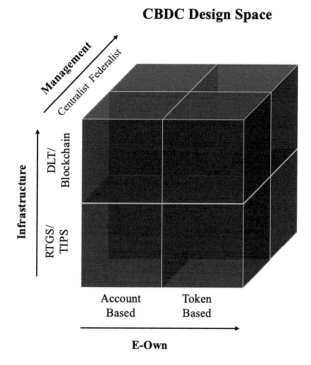

utilise TIPS as an example when referring to RTGS systems. However, this can be substituted for any other RTGS deployment without influencing the evaluation.

2.2.1 Instrument Features

Convertible In every potential version of a CBDC, regardless of the technology, or management scheme, convertibility with other forms of money in the economy and in particular those issued by the central bank can be established as a prerequisite. Central banks, in collaboration with the government, can, and likely will, establish CBDCs as a form of legal tender in the country of issuance. This would mean that both cash and deposits with the central bank will also be convertible for CBDCs and vice versa. While a convertibility at par might facilitate for undesirable events in times of stress, such as bank runs, and even be undesirable by commercial banks for profitability reasons [22], nevertheless, the technology options presented above allow for it. Whether it is implemented or not is strictly a policy decision.

Convenient The growing popularity of digital payments in the form of fintech, neobanks, e-money and cryptocurrencies has showcased their convenience over other forms of money such as cash or cheques. CBDCs regardless of the technological infrastructure, or other design choices, would offer comparable features and convenience.

Accepted and Available BIS' report establishes POS functionality and offline capabilities as main prerequisites for a CBDC to be considered "accepted and available". As we have established, CBDCs draw from advancements in digital payments and thus will likely adopt payments with NFC, QR codes and plastic. While the digital nature of a CBDC will necessitate communication with a remote database, to facilitate for offline transactions, this could be done asynchronously [13]. The Bank of Japan published a research paper [9] examining the options for offline transactions. The solutions proposed cover a wide range of mediums including an IC chip on a SIM card, PASMO/Suica cards otherwise electronic money and railway and transport tickets and finally debit cards with built-in keypads and screens. Visa has also published a paper [12] outlining how offline transactions utilise secure elements in mobile devices. The options presented above do not rely on the technological infrastructure as the components responsible (smartphone, servers, etc.) would be most likely shared among any CBDC deployment. Thus, offline CBDC transactions are possible and likely.

Low Cost When it comes to costs, those largely depend on the infrastructure as well as management scheme of a CBDC. In terms of the latter, as centralist schemes are entirely managed by the central bank, the costs of research, development and management of the infrastructure and end-customer relationships would be incurred by the central bank alone. This is in contrast to federalist schemes that involve entities besides the central bank. In those, the costs would be shared between the central bank and the various participating entities. Regarding the infrastructure, the operational costs of CBDC deployed existing RTGS infrastructures that would be comparable to existing systems. On the other hand, DLT solutions would, at least initially, necessitate costs in the form of research and deployment. Whether the central banks can afford not to pass those costs to the end consumers is debatable.

2.2.2 System Features

Secure BIS' report considers resistance to cyberattacks and counterfeiting as primary components of security. While cyber incidents involving both RTGS systems and blockchains have taken place in the past, indicatively [11, 30], both RTGS systems and DLT/blockchain have proven overall cyber-resilient and cyber-efficient against counterfeiting. Financial institutions employ high levels of security to protect information and funds, and the nature of the immediate settlement makes data vulnerable for a briefer time window. Blockchains, by employing a novel form of distributed consensus and protocol rules, ensure the validity of value transfers and mitigate instances of double spent. While some of those benefits are lost in networks with trusted entities, a proof-of-authority consensus mechanism, involving reputable entities that can be held accountable for wrongdoing, would offer comparable security to RTGS deployments.

Instant In RTGS systems such as Europe's TIPS, transactions are almost "instant", with 99 per cent of them settling in under 5 s [37]. In blockchains, most of the inefficiencies stem from the use of novel forms of consensus mechanisms and artificial limits to block size that result in limited processing capacity. While the above serve a purpose in networks comprised of mutually distrusting nodes, in the case of CBDCs, their use of blockchain technology will be different. Specifically, with trusted entities, most of the limitations presented above would be redundant; thus, transaction times will be comparable to that of existing RTGS schemes.

Resilient BIS' report defines resilience as the extreme resistance to operational failures and disruptions such as natural disasters and electrical outages. Both RTGS systems and DLT/blockchains utilise geographically distributed architectures that ensure data integrity in case of disruption. Moreover, RTGS systems such as TIPS operate on a continuous basis with recovery times in case of disaster limited to a maximum of 15 min [37].

Available Availability is defined as the capacity of a system to process transactions 24 h per day, 7 days per week, 365 days per year. Like DLT/blockchain, RTGS systems operate on a continuous automated basis, beyond business hours and without the need for human interference. Indicatively, the longest running blockchain network, Bitcoin, boasts an uptime of 99. 98% with only two downtime events, aggregately amounting to 15 h of downtime.

Throughput Intuitively, throughput translates to the number of transactions that can be processed in a given timeframe. RTGS systems, such as TIPS, can process north of 40 million transactions per day with a maximum capacity of 2000 transactions per second. Moreover, their usability, speed and efficiency can also serve as the platform for the deployment of a CBDC [37]. The throughput of many decentralised deployments is much more limited. Indicatively, the processing capacity of Bitcoin is approximately 4.5 transactions per second. However, it would be faulty to assume that this limitation is inherent for all DLT/blockchain-based systems. In fact, for Bitcoin and many other decentralised cryptocurrencies, this limited capacity is a conscious design choice in the blockchain trilemma. Simply put, the blockchain trilemma is "an oversimplification of the multidimensional mutually exclusive choices made by developers, users, and markets alike", with Vitalik Buterin defining the trilemma as the balancing act between decentralisation, security and scalability [17]. Simply put, many decentralised networks favour security and decentralisation over scalability and, as a result, have a lower processing capacity. Proof-of-authority CBDC deployments utilising DLT/blockchain can, and likely will, favour scalability and security at the expense of decentralisation, to achieve processing capacity comparable to RTGS deployments.

Scalable BIS' report defines scalability as the ability of a system to "accommodate the potential for large future volumes". A plethora of schemes have been proposed for scaling both RTGS and DLT/blockchain systems. Indicatively, in a recent report, the Bank of Italy showcases how existing RTGS systems such as TIPS can be succeeded by "by a distributed, horizontally scalable system" that will facilitate for greater processing capacity [2]. Moreover, recent advancements leading up to

Ethereum 2.0, such as *sharding*, present universally applicable solutions for scaling blockchain-based systems [21].

Interoperable "Interoperability refers to the technical ability of two or more disparate systems to exchange information in a manner that maintains the 'state and uniqueness' of the information being exchanged" [38]. RTGS systems are widely available and interoperable. In Europe, TIPS is widely available in 56 countries and integrates with domestic payment systems and procedures. Moreover, it also offers the ability to settle transactions in foreign currency. Besides the traditional financial system, solutions have been proposed for interoperability of RTGS systems with deployments from the private sector through a decentralised network of oracles [38]. For DLT/blockchain systems, a plethora of interoperable solutions have been proposed over the years [17].

Flexible and Adaptable BIS' report defines flexibility and adaptability as the ability of a CBDC system to be able to change according to policy imperatives. Given the modularity of RTGS and DLT/blockchain systems, as evident by the applications that they enable, they are likely to be adaptable to changing condition and policy imperatives.

Our analysis indicates that for the most favourable options for CBDCs, the dichotomy between accounts or token-based E-Own and the RTGS and DLT/blockchain infrastructure is redundant, as the choices do not represent material differences. This comes as an extension to recent bibliography from the Bank of England and Riksbank, which underline the insignificance of the account versus token debate for privacy [13, 20]. Moreover, the convergence of CBDC deployments on an RTGS versus a DLT/blockchain infrastructure in how they satisfy the necessary features relies on the permissioned nature of the deployment. As put by the Bitcoin white paper, "the main benefits are lost if a trusted third party is still required", and hence DLT/blockchain solutions more closely resemble their centralised counterparts.

Acknowledgments Part of this work has been carried out in the H2020 INFINITECH project, which has received funding from the European Union's Horizon 2020 Research and Innovation Programme under Grant Agreement No. 856632.

References

1. Adrian. (2019). *Stablecoins, Central Bank digital currencies, and cross-border payments: A new look at the International Monetary System.* IMF. Available at: https://www.imf.org/en/News/Articles/2019/05/13/sp051419-stablecoins-central-bank-digital-currencies-and-cross-border-payments. Accessed 14 June 2021.
2. Arcese, M., Giulio, D. D., & Lasorella, V. (2021). *Real-time gross settlement systems: Breaking the wall of scalability and high availability, Mercati, infrastrutture, sistemi di pagamento* (Markets, Infrastructures, Payment Systems). 2. Bank of Italy, Directorate General for Markets and Payment System. Available at: https://ideas.repec.org/p/bdi/wpmisp/mip_002_21.html. Accessed 16 June 2021.

3. Auer, R., & Böhme, R. (2020). *The technology of retail central bank digital currency.* www.bis.org. [online] Available at: https://www.bis.org/publ/qtrpdf/r_qt2003j.htm.
4. Bank of International Settlements (Ed.). (1996). *Implications for central banks of the development of electronic money.* Basle.
5. Bank of International Settlements. (2017). *Central bank cryptocurrencies* (p. 16).
6. Bank of International Settlements. (2019). *Investigating the impact of global stablecoins* (p. 37).
7. Bank of International Settlements. (2021). *Ready, steady, go? Results of the third BIS survey on central bank digital currency.* Available at: https://www.bis.org/publ/bppdf/bispap114.pdf. Accessed 9 June 2021.
8. Bank of International Settlements and Committee on Payments and Market Infrastructures. (2015). *Digital currencies.*
9. Bank of Japan. (2020). 中デジタル通が金同等の能を持つための技的(p. 26).
10. Bank of Thailand. (2021). BOT_RetailCBDCPaper.pdf.
11. Carnegie Endowment for International Peace. (2017). *Timeline of cyber incidents involving financial institutions.* Carnegie Endowment for International Peace. Available at: https://carnegieendowment.org/specialprojects/protectingfinancialstability/timeline. Accessed 15 June 2021.
12. Christodorescu, M., et al. (2020). Towards a two-tier hierarchical infrastructure: An offline payment system for Central Bank digital currencies. *arXiv:2012.08003* [cs]. Available at: http://arxiv.org/abs/2012.08003. Accessed 15 June 2021.
13. Claussen, C.-A., Armelius, H., & Hull, I. (2021). *On the possibility of a cash-like CBDC* (p. 15).
14. Cœuré, B., et al. (2020), *Central bank digital currencies: foundational principles and core features report no. 1 in a series of collaborations from a group of central banks.* Available at: https://www.bis.org/publ/othp33.pdf. Accessed 7 June 2021.
15. Diem Association. (2020). White Paper | Diem Association. Available at: https://www.diem.com/en-us/white-paper/. Accessed 14 June 2021.
16. Diez de los Rios, A., & Zhu, Y. (2020). *CBDC and monetary sovereignty.* Bank of Canada. https://doi.org/10.34989/san-2020-5
17. Dionysopoulos, L. (2021). Big data and artificial intelligence in digital finance. *INFINITECH OA Book.*
18. EC. (2019). *Joint statement by the Council and the Commission on "stablecoins."* [online] www.consilium.europa.eu. Available at: https://www.consilium.europa.eu/en/press/press-releases/2019/12/05/joint-statement-bythe-council-and-the-commission-on-stablecoins/#:~:text=12%3A54-. Accessed 21 Feb 2022.
19. ECB. (2020). *Report on a digital euro.* [online] Available at: https://www.ecb.europa.eu/pub/pdf/other/Report_on_a_digital_euro~4d7268b458.en.pdf.
20. England, B. (2020). *Discussion paper – Central Bank digital currency: Opportunities, challenges and design* (p. 57).
21. Ethereum. (2020). *Shard chains.* ethereum.org. Available at: https://ethereum.org. Accessed 16 June 2021.
22. EU Blockchain Observatory & Forum. (2021). *Central Bank digital currencies and a Euro for the future.*
23. European Central Bank. (2020). *Report on a digital Euro* (p. 55).
24. Holmes, A. (2019). The biggest tech scandals of the 2010s, from NSA spying to Boeing's deadly crashes to WeWork. *Business Insider.* Available at: https://www.businessinsider.com/biggest-tech-scandals-2010s-facebook-google-apple-theranos-wework-nsa-2019-10. Accessed 17 June 2021.
25. Insights, L. (2020) SIBOS 2020: Chinese central bank says 1.1 billion digital yuan paid so far. *Ledger Insights – enterprise blockchain.* Available at: https://www.ledgerinsights.com/sibos-2020-chinese-central-bank-digital-yuan-currency-cbdc/. Accessed 14 June 2021.
26. Kharpal, A. (2021). El Salvador becomes first country to adopt bitcoin as legal tender after passing law. *CNBC.* Available at: https://www.cnbc.com/2021/06/09/el-salvador-proposes-law-to-make-bitcoin-legal-tender.html. Accessed 16 June 2021.

27. Kiersz, A. (2019). 10 simple charts that show the wild ways America has changed since 2010. *Business Insider*. Available at: https://www.businessinsider.com/charts-show-how-americas-economy-and-population-changed-since-2010. Accessed 17 June 2021.
28. Koranyi, T. W., & Balazs. (2019). Facebook's Libra cryptocurrency faces new hurdle from G7 nations. *Reuters*, 18 October. Available at: https://www.reuters.com/article/us-imf-worldbank-facebook-idUSKBN1WW33B. Accessed 14 June 2021.
29. Li, Y. (2019). This is now the longest US economic expansion in history. *CNBC*. Available at: https://www.cnbc.com/2019/07/02/this-is-now-the-longest-us-economic-expansion-in-history.html. Accessed 17 June 2021.
30. MIT Technology Review. (2019). Once hailed as unhackable, blockchains are now getting hacked. *MIT Technology Review*. Available at: https://www.technologyreview.com/2019/02/19/239592/once-hailed-as-unhackable-blockchains-are-now-getting-hacked/. Accessed 15 June 2021.
31. Simmons, R., Dini, P., Culkin, N., & Littera, G. (2021). Crisis and the role of money in the real and financial economies—An innovative approach to monetary stimulus. *Journal of Risk and Financial Management, 14*(3), 129. Available at: https://www.mdpi.com/1911-8074/14/3/129 Accessed 21 Feb. 2022.
32. Soergel. (2020). Dow plunges to biggest loss in history on Coronavirus fears | Economy | US News, US News & World Report. Available at: www.usnews.com/news/economy/articles/2020-02-27/dow-plunges-to-biggest-loss-in-history-on-coronavirus-fears. Accessed 17 June 2021.
33. Statista. (2020). Number of IPOs in the U.S. 1999–2020. *Statista*. Available at: https://www.statista.com/statistics/270290/number-of-ipos-in-the-us-since-1999/. Accessed 17 June 2021.
34. Sveriges Riksbank. (2017). *The Riksbank's e-krona project*. S E (p. 44).
35. The World Bank. (2021). GDP growth (annual %) | Data. Available at: https://data.worldbank.org/indicator/NY.GDP.MKTP.KD.ZG?end=2019&start=1970&view=chart. Accessed 17 June 2021.
36. U.S. Bureau of Labor Statistics. (2017). *Below trend: the U.S. productivity slowdown since the Great Recession: Beyond the numbers*. U.S. Bureau of Labor Statistics. Available at: https://www.bls.gov/opub/btn/volume-6/below-trend-the-us-productivity-slowdown-since-the-great-recession.htm. Accessed 17 June 2021.
37. Visco, I. (2020). *The role of TIPS for the future payments landscape Speech by Ignazio Visco Governor of the Bank of Italy Virtual conference "Future of Payments in Europe."* [online] Available at: https://www.bis.org/review/r201130c.pdf.
38. World Economic Forum. (2020). WEF_Interoperability_C4IR_Smart_Contracts_Project_2020 (1).pdf. Available at: http://www3.weforum.org/docs/WEF_Interoperability_C4IR_Smart_Contracts_Project_2020.pdf

Chapter 6
Historic Overview and Future Outlook of Blockchain Interoperability

Lambis Dionysopoulos

1 Multidimensional Mutually Exclusive Choices as the Source of Blockchain Limitations

The limitations of blockchains have been repeatedly highlighted by some of the most accredited members of the decentralized community and summarized by the popular term *blockchain trilemma* [1]. While this *trilemma* is an oversimplification of the multidimensional mutually exclusive choices made by developers, users, and markets alike, it reveals, in an intuitive manner, why dissimilar networks must necessarily exist. Various iterations of this trilemma have been proposed: Vitalik Buterin views the blockchain trilemma as the balancing act between *decentralization*, *security*, and *scalability* [1], while Trent McConaghy considers *decentralization*, *consistency*, and *scalability* part of the trifecta [2]. For the purposes of this chapter, we will adopt the former approach. As the blockchain trilemma suggests, blockchains can only satisfy two of three trilemma dimensions and thus by definition can be one of the following: (1) *secure* and *scalable* but lack *decentralization*, (2) *scalable* and *decentralized* but lack *security*, and (3) *secure* and *decentralized* but lack *scalability* [1]. The various attempts at striking the perfect balance in the trilemma along with the ever-growing use cases for blockchains effectively translated in the creation of entirely new systems and cryptocurrencies, in an attempt to satisfy additional requirements. This further aggravated the problem of incompatibility between systems and in itself gave birth to a second type of dilemma: Choosing an innovative, dormant blockchain enables new technologies and state-of-the-art features; however, the risk of software bugs, security breaches, and potential loss of funds is considerably higher when compared to that of an

L. Dionysopoulos (✉)
Institute for the Future, University of Nicosia, Nicosia, Cyprus
e-mail: dionysopoulos.c@unic.ac.cy

© The Author(s) 2022
J. Soldatos, D. Kyriazis (eds.), *Big Data and Artificial Intelligence in Digital Finance*,
https://doi.org/10.1007/978-3-030-94590-9_6

established system. On the contrary, mature blockchains reduce this risk but at the same time limit access to innovative novel features. Blockchain interoperability is a direct response to those dilemmas, and an attempt at capturing the upsides that the technology has to offer, while minimizing the downsides.

2 First Attempts at Interoperability

2.1 Anchoring

One of the first implementations in blockchain interoperability came in the form of *anchoring*. Anchoring describes the process of storing data from one blockchain to another. This approach is usually employed by chains that adopt a *faster* and more *efficient model* that want to benefit from the inherent security of a less flexible system, without compromising on performance. In practice, this is done by producing a unique fingerprint of all the information deemed meaningful at determining one system's current state and storing it on the *primary anchor* (the less efficient more robust chain) in the form of a *hash*. In this sense, the primary anchor serves a trusted immutable timestamp log that can be used to verify the true state of the secondary chain. The hash can be generated in a number of different ways, with the most common being that of a Merkle tree[1] with the Merkle root serving as the anchor. As a rule of thumb, hashed data from the secondary chain is stored in the primary chain in periodic intervals to represent its updated state. Systems that employ proof of work are preferred as primary anchors for their inherent connection to physics, vast deployment, and proven reliability. While immutability is difficult to measure precisely, with *anchors*, efficient chains can achieve an otherwise unlikely degree of security.

2.2 Pegged Sidechains

Another approach that facilitated communication between otherwise isolated systems came in the form of a layer 2 solution coined *pegged sidechains*. Pegged sidechain is a technology that allows ledger-based assets of one blockchain to be transferred and used freely in a separate blockchain and even moved back to the original if necessary [3]. The secondary chain, or *sidechain*, is fully independent meaning that its potential compromise will have no impact on the main or *parent* chain. One early implementation that didn't allow for such flexibility was that of a *one-way pegged chain*. This term refers to the practice of destroying assets

---

[1] A tree structure of data in which every leaf node is labelled with the cryptographic hash of a data block and every non-leaf node is labelled with the hash of the labels of its child nodes.

in a publicly recognizable way on the parent chain which, when detected by the sidechain, would initiate the creation of new assets, effectively "transferring" them. With the introduction of the *symmetric two-way peg*, the "destruction" of assets is no longer necessary. Instead, they are sent to a special output on the parent chain that can only be unlocked by a simple payment verification (*SPV*) *proof of possession* [4] on the sidechain. Once locked on the parent chain, assets are allowed to move freely on the sidechain and can always be sent to an SPV-locked output so as to produce SPV proof and unlock funds on the parent chain. In the case of an asymmetric *two-way peg*, SPVs are no longer needed as users of the sidechain are simultaneously full validators of the parent chain and, thus, are at all times aware of its state.

2.3 Cross-Chain Atomic Swaps

A promising development in the space of blockchain interoperability is *cross-chain atomic swap*. Cross-chain atomic swaps allow peers to exchange different ledger-based assets directly, without the need of a custodian third party or trusted intermediary. Those exchanges are *atomic* in the sense that they completely either occur as described or have no effect at all. This is essential for the security of the exchange, as it ensures that a scenario in which – either through manipulation or human error – a single entity that is counterparty to the transaction cannot control both assets at the same time. To initiate an atomic swap, funds from two chains are initially deposited in *hashed time-locked smart contracts* (HTLCs) that can be publicly inspected on the blockchain. Those contracts can only be unlocked through a special key called a *preimage*, a combination of a key and a code. The code is set by the initiator of the swap and is initially kept secret, while the key part of the preimage corresponds to the keys held by the two peers enmeshed in the swap. If the initiator deems funds to be deposited as agreed, they publish the secret code, unlocking assets for both parties, and thus the cross-chain atomic swap occurs. At any time and for any reason, any of the two parties can walk away from the exchange, effectively canceling it. HTLCs are programmed to expire when enough time elapses, returning funds to the original owner's address, and by doing so they fulfill their premise at an atomic exchange. While the proposition of fast and nearly feeless asset exchanges between blockchains, without the need for a trusted intermediary, seems attainable, in practice, such deployments face many obstacles. As atomic swaps are built on the technologies of time and hash locks, they can only be employed by chains that adopt both. At the same time, assets that are exchanged must have the same *hash algorithm*. Arguably, the most important limitation toward mainstream adoption is that the initiation of cross-chain atomic swaps currently involves prohibitive levels of proficiency with the aforementioned technologies.

2.4   Solution Design

Technical incompatibilities are not exclusively responsible for the fragmentation of the blockchain space. As with all dormant technologies, the pursue of "all possible roads," the urge to discover the limitations of the technology, the sheer curiosity of human nature, and the notion that the "next big thing" lies in the next line of code contribute to the infinite complexity that characterizes today's blockchain ecosystem. Lacking a *focal point*, this complexity is just *noise*, serving as an obstacle to greater adoption and deterring end users from utilizing blockchains to achieve their goals. Conversely, multiple benefits could be obtained if developers were to utilize interoperable solutions to "tame" the chaotic blockchain space. We deem that the technology available currently is enough to employ (1) a user-friendly, zero-fee, multicurrency micropayment network, (2) a publicly verifiable archive of critical information on private blockchains, and (3) "modules" that extend the functionality of less flexible systems.

To begin with, given a wide enough employment and sufficient liquidity, atomic swaps can allow any cryptocurrency to "transmogrify" itself into any other almost instantaneously. This could in practice eliminate the high switching costs imposed by incompatible technologies and maintained by centralized exchanges. For most use cases, save that of exchanging cryptocurrencies for fiat, such implementation might render those centralized exchanges unnecessary. At the same time, exchanges that employ questionable practices or high fees will be forcefully discontinued. This could further add to the security and make the blockchain space more attractive overall.

In a similar manner, atomic swaps when used in conjunction with a *Lightning Network* [5] and other layer 2 solutions could reform payments, by making them cross-chain and extremely efficient, economic, anonymous, or secure depending on the user needs. Wallets that utilize both technologies can, by looking at a list of available cryptocurrencies, determine the one that will allow for the fastest possible transaction, with the lowest possible fees, effectively creating an almost zero-fee multicurrency micropayment network. Similarly, users can choose to optimize their transaction for anonymity, by selecting paths that obscure the origin of sent coins through *mixing*.[2] When transactions require more security or finality, a blockchain with large amounts of computational resources dedicated to its upkeep could be preferred. Lastly, the use of atomic swaps has additional implications for the use of cryptocurrency as a medium of exchange between customers and retailers. Even with the current state of the technology, customers would be able to pay in any cryptocurrency they desire, with retailers receiving a *stablecoin*, ensuring that they are protected against extreme price fluctuation. At the same time, by bypassing third-party payment verification services that charge fees, businesses save money and – if they so choose – can pass those savings on to the consumer.

---

[2] The practice of mixing one's funds with other people's money, intending to confuse the trail back to the funds' original source.

In conjunction with the above, anchoring can be leveraged in creating a publicly verifiable archive of blockchain checkpoints. Businesses that use blockchains in a private or consortium setting can publish information for auditing and transparency purposes. By the use of the same technology, blockchains that aim at being extremely efficient could mitigate reorganizations and achieve a relatively higher level of finality by "checkpointing" their latest block on a public blockchain and then building the next one on top of that. Lastly, layer 2 solutions with pegged sidechains can be utilized in extending the functionality of less flexible main chains. Projects like BitPay's ChainDb builds a peer-to-peer database system on top of the Bitcoin blockchain, treating it like a secure database [6]. Counterparty brings user-created tradable currencies, a decentralized exchange, and most importantly the ability to run Ethereum smart contracts and decentralized application on top of Bitcoin through the counterparty protocol [7]. Through the use of pegged sidechains and pegged sidechains on top of sidechains, virtually any feature available in the blockchain space can be added to almost any chain, naturally with questionable usability.

3   Later Attempts at Interoperability

Even when utilizing present interoperable solutions, the limitations of the decentralized space have begun to show [8]. Many projects, with the most notable being Polkadot, Cosmos, and Interledger, are working toward building an Internet-like network infrastructure that will combat present limitations by allowing any blockchain to seamlessly communicate with another in an Internet-like network infrastructure. In this final part of the chapter, we provide a brief overview of the aforementioned projects and present an idealistic solution design, disregarding the dormant stage of the technology and present limitations.

3.1   *Polkadot*

The Polkadot protocol was introduced in 2017 and was followed by a successful initial coin offering (ICO) that raised more than 145$ million. Polkadot is one of the most ambitious interoperable projects and was originally conceived by the co-founder of Ethereum and creator of the Solidity programming language, Gavin Wood. It aims at alleviating the primary pain points [9] of current blockchain implementations with the use of a heterogeneous multichain framework. If successful, this would allow for seamless cross-chain interaction, as smart contracts, applications, value, and data will be able to flow freely [10]. Additionally, networks connected with Polkadot will be able to make use of each other services and harness their unique advantages with near-native efficiency. To achieve this, the core of the Polkadot structure consists of three main elements, the *relay chain*, the *parachain*, and the *bridge chain*. The relay chain serves as the main chain or hub, where all parachains connect. It is responsible for coordinating consensus, providing pooled security for the network, and transferring data between parachains. A parachain is any blockchain or other data structures, public or permissioned, that connect to the relay chain. For scaling purposes, they are responsible for processing their own transactions but are secured by the network consensus. The last main element is that of the bridge chain, which is responsible for connecting completely sovereign blockchains that do not comply with Polkadot's governance protocols to the relay chain. One such example is the Ethereum blockchain. To support the network of connected chains described above, a structure of adjacent roles is put in place. *Validators* are responsible for submitting valid finalized blocks to the relay chain. To qualify, they are required to *stake*[3] a significant *bond*[4] in the form of the native DOT token. They receive candidate blocks from *collators* and are approved by *nominators*. Nominators are parties that hold a significant stake in the network and are responsible for electing trustworthy validators. Collators, on the other hand, are essentially the validators of each individual parachain. Lastly, *fishermen* are responsible for seeking out malicious behavior to report to validators in exchange for a one-off reward. Bad actors have their stakes *slashed*,[5] and parts of their assets are used to fund the bounty-like rewards given to fishermen.

3.2 Cosmos

Cosmos is an upcoming Tendermint-based [11] framework, which similar to Polkadot aims at standardizing cross-blockchain communication for interoperable purposes. It raised more than 4870 Bitcoin in 2017. Building on the foundation we laid in the previous paragraph, we can easily interpret the various elements of Cosmos. *Hub* is the chosen name for the main chain of the network, and it serves an almost identical role to that of the relay chain in Polkadot. The Hub is built on top of a Tendermint engine, which is comprised of two main parts, the Tendermint core, which offers a BFT proof-of-stake consensus engine, and an ABCI (application blockchain interface),[6] which replicates *dApps*[7] deployed in multiple programming languages. Similar to Polkadot, Cosmos utilizes parachains, called *zones*, which connect to the Hub. Validators commit blocks originating from zones

---

[3] The process of holding funds in a wallet to support operations of a blockchain network and receive rewards as an exchange.

[4] As in the financial instrument.

[5] Reduced as a punishment for misbehavior.

[6] An interface that defines the boundary between the replication engine (the blockchain) and the state machine (the application). Using a socket protocol, a consensus engine running in one process can manage an application state running in another.

[7] Abbreviation for decentralized application. A backend code running on a decentralized peer-to-peer network.

to the Hub. Lastly, a native digital currency called ATOM serves as a license for holders to vote, validate, and delegate other validators. ATOMs are also used for antispam protection, in a similar way to Ethereum's *Gas*, and slashed in case of bad acting. Cosmos aims at bringing together cryptocurrencies by offering a more robust, scalable, and faster model for distributed asset exchange compared that is possible with a cross-chain atomic swap. At the same time, it aims at solving Ethereum's scaling problems, by providing faster commit times by utilizing its Tendermint consensus and having Ethereum contracts run on different zones in a form of sharding [1].

3.3   Interledger

Similarly, Interledger is an open protocol suite that aims at enabling seamless exchange of value, across different payment networks or ledgers. It is a revised and open-source implementation of the cross-chain atomic swap protocol. Interledger positions itself as an "internet of money" and intents to route packets of value similar to how packets of information are routed on the Internet. To achieve this, it operates on a stack of four layers, (a) the *application layer*, which is responsible for coordinating the atomic swap sender and destination addresses; (b) the *transport layer*, which serves the function of an end-to-end protocol between the sender and receiver of value; (c) the *Interledger layer*, which actualizes the transaction data; and, lastly, (d) the *ledger layer*, which is used for payment settlement. At the same time, peers on the Interledger network are nodes that can have any or all of the following roles: *senders*, *routers*, or *receivers*. Senders are essentially the initiators of a transfer of value, while receivers are the receivers of that. Routers, on the other hand, serve as intermediaries in the transaction and are responsible for applying currency exchange and forwarding of the packets of value.

3.4   Idealistic Solution Design

Disregarding any present limitations, we speculate that blockchain interoperability will facilitate the frictionless data flow between systems. An exhaustive categorization of all possible derivate applications is impossible, but we opt to highlight three specific use cases that we feel encapsulate the magnitude of changes that one can expect. Blockchain interoperability could (1) allow for a state-of-the-art, fully automated, and multi-asset financial system accessible anywhere and by anyone in the world, (2) provide a new model for consumer-business relationships by rendering traditional Know Your Customer practices and the monthly subscription scheme obsolete, and (3) enable efficient, scalable, and secure IoT and AI applications. To start, completely frictionless flow of information between blockchains when coupled with existing *digital identity* protocols as those proposed by *Consensys*,[8] *Bitnation*[9], or *Evernym*[10] can lead to a state where any application connected the interoperable blockchain network can be aware of the identity of anyone. To facilitate for privacy, data minimization techniques such as *zero-knowledge proofs*[11] can be utilized. This could enable individuals to verifiably prove their identity, without necessarily revealing sensitive information. In the context of an online service, this would mean that a traditional registration with the input of an individual's personal data will no longer be necessary. By making payments efficient, new opportunities for alternative business models emerge. As an example, the prominent scheme of monthly subscriptions and the accompanying subscription fatigue can be replaced with a pay-as-you-use model, which has the potential to yield multiple benefits for consumers and businesses alike. Up until now, subscription services operated under educated assumptions about monthly use of their services, reflected in a monthly fee. This approach necessarily leads to overcharging or undercharging of individuals as it is not tied to use. Additionally, and due to the periodic nature of payments, it exposes companies to credit risk. By making transactions efficient, a network of interoperable blockchains could allow for constant streaming of funds, as long the user streams video and audio or in other terms uses a service. This intermediate network ensures that no party can "cheat." By the same principles, the use of otherwise prohibitively expensive enterprise-grade programs could become more widely accessible. Users will be able to pay only for their use instead of the full price, and businesses would attract otherwise unattainable customers.

By utilizing interoperable blockchains, modern cryptocurrency wallets could extend their functionality beyond what is possible today. To begin, the sheer efficiency of cryptocurrency exchange will allow for any user or application to use any token that best fulfills their needs as it could be transmogrified to any other instantly and for free. At the same time, wallets could seamlessly make use of the growing network of decentralized finance (DeFi) to provide multiple benefits for their users. Deposited funds could automatically be lent out through a network of services similar to Compound[12] where the on-wallet software would determine the combination that will yield the highest returns. This will allow for asset holders to enjoy industry's high interest rates, with immediate access to their funds. At the same time, borrowers could have instant access to liquidity on fair and transparent terms. Taking this above concept a step further, funds could be automatically hedged or stored based on behavioral patterns. On-wallet software could monitor the spending habits of the financial profile of the user and make decisions to their best interest. For example, a wallet could determine that the best use of its funds is to be

---

[8] Consensys Website.

[9] Bitnation Website.

[10] Evernym Website.

[11] A method by which one party (the prover) can prove to another party (the verifier) that they know a value x, without conveying any information apart from the fact that they know the value x.

[12] Compound Finance Website.

traded on one of the many token sets provided by the Set Protocol[13] or invested long term in Bitcoin. Additionally, the growing popularity tokenization and non-fungible tokens (NFTs) [12] could enable additional novel applications. For example, a user could pay for their morning coffee with one-millionth of their tokenized home value or even opt to use part of it as collateral for a loan, if they so choose.

Artificial intelligence (AI) is another transformable technology. Interoperable blockchains and Internet-native money can be utilized in making AI efficient, secure, and scalable. To start, AI built on top of interoperable blockchains will be able to exchange information and learning models for free or even in exchange for digital assets. In this machine-to-machine economy, thousands of machine learning agents could simultaneously share and improve their knowledge. On the topic of security, a network of interoperable chains can also be used to provide a truly *explainable AI*.[14] As humans are not aware of the inner workings of deep learning, there is no certainty as to what inputs result in what outputs. As a result, such systems are treated as *black boxes*. By offering an immutable ledger and precise access to information, blockchains can be used to record specific events and thus offer accountability and explanations in an unbiased manner. Finally, blockchain-backed AI could unleash the full potential of IoT devices [13]. Billions of connected devices around the world record our universe of data, serving as a nervous system for the a distributed on-chain *master brain* that would process this wealth of information.

Acknowledgments This work has been carried out in the H2020 INFINITECH project, which has received funding from the European Union's Horizon 2020 Research and Innovation Programme under Grant Agreement No. 856632.

References

1. Buterin, V.. github.com [Online]. Available: https://github.com/ethereum/wiki/wiki/Sharding-FAQ#this-sounds-like-theres-some-kind-of-scalability-trilemma-at-play-what-is-this-trilemma-and-can-we-break-through-it
2. McConaghy, T. (2016, July 10). Medium [Online]. Available: https://blog.bigchaindb.com/the-dcs-triangle-5ce0e9e0f1dc
3. Back, A., Corallo, M., Dashjr, L., Friedenbach, M., Maxwell, G., Miller, A., Poelstra, A., Timón, J., & Wuille, P. (2014, October 22). http://kevinriggen.com/ [Online]. Available: http://kevinriggen.com/files/sidechains.pdf
4. Rosenbaum, K. (2016, August 30). popeller.io [Online]. Available: https://popeller.io/index.php/2016/08/30/spv-proofs-in-sidechains/
5. Joseph Poon, T. D. (2016, January 14). lightning.network [Online]. Available: https://lightning.network/lightning-network-paper.pdf
6. Bitpay. bitpay(dot)com [Online]. Available: https://bitpay.com/chaindb.pdf

---

[13] TokenSets Website.

[14] On explainable AI.

7. counterparty. counterparty(dot)io. [Online]. Available: https://counterparty.io/docs/assets/
8. Hertig, A. (2017, December 4). Coindesk(dot)com. [Online]. Available: https://www.coindesk.com/loveable-digital-kittens-clogging-ethereums-blockchain
9. WEB3 Foundation. polkadot(dot)network [Online]. Available: https://polkadot.network/Polkadot-lightpaper.pdf
10. Wood, G. (2017). POLKADOT: Vision for a heterogeneous multi-chain framework. [Online]. Available: https://polkadot.network/PolkaDotPaper.pdf
11. Kwon, J. (2014). tendermint(dot)com. [Online]. Available: https://tendermint.com/static/docs/tendermint.pdf
12. Singhal, S. (2019, August 9). Hackernoon(dot)com. [Online]. Available: https://hackernoon.com/what-can-be-tokenized-the-tokenization-of-everything-mw1ay3bk7
13. Lin, X., Li, J., Wu, J., Liang, H., & Yang, W. (2019). Making knowledge tradable in edge-AI enabled IoT: A consortium blockchain-based efficient and incentive approach. *IEEE Transactions on Industrial Informatics, 15*(12), 6367–6378.
14. Alex Tapscott, D. T. (2017, March 1). How blockchain is changing finance. *Harvard Business Review, 4–5.*
15. Treleaven, P., Brown, R. G., & Yang, D. (2017). Blockchain technology in finance. *Computer, 50*(9), 14–17.
16. Clare Sullivan, E. B. (2017). E-residency and blockchain. *Computer Law & Security Report.*
17. Bhowmik, T. F. D. (2017). The multimedia blockchain: A distributed and tamper-proof media transaction framework. In *22nd international conference on Digital Signal Processing (DSP).*
18. Wang, J., Wang, S., Guo, J., Du, Y., Cheng, S., & Li, X. (2019). A summary of research on blockchain in the field of intellectual property. *Procedia Computer Science, 147,* 191–197.
19. Mettler, M. (2016). Blockchain technology in healthcare: The revolution starts here. In *IEEE 18th international conference on e-Health Networking, Applications and Services (Healthcom).*
20. Loeber, J. (2018, January 28). *Fragmentation in cryptocurrencies* [Online]. Available: https://johnloeber.com/w/fragmentation.html
21. Bhowmik, T. F. D. (2017). The multimedia blockchain: A distributed and tamper-proof media transaction framework. In *22nd international conference on Digital Signal Processing (DSP).*
22. Bitpay. Bitpay.com. [Online]. Available: https://bitpay.com/chaindb.pdf

Chapter 7
Efficient and Accelerated KYC Using Blockchain Technologies

Nikolaos Kapsoulis, Antonis Litke, and John Soldatos

1 Introduction

In state-of-the-art financial relations, Know Your Customer (KYC) and Know Your Business (KYB) policies expect that customer parties, either individuals or entire corporations, endeavor verification of their identity. Particularly, a KYC/KYB mechanism ensures that the identification and verification of a client occurs against national and international regulations and laws set by governments, commissions, central banks, and financial associations. Each financial organization is able to estimate the risks involved with sustaining a new business–customer partnership. Within the wider Finance field of Anti-Money Laundering (AML) procedures, every financial institution is obliged to establish KYC and KYB operations at the time they register a new customer. As both the customer profile information and the relevant laws and rules are subject to changes over time, the updates and maintenance of the data become more complicated. Furthermore, the adopted centralized systems are exposed to new-generation risks of data protection and cybersecurity that form cheaper to launch relevant attacks led by more sophisticated adversaries year by year [1].

Blockchain technology and particularly permissioned blockchain networks are capable of providing security to the KYC and KYB processes through decentralization [2, 3]. The concept of decentralization mainly exploits the idea that the

N. Kapsoulis (✉) · A. Litke
INNOV-ACTS Limited, Nicosia, Cyprus
e-mail: nkapsoulis@innov-acts.com; alitke@innov-acts.com

J. Soldatos
INNOV-ACTS LIMITED, Nicosia, Cyprus

University of Glasgow, Glasgow, UK
e-mail: jsoldat@innov-acts.com

© The Author(s) 2022
J. Soldatos, D. Kyriazis (eds.), *Big Data and Artificial Intelligence in Digital Finance*,
https://doi.org/10.1007/978-3-030-94590-9_7

information is replicated across all network nodes. In this context, the information integrity cannot be harmed by sabotaging one or more nodes, and thus, a single point of failure is avoided. In particular, the permissioned blockchain technology isolates the sensitive information inside a dedicated private network where only privileged parties can access it. Particularly, every party is accepted into the network by an invitation from inside that enables the participant to engage in the blockchain activities. The customer information is kept safe on the private ledger where data transparency is offered to the privileged group of the legal network participants. Both the clients and the organizations are able to perform CRUD operations on the data (create, read, update, and delete) under pre-configured access control policies. For instance, the various features of permissioned blockchains enable applications of different policies that are able to separate legal parties into a higher privacy network running inside the initially defined private one. Improved privacy control and data immutability rule inside the aforementioned technological scenario, while they ensure legitimate customer data protection and management together with proper administration of the data by financial enterprises [4, 5].

2   Architecture

The implemented KYC/KYB blockchain solution resolves the aforementioned issues of the industry by exploiting blockchain technology as the underlying infrastructure. In Fig. 7.1, the high-level architecture of the solution is depicted.

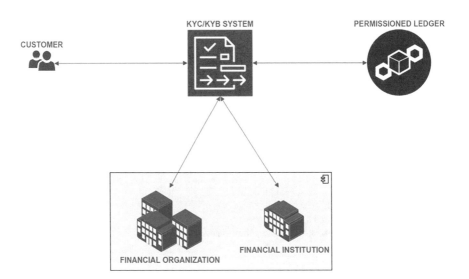

Fig. 7.1 High-level architecture of KYC/KYB solution

In general, the customer participant being either an individual or an entire organization needs to acquire financial services that are offered by financial institutions or organizations. For the completion of the relevant transaction, the financial institution requests that the customer identity information is documented specifically in KYC/KYB data format. Thus, after the data is legally verified, the customer participant uploads their KYC/KYB documentation to the KYC/KYB system that interacts with the permissioned ledger and stores the information on-chain. When a financial organization requires to initiate business relations with the same customer, the latter's KYC/KYB information is easily and rapidly obtained through the access rights provided by the first financial institution since business partnership is already arranged with the customer. It is important to clarify that initially the customer has declared their consent in sharing KYC/KYB information among privileged participants of the blockchain network, while the privacy of the information is guaranteed. In this system, data security and efficient data management and maintenance govern since the united underlying blockchain infrastructure with common data structures offers stability, security, sustainability, and high transactions per second.

3   Use Case Scenarios

The design specifications of blockchain applications are tightly coupled with the definition of the underlying business intelligence. For each use case scenario, as they are analyzed at a later stage in the current section, the respective business objects prompt the corresponding developments and formulate the relevant blockchain submissions. The following sections provide the descriptions of the scenarios that are encapsulated in the solution by elaborating on the relevant details. The entire INFINITECH on-chain solution serves in principle the use case scenarios analyzed below.

Use Case Scenario I: Assertion of Identification Documents
In scenario I, the customer asserts their identity documents to the corresponding financial organization through the KYC/KYB on-chain process. In particular, a customer representing either an individual or a corporation acknowledges their KYC or KYB information in order to initiate business relations with the finance participants of the permissioned blockchain network. In this context, each enterprise participant ensures the legitimacy of their customer, and thus, new business relationships are established. The scenario analysis is stated in Table 7.1.

Use Case Scenario II: Read Access for KYC/KYB Information
In scenario II, a financial organization as a legal network participant has read access to customer KYC/KYB documents information (Table 7.2). In particular, inside the permissioned blockchain network, each party is able to view the corresponding ledgers with the submitted client information. In this context, each financial organization has read access to this data stored on-chain, and by initiating a simple

Table 7.1 Scenario I: Assertion of identification documents

Stakeholders involved	Customer
Pre-conditions	A private blockchain network is set up
Post-conditions	The customer's KYC/KYB information is stored on-chain
Data attributes	The customer's KYC/KYB documents
Normal flow	I. The customer uploads their KYC/KYB documentation on the blockchain ledger
	II. The documentation information is successfully stored on-chain
Pass metrics	The documentation information is successfully stored on-chain
Fail metrics	There is no customer to upload their KYC/KYB documents

Table 7.2 Scenario II: Read access for KYC/KYB information

Stakeholders involved	Financial organization
Pre-conditions	The financial organization is a legal participant of the network
Post-conditions	The financial organization has read access control over the requested KYC/KYB information
Data attributes	I. The financial organization's network attributes
	II. The requested KYC/KYB documents
Normal flow	I. The financial organization requests for read access of a specific customer's KYC/KYB documentation
	II. The requested documentation information access is successfully granted
Pass metrics	The requested documentation information is successfully granted
Fail metrics	The financial organization is not eligible to access the requested documentation information

read access request, the financial organization fetches the information of a customer they are interested in. Additionally, upon using the system, each customer approves that their data may be accessed by the financial organization that they will initiate business relations with.

Use Case Scenario III: Sharing of Customer KYC and KYB Documents
In scenario III, financial organization **A** shares the KYC/KYB document information of customer **B** with the financial organization **C** through the secure and private blockchain network. In particular, each of the financial organizations participating in the permissioned network is eligible for accessing the data stored on-chain. However, depending on the different access control rights granted by the time of joining the blockchain network, there exists the case where different organizations are qualified to access different groups of data. In this context, together with the initial consent given by the client, a financial organization may grant access to another organization or institution for a specific KYC/KYB documentation (Table 7.3).

In the previous sections, all the relevant use cases of the designed blockchain solution are documented. For each of the scenarios, the related sequence diagrams are produced and analyzed in the following section.

Table 7.3 Scenario III: Sharing of customer KYC and KYB documents

Stakeholders involved	Financial organizations
Pre-conditions	I. The financial organizations **A** and **C** are legal participants of the network
	II. The requested customer **B** has submitted their KYC/KYB document information
Post-conditions	The financial organization **C** has read access control over the requested KYC/KYB information
Data attributes	I. The financial organizations' **A** and **C** network attributes
	II. The requested KYC/KYB documents
Normal flow	I. The financial organization **A** requests for sharing of a specific customer's KYC/KYB documentation with a financial organization **C**
	II. The requested documentation information access is successfully granted
Pass metrics	The requested documentation information is successfully granted
Fail metrics	The financial organization **A** is not eligible to share the requested documentation information

4 Sequence Diagrams

The delivered use case scenarios of the previous sections correspond to specific UML sequence diagrams that are elaborated on in this section. The interactions between the different stakeholders are depicted in detail along with the components of the designed solution and their interoperability.

Scenario I Sequence Diagram
In scenario I, *Assertion of customer identity documents*, a customer provides their documentation information in order to be eligible for business partnerships with the participating financial organizations. Particularly, a customer actor, either an individual or an entire enterprise, acknowledges the requested KYC or KYB documents in the KYC/KYB system by uploading them on the ledger. This action is translated to an internal request that propagates the information inside the blockchain network, through the different blockchain components. The entire procedure is illustrated in Fig. 7.2, and the individually depicted Blockchain Modules are explained in Chap. 1, *A Reference Architecture for Big data Systems and AI Applications in the Finance Sector (INFINITECH-RA)*, while their corresponding development exploitation on behalf of the current solution is analyzed in the dedicated implementation section. Finally, the data is safely stored on the INFINITECH private ledger, and afterward, further actions and use case scenarios are enabled to emerge and execute, i.e., use case scenarios II and III.

Scenario II Sequence Diagram
In scenario II, *Read access for KYC/KYB information*, the blockchain network parties are constituted by the financial organizations that are able to inspect the

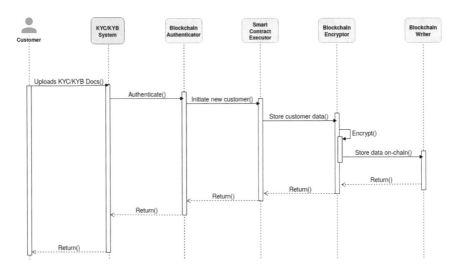

Fig. 7.2 Scenario I sequence diagram

KYC/KYB documentation information of the clients. Through the underlying private and secure blockchain network, the different organizations obtain access to a customer's KYC/KYB submitted data and are able to estimate the risk of establishing business relationships with them. As stated, the customer data submission is already accompanied by the customer's consent of the KYC/KYB information exploited by the legal parties of the network, i.e., the financial organizations. As in Fig. 7.3, through the sequential execution of the blockchain components, the read access is propagated to the financial organization that requested it.

Scenario III Sequence Diagram
In scenario III, *Sharing of customer KYC/KYB documents*, sharing of customer **B** information among participant organizations **A** and **C** takes place. Particularly, organization **C** obtains customer's **B** information through the cooperation of organization **A**. Since the customer data already exists inside the secure and private blockchain ledger, organization **C** requests it indirectly. Organization **A** responds to the request by granting read access of customer **B** information to organization **C**. In such an efficient data sharing system, the customer and the financial organizations benefit from the underlying technological features and infrastructures since, for instance, the customer avoids reentering their KYC/KYB data to a different system of a different organization, while the financial organizations rapidly obtain customer information and make decisions upon new business relations establishment (Fig. 7.4).

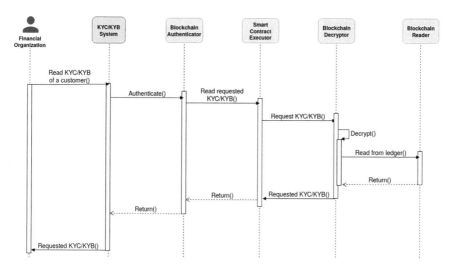

Fig. 7.3 Scenario II sequence diagram

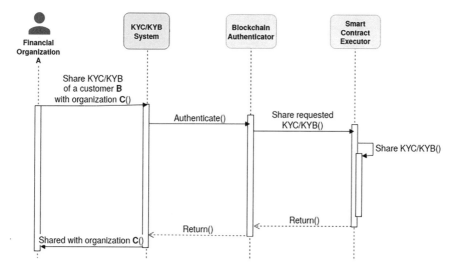

Fig. 7.4 Scenario III sequence diagram

5 Implementation Solution

This section explains the particulars of the implementation of the INFINITECH
Know Your Customer/Know Your Business On-Chain Solution. The designed and
developed solution has several important parts that are underlined below in order to
present in a more clear and coherent way the built framework and the technological
manifestation of the presented concept architecture and use case scenarios.

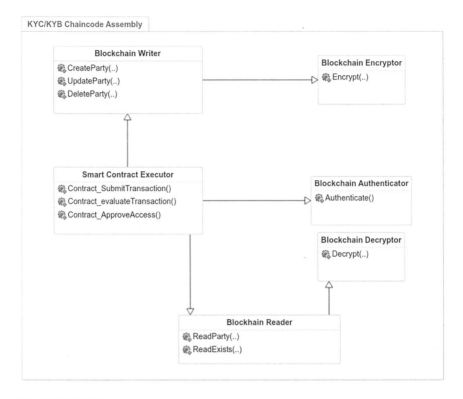

Fig. 7.5 KYC/KYB chaincode structure

Chaincode Structure

In Fig. 7.5, the KYC/KYB chaincode structure is depicted at a component level. As stated, the depicted Blockchain Modules are explained in Chap. 1, *A Reference Architecture for Big data Systems and AI Applications in the Finance Sector (INFINITECH-RA)*, though there is a following individual short description for consistency.

The *Blockchain Writer* component undertakes the responsibility to submit new transactions to the blockchain ledger. In the Know Your Customer/Know Your Business Use Case, the *Blockchain Writer* component refers to the transactions submitted for registering a new customer on the private ledger, updating an existing one and withdrawing the KYC/KYB information of an existent participant.

The *Blockchain Reader* component is used in order to read the ledger state and fetch a particular party's KYC or KYB submitted information.

The *Smart Contract Executor* component's main purpose is to consolidate the business intelligence of the defined use case and execute the chaincodes on the blockchain ledger.

The *Blockchain Authenticator* component is responsible for performing the authentication of the blockchain network user in order to grant access to a specific channel of the blockchain network.

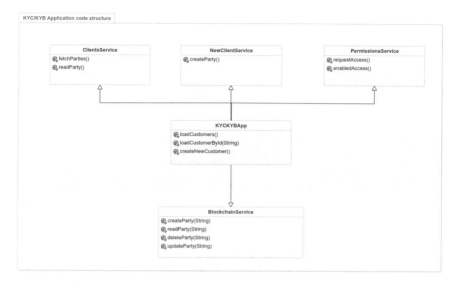

Fig. 7.6 KYC/KYB application code structure

The *Blockchain Encryptor* component performs the encryption of the related data that are involved and produced within the smart contract execution.

The *Blockchain Decryptor* component performs the decryption of the data encrypted by the *Blockchain Encryptor* component.

Web Application

With regard to the KYC/KYB web application, the following analysis presents the user interface (UI) framework essential services and their functions that are hitherto developed for the user interaction with the on-chain part of the solution. In Fig. 7.6, the application code structure is illustrated.

The *Clients Service* is responsible for retrieving the requested KYC or KYB information from the blockchain ledger and for delivering it on the web interface of the participant channel organization.

The *New Client Service* is responsible for delivering the KYC or KYB data of a newly inserted client on the specified blockchain end points in order to be submitted on the ledger.

The *Permissions Service* is responsible for switching the access control state of a financial organization.

The *KYC/KYB App* constitutes the main user web interface end point through which the end-user can initiate all the permitted actions and navigate into the application.

The *Blockchain Service* is responsible for submitting new use case related data on the blockchain ledger after triggered by the web user interface. The functionalities enable immediately the corresponding chaincode functions.

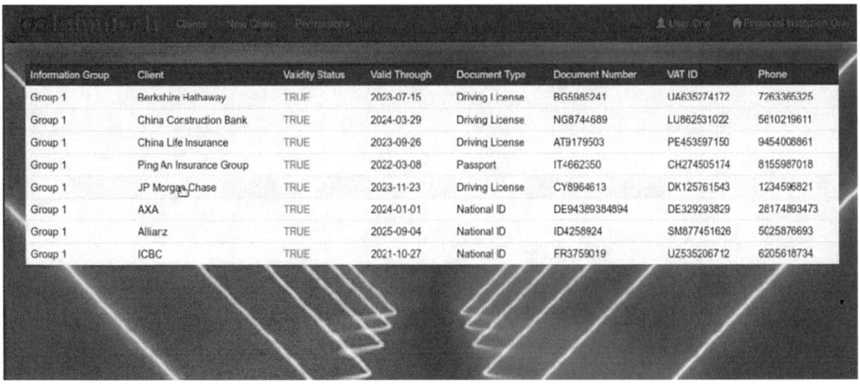

Fig. 7.7 Clients view

Fig. 7.8 New client view

Navigation

With regard to the web application navigation routes by a web end-user, the following illustrations depict the relevant views' information accompanied by their descriptions. In general, the last part of the implementation showcases in a practical way the important parts of the user interface framework that is hitherto developed for the user interaction with the blockchain ledger. In Fig. 7.7, the specific *Clients* view displays the KYC/KYB information of the on-chain submitted participants as seen from the account of *User One* of *Financial Institution One*. In every *Clients* view, the ledger information that the specific financial institution user is allowed to read is depicted as defined from the underlying blockchain mechanisms. In Fig. 7.8, the *New Client* view depicts the user interface (UI) that a financial institution user exploits in order to register the KYC or KYB information of a new client. Upon right form completion, the new client is being inserted and the submission sends the data to the ledger. In Fig. 7.9, the *Permissions* view displays the control access status

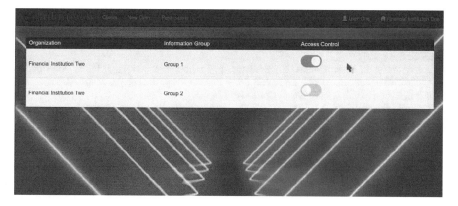

Fig. 7.9 Permissions view

of a financial institution. Every request for access from other financial institutions is depicted in this view, while the financial institution user is able to permit or forbid access.

6   Conclusions and Future Works

In this chapter, an efficient and accelerated KYC process exploiting the underlying blockchain technology is analyzed using the *INFINITECH KYC/KYB On-Chain Solution*. The state-of-the-art financial sector is greatly benefited by the solution since the addressing of various technological issues including trust, immutability, secure data sharing, and confidential transactions. Such blockchain approaches are meant to disrupt the financial scene as promoting the adoption of new secure enterprise data networks. For the current solution, future plans are being under consideration for integrating with the technological results presented in Chap. 8.

Acknowledgments The research leading to the results presented in this chapter has received funding from the European Union's funded Project INFINITECH under grant agreement no: 856632.

References

1. Polyviou, A., Velanas, P., & Soldatos, J. (2019). *Proceedings, 28*(1). https://doi.org/10.3390/proceedings2019028007. https://www.mdpi.com/2504-3900/28/1/7
2. Kapsoulis, N., Psychas, A., Palaiokrassas, G., Marinakis, A., Litke, A., & Varvarigou, T. (2020). *Future Internet, 12*(2). https://doi.org/10.3390/fi12020041. https://www.mdpi.com/1999-5903/12/2/41

3. Bhaskaran, K., Ilfrich, P., Liffman, D., Vecchiola, C., Jayachandran, P., Kumar, A., Lim, F., Nandakumar, K., Qin, Z., Ramakrishna, V., Teo, E. G., & Suen, C. H. (2018). *2018 IEEE International Conference on Cloud Engineering (IC2E)* (pp. 385–391). https://doi.org/10.1109/IC2E.2018.00073
4. Karagiannis, I., Mavrogiannis, K., Soldatos, J., Drakoulis, D., Troiano, E., & Polyviou, A. (2020). *Computer Security - ESORICS 2019 International Workshops, IOSec, MSTEC, and FINSEC, Revised Selected Papers*, Germany.
5. Norvill, R., Steichen, M., Shbair, W. M., & State, R. (2019). *2019 IEEE International Conference on Blockchain and Cryptocurrency (ICBC)* (pp. 9–10). https://doi.org/10.1109/BLOC.2019.8751480

Chapter 8
Leveraging Management of Customers' Consent Exploiting the Benefits of Blockchain Technology Towards Secure Data Sharing

Dimitris Miltiadou, Stamatis Pitsios, Spyros Kasdaglis, Dimitrios Spyropoulos, Georgios Misiakoulis, Fotis Kossiaras, Inna Skarbovsky, Fabiana Fournier, Nikolaos Kapsoulis, John Soldatos, and Konstantinos Perakis

1 Introduction

The banking sector is currently undergoing a major transformation driven by the new revised payment service directive (PSD2) which could act as a catalyst for the innovation in the new wave of financial services. It is clear that the introduction of PSD2 is reshaping the banking sector in a way that has not been seen before in the specific sector posing challenges to the banks that are seeking for solutions to abide by this new regulation while at the same time to leverage the opportunities offered to strengthen their position.

PSD2 was initially introduced in 2017 and entered into force in December 2020. It was introduced as an amendment of the previously established in 2007 payment service directive (PSD) and is a European legislation composed of a set of laws

D. Miltiadou (✉) · S. Pitsios · S. Kasdaglis · D. Spyropoulos · G. Misiakoulis · F. Kossiaras
K. Perakis
UBITECH, Chalandri, Greece
e-mail: dmiltiadou@ubitech.eu; spitsios@ubitech.eu; skasdaglis@ubitech.eu; dspyropoulos@ubitech.eu; gmisiakoulis@ubitech.eu; fkossiaras@ubitech.eu; kperakis@ubitech.eu

I. Skarbovsky · F. Fournier
IBM ISRAEL – SCIENCE AND TECHNOLOGY LTD, Haifa, Israel
e-mail: INNA@il.ibm.com; fabiana@il.ibm.com

N. Kapsoulis
INNOV-ACTS Limited, Nicosia, Cyprus
e-mail: nkapsoulis@innov-acts.com

J. Soldatos
IINNOV-ACTS LIMITED, Nicosia, Cyprus

University of Glasgow, Glasgow, UK
e-mail: jsoldat@innov-acts.com

and regulations for electronic payment services in the European Union (EU) and the European Economic Area (EEA) [1]. The main objectives of PSD2 are built around the following main pillars: (a) the establishment of a single, more integrated and efficient payment market across the European Union, (b) the establishment of more secure and safe electronic payments, (c) the protection of the customers of the electronic payment services against possible financial frauds and (d) the boost of innovation in the banking services with the adaptation of new novel technologies.

One of the fundamental changes introduced by PSD2 [2], along with the employment of strong customer authentications and the requirement of the authorised and registered by the European Banking Authority (EBA) payment license, is the requirement to open the access to the financial (or banking) data, currently maintained by the banks, to third-party providers (TPPs). The rationale of this change is to enable banks and their customers to benefit from the usage of third-party APIs towards novel financial services that will foster the innovation in the banking sector. The opening of the payment services of the banks to other TPPs sets the basis for what is referenced in the banking sector as open banking. The initiative of open banking generally refers to the ability for banking customers to authorise third parties to access their bank account data either to collect account information or to initiate payments, as stated on a recent report by KPMG [3]. Through PSD2, the European Commission aims to boost the innovation, transparency and security in the single European payment market [4]. Moreover, PSD2 will support the collaboration between banks and fintech innovative institutions towards the realisation of disruptive business models and new financial services which their customers can leverage.

Open banking imposes that banks have to share personal, and in most cases even sensitive, data of their customers with TPPs. However, besides PSD2, the European Commission has enforced another important directive, the General Data Protection Regulation (GDPR) [5]. GDPR constitutes a European law related to data protection and data privacy imposed in EU and EEA. GDPR was put into effect on May 25, 2018, and imposes a set of strict privacy requirements for the collection, processing and retention of personal information. The primary aim of GDPR is to provide the person whose data is processed – which is referenced in the regulation as the data subject – the control over his/her personal data. It provides the right to data subjects to provide (or withdraw) to their consent for third parties to legitimately access and process their personal data, among others. Furthermore, GDPR imposes strict security and privacy-preserving requirements to the ones collecting and/or processing the data of the data subjects.

As both regulations, PSD2 and GDPR, come into force, it is imperative that the right balance between innovation and data protection is defined and assured by the banks. The reason for this is that, on the one hand, banks are obliged to share the personal and financial data of their customers to the TPPs, while at the same time, on the other hand, banks are responsible for the security of the personal data collected from their customers. To this end, while PSD2 is reshaping the banking sector and is supporting the open banking initiative towards the development and provision of novel financial services, the access to these services that involve personal data must be performed in a GDPR-compliant manner.

Fig. 8.1 Consent as the bridge between PSD2 and GDPR towards innovation

Hence, in order to properly harness the opportunities offered by PSD2, several aspects need to be taken into consideration. Nevertheless, as stated in a recent report from Ernst & Young, when properly implemented in harmony, PSD2 and GDPR enable banks to better protect and serve consumers, to move beyond compliance and to seize new opportunities for growth [6]. In addition to this, PSD2 constitutes a major opportunity for banks to further strengthen their business models, to introduce new capabilities to their customers and at the same time to become savvier about their customer's patterns of behaviour [3]. As reported by Deloitte, open banking will expand traditional banking data flows, placing the customer at its core and in control of their banking data, including their personal information [4].

Data sharing is the core aspect of open banking. Data sharing drives innovation and in most cases increases the effectiveness and efficiency of existing services, reduces costs and strengthens the relationships of businesses with their clients. However, in the case of open banking as it also involves sharing of personal data, strong requirements related to data protection exist, as explained above. While being two distinct regulations, the bridge between PSD2 and GDPR lays on the consent of the customer of the bank to share his/her data. According to GDPR [5], consent is one of the six legal grounds of lawful processing of personal data, and it needs to be given by a statement or by a clear affirmative action (Fig. 8.1).

Hence, the need for robust and efficient consent management becomes evident. Consent management is the act or process of managing consents from the data subject (or customers in the case of banks) for processing and/or sharing their personal data. And as the PSD2 is built around digital innovation and integration, digital consent management is considered the proper way to tackle the challenges faced for banks towards the compliance with PSD2 and GDPR. The Open Banking Working Group of Euro Banking Association (EBA) reports that digital consent management lies at the heart of any possible solution for the challenges ahead [7]. Digital consent is enabling the banks and/or third-party providers to leverage the opportunities offered by PSD2 in a GDPR-compliant manner. However, the consent management constitutes a process with multiple aspects which shall be taken into consideration during the design and implementation phase.

2   Consent Management for Financial Services

Consent is defined in Article 4 of GDPR as "any freely given, specific, informed and unambiguous indication of a data subject's wishes by which he or she, by a statement or by clear affirmative action, signifies agreement to the processing of personal data relating to him or her". Furthermore, GDPR requires the explicit consent of the customer in order to process and/or share his/her data. Additionally, the conditions of a consent are strictly defined under Article 7 of GDPR and include, among others, the exact definition of the context of the given consent, the explicit and distinguishable way of the consent gathering as well as the right of the customer to withdraw his/her consent at any time. PSD2 also clearly states that TPPs shall access, process and retain only the personal data that is necessary for the provision of their payment services and only with the "explicit consent" of the payment service user.

Hence, the main characteristics of consent can be grouped as follows [8]:

- The consent must be given as real choice and under the control of the customers. Any element that prevents the customers to exercise their free, such as inappropriate pressure or influence, invalidates the consent. Customer must be able to refuse their consent.
- Consent must be explicit, meaning that the customer must provide consent with affirmative action and the processing purpose must be thoroughly and granularly described to the customer and be distinguishable from other matters.
- The customer must be able to withdraw the consent at any point without negative consequences for him/her.
- In the case where multiple processing operations for more than one purpose, customers must be able freely to choose which purpose they accept rather than having to consent to a bundle of processing requests.
- Consent can be issued in a written and signed form, as well as in an electronic form using an electronic signature or a scanned document carrying the signature of the customer.

In addition to this, a consent must contain at least the following information [8] (Fig. 8.2):

- The data controller's identity
- The purpose of each of the processing operations for which consent is sought
- What (type of) data will be collected and used
- The existence of the right to withdraw consent
- Information about the use of the data for automated decision-making

Consent management constitutes the process of managing consents from the customers for processing their personal data. The effective consent management enables tracking, monitoring and management of the personal data lifecycle from the moment of opt-in to the data erase in GDPR-compliant manner. The core aim of this process is to facilitate and improve the customer's control over their personal

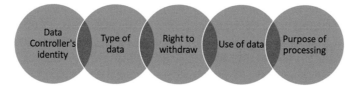

Fig. 8.2 Core content of a valid consent

Fig. 8.3 Consent
management lifecycle

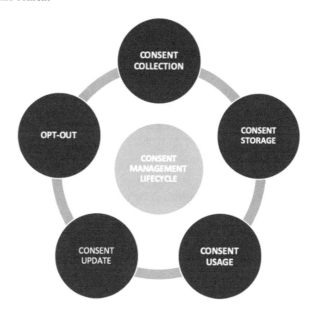

data enabling their right to provide (or withdraw) consent which allows authorised parties to access their personal and business information and its immediate effect. As it enables granular permission consent, it constitutes a key enabler of trust which is vital to maximise data sharing and assure that customers are comfortable with sharing data.

Consent management is composed of the following distinct phases (Fig. 8.3):

- *Consent Collection*: During this first phase, the consent of the customer (data subject per the GDPR terminology) is formulated. The terms of the consent are strictly defined during this stage. The formulated consent includes at least the information described in the previous paragraph in a clear and unambiguous manner.
- *Consent Storage*: During this second phase, the formulated consent is stored and securely maintained into a single source of truth which is usually a secure data storage modality. To this end, it is imperative that all the necessary measures are undertaken to assure the preservation of the security, privacy and integrity of the stored information.

- *Consent Usage*: During this third phase, the stored consents are effectively leveraged to formulate the decisions related to data access and processing. Hence, the stored consents are provided as input into the proper data access control mechanism in order to evaluate any request to access the personal data of the customer, thus regulating the access and preventing their unauthorised disclosure. The terms of the consent are the holding a critical role in this process. It is very important that they are solid and clear, as well as their integrity is assured.
- *Consent Update*: During this forth phase, it is safeguarded that the customer is able to update the terms of the consent and at the same time that the latest terms of the consent are maintained in a robust and transparent manner. This phase also introduces the requirement for storing the evolution of each consent during its existence.
- *Opt-Out*: During this last phase, the right of the customer to withdraw his/her consent, which constitutes a fundamental right of the data subject per GDPR, is ensured. Hence, at any point, the customer has the legitimate right to withdraw the consent he/she has provided to third parties to access and process their personal data.

A consent management system is the software that facilitates the execution of digital consent management by effectively handling all the operations performed during the complete consent management lifecycle. It acts as the mediator between the involved parties, enabling the formulation, as well as the update or a withdrawal, of a consent and finally its storage into a secure data storage modality.

3 Related Work

The importance of consent management is evident from the large research efforts and the broad number of diverse approaches and solutions that have been proposed in various domains. All the proposed approaches have the safeguarding of individual's privacy as their core requirement; however, the proposed solutions are handling the specific requirement in many different ways. It is also acknowledgeable that although some generic and cross-domain solutions exist, the majority of them are tailored to the needs of a specific domain. It is also evident from literature that IoT and health domains are the most widely explored research areas on this topic.

Taking into consideration the requirement for the existence of valid consent, a web standard protocol called User-Managed Access (UMA) is proposed [9] in an attempt to enable digital applications to offer the required consent management functionalities in a domain-agnostic manner. UMA enables the individuals to perform authorisation control over their personal data access and usage in a unified manner. Nevertheless, this generic solution is lacking in terms of data access auditing, access revocation and the right to be forgotten. Leveraging blockchain technology and smart contract is an additional proposed solution [10] for efficient and effective consent management. Another domain-agnostic approach is based on blockchain technology and smart contracts [11]; however, it is more focused on data accountability and data provenance. The solution encapsulates the user's consent into a smart contract but is focused on the data subject and data controller aspects while lacking support of the data processor aspects in case where data controller and data processors are different entities.

When it comes to the health domain, multiple approaches and solutions have been proposed, as expected due to the nature of health data which are characterised as sensitive data per the GDPR. The generic consent management architecture (CMA) is proposed [12] as an extension of UMA in combination with the MyData [13] approach. It enables secure data transactions with a context-driven authorisation of multisourced data for secure access of health services maintaining personal data ownership and control. Nevertheless, the proposed architecture is still defined at a general level, and further work is still needed, while some of the utilised technologies require further development. Another approach for consent-based data sharing is also proposed via a web-based solution and video files [14]. In this approach for genomic data sharing, a single one page consent form is introduced on a web application, supplemented with online video that informs the user for the key risks and benefits of data sharing. Nevertheless, this approach does not effectively cover the complete consent management lifecycle. For mobile applications, another initiative is the participant-centred consent (PCC) toolkit [15] whose purpose is to facilitate the collection of informed consents from research participants via mobile. Researchers have also proposed several consent management solutions which are based on the blockchain technology and enable health data sharing. A domain-specific solution is proposed with a blockchain-based data sharing consent model that enables control of the individuals' health data [16]. The proposed solution offers – with the help of smart contracts – dynamic consent definition over health data with the Data Use Ontology (DUO), as well as their search and query via the Automatable Discovery and Access Matrix (ADA-M) ontology. An additional blockchain-based solution for consent management utilises smart contracts for interorganisational health data sharing [17]. It supports searching for patients in the framework that match certain criteria; however, it is based on open and public blockchain networks. Furthermore, several blockchain-based solutions have been proposed in literature [18–20] for the effective data sharing of EHRs and health data in a secured manner.

In addition to the health domain, for the IoT domain, a large number of solutions for consent management is also offered. ADVOCATE [21] platform is providing a user-centric solution that effectively implements the consent management lifecycle for secure access of personal data originating from IoT devices. The solution offers consent management and data disposal policy definition; however, it is domain-specific to the IoT ecosystem. Another solution offers the ability to set their privacy preferences of their IoT devices and safeguards that their data are only transmitted based on these preferences [22]. In this solution, blockchain is used to store and protect these preferences; however, this solution can be considered as a GDPR-compliant solution. A lightweight privacy-preserving solution for consent management is also proposed [23] with cryptographic consents being issued by the individuals. The specific proposal supports multiple security features, such as untraceability and pseudonymity, exploiting the hierarchical identity-based signature (HIBS). Nevertheless, the proposed approach partially addresses the requirements of GDPR. Additionally, several solutions are proposed for specific IoT device-type only, such as Cooperative Intelligent Transport Systems [24], smart homes [25], medical devices [26, 27] and smart watches [28].

In the finance domain, despite the fact that the requirement for consent management is crucial for financial institutions, the list of available solutions is still limited. A solution built around the double-blind data sharing on blockchain has been proposed [29]. The specific solution is focused on establishing a Know Your Customer (KYC) application through which the consents are formulated supporting the dynamic definition of the consent terms with respect to the data usage. Nevertheless, the specific solution is not supporting the dynamic definition of the purpose of the data usage in the formulated consents. A new blockchain-based data privacy framework that is combined with the nudge theory is also proposed [30]. The specific solution offers a data privacy classification method according to the characteristics of financial data, as well as a new collaborative filtering-based model and a confirmation data disclosure scheme for customer strategies based on the nudge theory. However, the specific approach requires a layered architecture for the designed application that is based on hybrid blockchain, a combination of public and private blockchain which is not usually embraced by financial institutions that operate under strict regulations.

4   Methodology

Within the INFINITECH project, the consortium partners seek to exploit the open banking opportunity, aspiring to examine how open banking and internal banking data sources can be effectively combined in order to gain useful insights and extract knowledge on the customer's patterns of behaviour from a range of data analysis methods.

To this end, the consortium partners were engaged into a series of internal roundtable discussions and brainstorming sessions, bridging the expertise within the consortium in the banking sector services and the existing regulatory framework in the banking sector with the expertise in the design and delivery in technological solutions in order to formulate a novel solution.

The scope of these interactive sessions can be grouped into the following axes:

- To identify the needs of the banking sector with regard to the consent management which could act as the key ingredient of an ecosystem of innovative financial services
- To define the proper consent management process which will ensure the privacy preservation and trustworthiness for data sharing and analysis of financial information between banks, their customers and TPPs

- To define the proper framework which supports customer-centric data services that involve both data acquisition and collaborative data sharing where customers maintain complete control over the management, usage and sharing of their own banking data

As mentioned before, there is a clear need for a solution that bridges the gap between the two distinct regulations, namely, the PSD2 and GDPR regulations. The consent of the bank's customers to share their financial information is considered a hard requirement imposed by these regulations. Hence, it is imperative that banks seeking to leverage the opportunities of open banking design a data-sharing framework which is built around their customer's consent to provide access to their data to any interested party outside their organisation. Within this context, banks shall evaluate how these data are shared and how this proposition is aligned with the requirements and needs of the fintech ecosystem. The data-sharing framework shall take into consideration how these parameters are directly depicted into each consent agreement provided by their customers. As it is clearly stated in GDPR article 7 [5], the consent of the data subject (in this case, the customer of the bank) must contain, in clearly and in easily understandable terms, the information about what they are consenting and, above all, the use of data cannot go beyond what is specified in the consent agreement. Hence, the consent and its terms shall define not only which data can be shared but also when and under which conditions.

In addition to this, a crucial aspect in every data-sharing approach is the security measures that are employed in order to ensure the proper access control besides the security of data at rest. Security is the heart of any data-sharing framework. It constitutes one of the major concerns of any data provider and one of the core aspects that can lower the barriers of adoption of any data-sharing framework. In our case, the proper data access control should be based on the terms of the consent as provided by the customer of the bank. The terms of the consents should be translated into the proper data access policies which prevent unauthorised disclosure to personal data and clearly define what personal data can be accessed and by whom, as well as when and for how long can this personal data be accessed.

Finally, banks need a framework that is, on the one hand, capable of offering custodian services and, on the other hand, ensuring that customers maintain the control over the management, usage and sharing of their banking data. By placing the consent of a customer in the centre of the data-sharing framework, customers will remain in control of their data. This supports secure and controlled accessibility of the data, as the explicit consent of the customers will formulate the decision what data are accessed, who can access them and when this access is revoked. Through the use of consent, the data-sharing framework is enabling both data acquisition and collaborative data sharing to deliver added-value services, both being basic principles of PSD2, in a GDPR-compliant manner.

Towards the design of a novel consent management system that will empower the banks to leverage the PSD2 opportunities in a GDPR-compliant manner, the consortium analysed:

- The requirements elicited from the internal roundtable discussions and brainstorming sessions related to the PSD2 regulation and needs of the banking sector
- The requirements related to the legitimate formulation of a consent of a customer in accordance with the GDPR regulation
- The different phases and characteristics of each phase of the consent management lifecycle

From the analysis of these requirements and characteristics, it became clear that with regard to the consents and their validity period, two different approaches should be supported by the consent management system. The first approach supports the consents for which the validity period is a predefined period as set in the terms of the consent. This consent type is referred as "once off" consent. The expiration of the validity is translated as automatic revocation of the consent, and a new consent is required in order for the recipient to be able to access the customer's data. An example of the cases considered for this consent type is the customer's consent for sharing of KYC data between two financial institutions (banks) for the scenario of a loan origination/application or an account opening. The second approach supports the consents that have an infinite validity period, referred to as permanent (or "regular") consents. This consent type is only invalidated in the case where the consent is withdrawn. An example of the cases considered for this consent type is the peer-to-peer (including person-to-person, person-to-organisation, organisation-to-organisation cases) customer data-sharing consent in which a customer utilises an interface of a mobile application to select a set of its specific customer data (such as specific accounts, specific transactions or alerts) that they would like to share with an individual person or a group of persons.

In addition to the different consent types, the analysis of these requirements and characteristics has also driven the design of the corresponding use cases that should be implemented in order to ensure that all requirements are effectively addressed and that the designed consent management system will bring the added value to the banking sector's stakeholders. The defined use cases cover the complete lifecycle of consent management system, starting from the registration of the relevant users, the formulation of the consent and its storage to the retrieval of the stored consent in order to formulate the data access control decision.

The key stakeholders which are involved and are interacting with the proposed solution are:

- The internal financial institution (namely the bank) that collects and has access to its customer data
- The customer of internal financial institution (namely, customer of the bank) whose data are collected by the internal financial institution
- The external financial institution or peer (namely, third-party provider) that aspires to obtain the data of the customer of internal financial institution upon the formulation of a valid and legitimate consent that is formulated between the three parties

The consortium analysed also the latest state-of-the-art technologies which can be leveraged for the implementation of the aspired solution. The core aspect of any data-sharing framework is the employed security. However, traditional technologies have failed to become a key enabler of trust, due to multiple security/data tampering incidents as it is evidenced by the recent large-scale personal data breaches around the world [31]. Nevertheless, blockchain technology and its latest advancements appear as a compelling technology to overcome the underlying challenges around trusted data sharing by exploiting its attractive features such as immutability and decentralisation.

In the following paragraphs, the different use cases which are supported by the aspired solution are presented in detail.

4.1 User's Registration

The first step includes the registration of the different users in the consent management system. Different roles are assigned to the users based on the role in the consent formulation process. In order for a customer to be able to receive consent requests, it is mandatory that they are registered in the consent management system, creating their profile that will be used in order to receive consent requests. Their profile includes the minimum customer discovery and communication details required in order to receive a consent request, while it ensured that this information and any consequent private information are not disclosed to any interested party. In the same manner, the third-party provider registers in the consent management system in order to be able to initiate consent requests to a customer of the bank. The users of the bank, who are considered as the administrators of the consent management system, are already preregistered so as to facilitate the whole execution (Fig. 8.4).

4.2 Customer Receives a Request to Provide New Consent for Sharing His/Her Customer Data

The consent formulation is initiated from the third-party provider. In particular, the third-party provider issues a new consent request to the consent management system indicating the specific customer that accesses to his/her is requested along with the details of the requested customer data (i.e. specific data, period of usage). As a first step, the initial request is received and reviewed by the bank before it is pre-approved and received by the customer. Upon this approval, the customer receives a notification with all the required information for the consent request in a proper way that it allows them to review the request (Fig. 8.5).

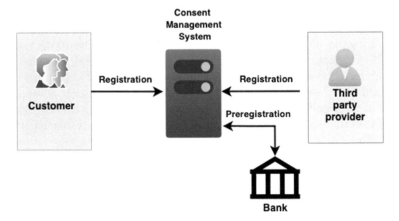

Fig. 8.4 User's registration

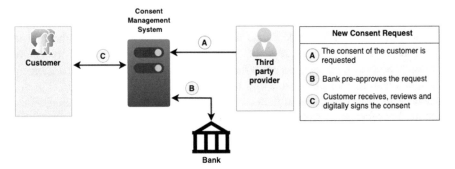

Fig. 8.5 New consent request

4.3   Definition of the Consent

In this step of the process, the customer is able to review, and possibly alter, the details and conditions of the consent request before he/she formulates his/her decision to provide the consent or deny the access to his/her personal data. In the case of approval, the final terms of the consent are defined by the customer, and it is submitted to the consent management system. The third-party provider is then informed and can approve the final terms of the consent, as set by the customer, or abandon the consent request. In the case of denial from the customer on the initial request, the request is blocked, and the third-party provider is informed for the completion of the process (Fig. 8.6).

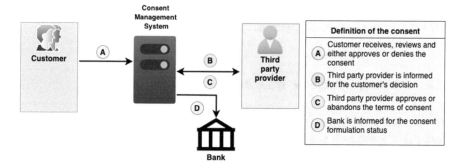

Fig. 8.6 Definition of the consent

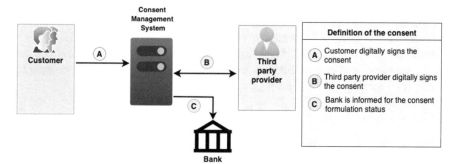

Fig. 8.7 Digital signing of the consent

4.4 Signing of the Consent by the Interested Parties

Once the terms of the consent have been formulated, the consent management system provides the formulated consent form to both parties (the customer and the third-party provider) in order to be digitally signed. The consent management system collects the digitally signed consent forms from both parties. At this point, the consent status is now set as "active" (Fig. 8.7).

4.5 Consent Form Is Stored in the Consent Management System

Once the digitally signed consent form is available, the consent management system is able to store this information in order to be used in the access control formulation process. In the case of the "once off" consent, where a specific validity period is defined, the consent management system is internally handling the validation of consent time period by creating and monitoring the specific timer in order to perform the validation of consent time period.

4.6 *Consent Update or Withdrawal*

In accordance with the GDPR regulation, the formulated consent form can be updated or withdrawn at any time by the customer side. On the one hand, the terms of the consent can be updated following the previously described steps (from the definition of the consent till the consent storage) in order to formulate the updated consent and maintain the consent status "active". In the case of the "once off" consent, the associated timer is restarted when the consent is updated. On the other hand, the consent can be withdrawn, which is internally translated into the update of the consent status to "withdrawn". In this case, if the withdrawn consent is a "once off" consent, the associated timer is stopped. For both cases, the consent management system ensures that the complete history of each consent is maintained for later usage (Figs. 8.8 and 8.9).

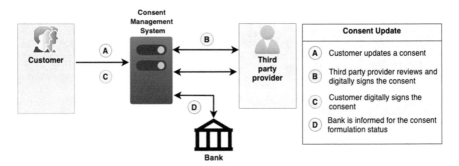

Fig. 8.8 Consent update

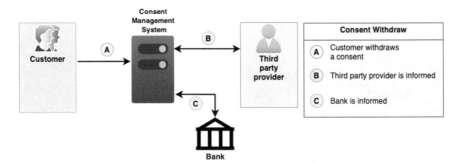

Fig. 8.9 Consent withdrawal

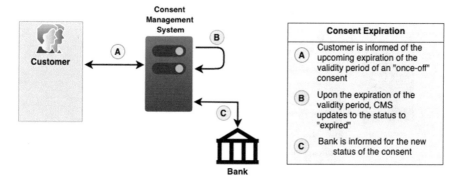

Fig. 8.10 Consent expiration

4.7 Expiration of the Validity Period

In the case of the "once off" consent, the validity period is set to a predefined time period. When consent is formulated, the consent management system creates and monitors the specific timer in order to perform the validation of consent time period. The consent management system informs the customer before the validity period of a "once off" consent expires in case he/she would like to initiate an update before the validity period expires. Once this timer is expired, the consent management system sets the status of the specific consent to "expired" (Fig. 8.10).

4.8 Access Control Based on the Consent Forms

The key aspect of the aspired data-sharing framework is the efficient and robust access control mechanism with the consent management system at its core. During each data access request to underlying data management system, the consent management system is consulted in order to validate the consent status between the requesting party and the customer whose data are requested. The data access control decision is formulated by the existing consents and their status, as well as the underlying terms of each consent (Fig. 8.11).

4.9 Retrieve Complete History of Consents

Another key aspect of the proposed solution is that the customer is able to be constantly informed of all the consents that are given to each specific recipient, as well as of the complete history of these consents. Furthermore, for each consent, all the different versions of the provided consent can be retrieved besides the latest

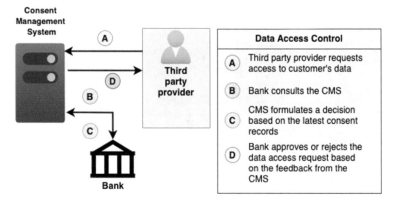

Fig. 8.11 Access control based on consents

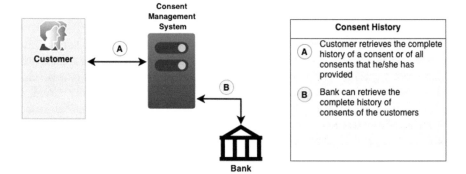

Fig. 8.12 Complete history of consents retrieval

one. In this sense, the customer shall be able to retrieve at any time their consent history per specific stakeholder or for all stakeholders, while the bank shall be to retrieve all the consents given by the customers for a stakeholder (Fig. 8.12).

5 The INFINITECH Consent Management System

The designed solution aims to enable the collaborative data sharing between customers, banks and other organisations, in order to facilitate the application of advanced analytics over particular datasets and intelligent support tools for better understanding of customers and their financial relationships among others, which is considered critically important in today's financial markets.

As explained, the development of such intelligence support tools is highly dependent on the customer's permission to share data. Hence, the requirement for a trusted and secure sharing mechanism of customer consent arises in order enable the development of new customer services that solve business problems, such as improved KYC processes and consequently AML, credit scoring and fraud detection services. A robust and solid consent management system that will support the granular permission consent is considered as a key enabler of trust which is vital to maximise data sharing and ensure customers are comfortable with sharing data.

Blockchain is a continuously growing, distributed, shared ledger of uniquely identified, linked transaction records organised in blocks that are sealed cryptographically with a digital fingerprint generated by a hashing function and are sequentially chained through a reference to their hash value [32]. In general, the blockchain technology is composed of multiple technologies related to cryptography, peer-to-peer networks, identity management, network security, transaction processing, (distributed) algorithms and more, which are all leveraged in order to formulate an immutable transaction ledger which is maintained by a distributed network of peer nodes formulating the blockchain network. The key characteristics of the blockchain technology are that it is decentralised, immutable, transparent, autonomous and open-sourced [33].

The blockchain technology has a set of key concepts that includes (a) the distributed ledger that is composed by blocks containing the transaction records, (b) the consensus model that is utilised in order to validate a transaction and to keep the ledger transactions synchronised across the blockchain network and (c) the smart contracts or chaincodes which are the trusted distributed applications that are deployed within the nodes of the blockchain network and encapsulate the business logic of the blockchain applications. The blockchain implementations can be characterised and grouped into two major high-level categories based on the permission model applied on the blockchain network, the *permissionless blockchain networks* which are open and publicly and anonymously accessible blockchain networks and the *permissioned blockchain networks* where only authorised users are able to maintain, read and access the underlying blockchain.

Our solution exploits the benefits of blockchain technology and specifically the permissioned blockchain that is considered as the appropriate candidate solution due to its security and trust characteristics. It is built around a blockchain application that implements a decentralised and robust consent management mechanism which facilitates the sharing of the customers' consent to exchange and utilise their customer data across different banking institutions. Built on top of the core offerings of the blockchain technology, namely, its decentralised nature and immutability, as well as the impossibility of ledger falsification, our approach ensures the integrity of customer data processing consents and their immutable versioning control through the use of the appropriate blockchain infrastructure. Hence, the blockchain-enabled consent management mechanism enables the financial institutions to effectively manage and share their customers' consents in a transparent and unambiguous manner.

Its key differentiating points, from the financial institutions' perspective, is the ability to inform the customer at any time:

- For any customer data that it managed upon their consent
- The latest status of their consent (active or revoked/withdrawn)
- The recipients (financial institutions or peers) of their customer data upon their consent
- The purpose (or even legal basis) and time period of their customer data sharing to the recipient (financial institutions or peers)

On the other hand, its key differentiating points, from the customer's perspective, are enables them to:

- Be constantly informed for all the requests for sharing their customer data
- Be able to activate or revoke their consents
- Be constantly aware of the active consents they have given to each specific recipient

Our solution exploits the benefits of the blockchain technology in order to assure the integrity of the formulated consents with the utilisation of its cryptographic techniques in combination with the usage of digital signatures. In addition to this, the formulated consent and their complete update history are stored in a secure and trusted manner within the blockchain infrastructure. Hence, the blockchain technology is leveraged to store this sensitive information, to apply immutable versioning control and to enable the retrieval of this information in an indisputable manner. With the use of blockchain technology, both the financial institutions and their customers are able to retrieve the latest untampered consent information, as well as any previous versions of the consent.

The solution is designed to hold the role of the middleware application between the data management applications of the financial institution, in which the secured APIs are available, and the third-party providers. Hence, the designed consent management system does not save or distribute any customer data. Being a middleware application, it provides the means to formulate a consent agreement and utilise the existing consent agreements to formulate an access control decision that should be propagated to the underlying data management applications through which the customer data are actually shared to the third-party provider.

The high-level architecture of our solution is depicted in Fig. 8.13. Our solution is composed by a set of core components, namely, the *consent management system*, the *file storage* and the *blockchain infrastructure*, which are effectively interacting via well-defined RESTful APIs. Following the analysis performed in the previous section, the main stakeholders that are involved are the internal financial institution (namely, the bank), the customer of internal financial institution (namely, customer of the bank) and the external financial institution or peer (namely, third-party provider).

The consent management system constitutes the mediator between the stakeholders and the underlying blockchain infrastructure. It receives and effectively handles the requests for the consent formulation from the external financial institution or peer to the internal financial institution and consequently the customer of the internal institution. The consent management system implements the complete consent

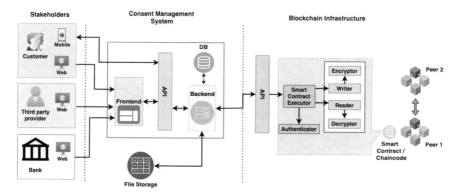

Fig. 8.13 High-level architecture of the blockchain-enabled consent management system

management lifecycle by leveraging the functionalities offered by the underlying blockchain infrastructure. To facilitate the interactions of the stakeholders with the consent management system, it offers both a Web application and a mobile application. The Web application is offered to all the involved stakeholders, in order to interact with the system on each step of the process. The Web application is composed of the core backend component that provides all the backend operations of the system, such as the processing of the input data, the retrieval of the requested information and all the interactions with the underlying blockchain infrastructure. The core backend component is supported by a local database instance in which only the operational data is maintained, such as the user management data or the data of a consent that is still under formulation. The graphical interface of the consent management system which provides all the functionalities of the Web application to the stakeholders is offered by the frontend component. The frontend component interacts with the backend via a set of APIs provided by the backend in order to perform all the requested activities by the stakeholders. In addition to the Web application, the customers of the bank are offered with the mobile application through which they are able to perform all the required operations in the same manner as they are offered by the web application.

The file storage component is providing the file repository that stores the finalised consent forms in digital format once all the steps of the process have been completed upon the acceptance of the details and terms of the consent by all the involved parties. It provides the required interface to the consent management system in order to store or retrieve the formulated consent forms upon need.

The blockchain infrastructure is providing the permissioned blockchain network of the solution. The designed blockchain network constitutes the cornerstone of the proposed solution and is based on Hyperledger Fabric[1]. In detail, the designed blockchain network is formulated by two peers (P1 and P2), which are owned by

---

[1] Hyperledger Fabric, https://www.hyperledger.org/use/fabric

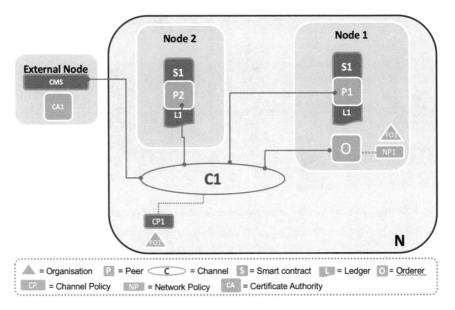

Fig. 8.14 Blockchain network of the solution

one single organisation (FO1) and are hosting their own copy of the distributed ledger (L1). Both peers are operating in one channel (C1) in which the smart contract/chaincode (S1) is deployed. In the designed blockchain network, a single order (O), which formulates the transactions into blocks and ensures the delivery of the blocks to the peers of the channel, is deployed. In this blockchain network, the consent management system is deployed on an external node which also holds the single certificate authority (CA1) that is utilised for the secure interaction with channel C1. Channel C1 is regulated by the underlying channel policy CP1 that ensures the isolation of the channel from external peers and other channels. Finally, the complete blockchain network is regulated by the network policy NP1 that is applied across the whole network with permissions that are determined prior to the network creation by organisation FO1. The interaction between the blockchain infrastructure and the backend of the consent management system is realised through the well-defined APIs provided by the smart contract/chaincode S1 (Fig. 8.14).

As explained before, the smart contract/chaincode is the trusted distributed application that encapsulates the business logic of solution. It contains the definition of the business objects for which the current and historical state will be maintained and updated through a set of functions which are also included in the chaincode. To facilitate the desired operations, the data schema that defines the core business objects has been defined, and it is depicted in Fig. 8.15. The definition was based on the consent receipt specification that is proposed by the Kantara Initiative [34] which has been adapted in terms of terminology in order to be aligned with the EU GDPR

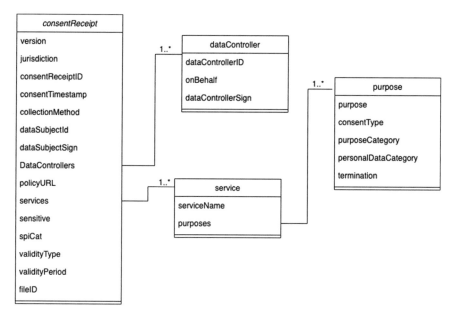

Fig. 8.15 Data schema

legislation. The designed data schema covers all the aspects of a consent, spanning from the generic information, such as the version number, the unique identifier and the timestamp of creation or modification, to more specific information such as the data subject information, the data controllers' information, the applied privacy policy, the service or group of services for which personal data are collected and the relevant purposes and of course the exact personal and/or sensitive data that are collected. Additionally, key information for the consent management lifecycle execution is included, such as the latest status of the consent, the validity type (permanent or once off) and the validity period in the case of once off consent.

The chaincode is directly built on top of the presented data schema with a set of functions which are conceptually organised into the following groups of functions:

- *Blockchain Reader*: The main purpose of this group is to enable the fetching of the requested data from the blockchain ledger. It provides the necessary functions to read the ledger state, fetch a particular consent by id, query the ledger state given different parameters to fetch a set of consents matching those parameters and get the consent history (history of all updates for particular consent data entity or data entities).
- *Blockchain Writer*: The main purpose of this group is to facilitate the submission of new transactions to the blockchain ledger. In particular, this group submits new transactions for a new consent, an updated consent or a withdrawn consent.

- *Smart Contract Executor*: The main purpose of this group is to encapsulate the business logic of the designed solution and execute the smart contracts on the blockchain ledger. In particular, this group invokes all the operations related to the read and write operations, leveraging the respective functions of the blockchain writer and blockchain reader.
- *Blockchain Authenticator*: The main purpose of this group is to perform the authentication of the blockchain network user in order to grant access to a specific channel of the blockchain network.
- *Blockchain Encryptor*: The main purpose of this group is to execute the encryption operations which are performed on the consent data prior to being inserted as new transactions to the blockchain ledger.
- *Blockchain Decryptor*: The main purpose of this group is to perform the decryption of the consent data which were encrypted by the *blockchain encryptor*.

5.1 Implemented Methods

In the following paragraphs, the implemented methods of the designed solution are presented. The basis of the implementation was the initial use cases, as well as the collected requirements, which were described in the methodology section. The implemented methods effectively cover all the aspects of the consent management lifecycle. On each method, the different interactions of the various components of the solution, as well as the interactions of the stakeholders with these components, are highlighted.

5.1.1 Definition of Consent

The formulation of the consent involves several steps from the initial consent request, as initiated by the third-party provider to the consent management system, to the actual consent formulation with the exact terms and validity period and its storage within the blockchain infrastructure. At first, the received consent request is pre-approved by the bank through the consent management system via the Web application. Once it is pre-approved by the bank again via the Web application, the request is received by the customer through the consent management system either via the Web application or via the mobile application. The customer reviews and defines the final terms of the consent and submits the digitally signed consent in the consent management system. The formulated consent is received by the third-party provider and is also digitally signed. Once the consent is digitally signed by both parties, the consent management system interacts with the blockchain infrastructure via the provided API, and a new transaction is created and inserted into the blockchain via the deployed chaincode with status "active". In the case of the "once off" consent, where a specific validity period is defined, the consent management system is internally handling the validation of consent time period

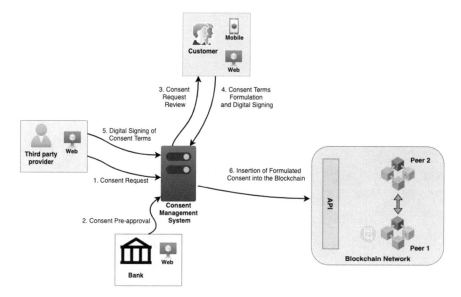

Fig. 8.16 The definition of consent method

by creating and monitoring the specific timer in order to perform the validation of consent time period (Fig. 8.16).

5.1.2 Consent Update or Withdrawal

The customer is able to retrieve and update or withdraw a consent at any time. At first, the selected consent is retrieved from the consent management system by interacting with the API of the blockchain infrastructure. At this point, the customer is presented with the existing terms of the consent and can modify its terms or withdraw it. In the consent update case, the updated terms are digitally signed by the customer and received by the consent management system. The involved third-party provider is notified and digitally signs the updated consent. Once the updated consent is digitally signed by both parties, the consent management system interacts again with the blockchain infrastructure via the provided API in order to initiate a new transaction and insert it into the blockchain via the deployed chaincode while the status remains "active". In the case of the "once off" consent, the associated timer is restarted when the consent is updated.

In the case of withdrawal, the consent management system interacts with the blockchain infrastructure via the provided API in order to "invalidate" the existing consent with the new transaction which sets the new status to "withdrawn". In the case of the "once off" consent, the consent management system also stops the respective timer. Every update or withdrawal of an existing consent introduces in

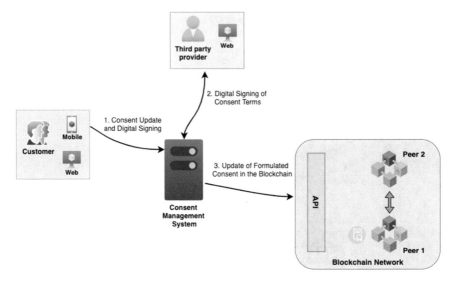

Fig. 8.17 The consent update method

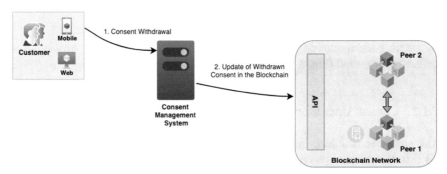

Fig. 8.18 The consent withdrawal method

a new version of the specific consent is appended into the complete history that is maintained within the blockchain infrastructure (Figs. 8.17 and 8.18).

5.1.3 Consent Expiration

The "once off" consent has a validity period as defined in the terms of the consent. Once a "once off" consent has been formulated, the consent management system creates and monitors the specific timer. At the point where the timer is about to expire, the consent management system informs the respective customer to initiate an update of the consent that will result in the timer renewal based on the terms of the updated consent. In the case where the timer expires, the consent management

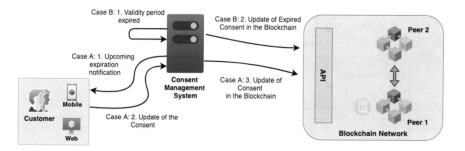

Fig. 8.19 The consent expiration method

system interacts with the blockchain infrastructure via the provided API in order to "invalidate" the existing consent with the new transaction which sets the new status to "expired" (Fig. 8.19).

5.1.4   Access Control

Once a data access request is received by the bank on their data management system from a third-party provider, the bank consults the consent management system via the Web application in order to verify and approve the specific data access request based on the terms and current status of the consent between the third party and the customer whose data are requested. To this end, the consent management system receives the information of the involved parties from the bank and initiates a query into the blockchain infrastructure via the provided API in order to retrieve the latest consent between them.

In the case where the consent status is "active" and the terms are met, the data access request can be approved, while in the case where the status is set to "withdrawn" and "expired" or a consent between the involved parties does not exist, the data access request should be denied (Fig. 8.20).

5.1.5   Complete History of Consents

The consent management system offers the retrieval and display of the latest consent information, as well as the complete history of the existing consents, at any point. The customer can retrieve the list of consents in a user-friendly way via the Web application or the mobile application. Depending on his/her selection, the customer can select and view a specific consent that has been provided to a third-party provider, as well as the complete of consents that he/she has provided to any third party provider. On the other hand, the bank is also able to retrieve a specific consent between a specific customer and a third-party provider, all the consents that a third

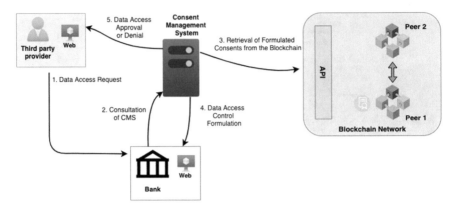

Fig. 8.20 The access control based on consent method

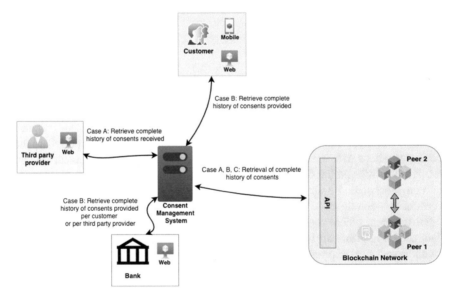

Fig. 8.21 The retrieval of the complete history of consent method

party has received by any of its customers and finally the complete list of consents provided by all his/her customer to any third-party provider.

To achieve this, the consent management system translates the request initiated by either the customer or the bank into a query with specific parameters which is executed into the blockchain infrastructure via the provided API in order to retrieve the requested information and display it to the requestor (Fig. 8.21).

6 Conclusions

The designed and implemented solution of the blockchain-empowered consent management system enables the sharing of customers' consent, thus facilitating the exchange and the utilisation of customer data, across different banking institutions. It enables the exploitation of the open banking opportunities towards the realisation of novel financial services through the collaborative data sharing in a PSD2- and GDPR-compliant manner.

Building directly on top of the key offering of the blockchain technologies that address the underlying challenges around trusted data sharing, the proposed consent management system effectively supports the employment of a trusted and secure sharing mechanism that is based on the customer consent to share his/her data. The provided solution constitutes a robust and solid consent management system that can act as a key enabler of trust of an ecosystem of innovative financial services in which customers are comfortable with sharing data since they are maintaining complete control over the management, usage and sharing of their own banking data.

Acknowledgments The research leading to the results presented in this chapter has received funding from the European Union's funded Project INFINITECH under Grant Agreement No. 856632.

References

1. European Commission – European Commission. (2021). *Payment services (PSD 2) – Directive (EU) 2015/2366* [online]. Available at: https://ec.europa.eu/info/law/payment-services-psd-2-directive-eu-2015-2366_en. Accessed 10 June 2021.
2. Tuononen, K. (2019). *The impact of PSD2 directive on the financial services industry*.
3. KPMG. (2021). *Open banking opens opportunities for greater customer* [online]. Available at: https://home.kpmg/ph/en/home/insights/2019/07/open-banking-opens-opportunities-for-greater-value.html. Accessed 8 June 2021.
4. Deloitte. (2018). *Open banking – Privacy at the epicentre* [online]. Deloitte. Available at: https://www2.deloitte.com/content/dam/Deloitte/au/Documents/financial-services/deloitte-au-fs-open-banking-privacy-epicentre-170718.pdf. Accessed 9 June 2021.
5. Official Journal of the European Union. (2016). *REGULATION (EU) 2016/679 OF THE EUROPEAN PARLIAMENT AND OF THE COUNCIL* [online]. Available at: https://eur-lex.europa.eu/legal-content/EN/TXT/HTML/?uri=CELEX:32016R0679&from=EN. Accessed 6 June 2021.
6. Ernst & Young Global Limited. (2019). *How banks can balance GDPR and PSD2* [online]. Available at: https://www.ey.com/en_lu/banking-capital-markets/how-banks-can-balance-gdpr-and-psd2. Accessed 18 June 2021.
7. Euro Banking Association, B2B Data Sharing: Digital Consent Management as a Driver for Data Opportunities. 2018.
8. European Data Protection Board. (2020). *Guidelines 05/2020 on consent under Regulation 2016/679* [online]. Available at: https://edpb.europa.eu/our-work-tools/our-documents/guidelines/guidelines-052020-consent-under-regulation-2016679_en. Accessed 5 June 2021.

9. Maler, E. (2015, May). Extending the power of consent with user-managed access: A standard architecture for asynchronous, centralizable, internet-scalable consent. In 2015 *IEEE security and privacy workshops* (pp. 175–179). IEEE.
10. Sağlam, R. B., Aslan, Ç.B., Li, S., Dickson, L., & Pogrebna, G. (2020, August). A data-driven analysis of blockchain systems' public online communications on GDPR. In 2020 *IEEE international conference on decentralized applications and infrastructures* (DAPPS) (pp. 22–31). IEEE.
11. Neisse, R., Steri, G., & Nai-Fovino, I. (2017, August). A blockchain-based approach for data accountability and provenance tracking. In *Proceedings of the 12th international conference on availability, reliability and security* (pp. 1–10).
12. Hyysalo, J., Hirvonsalo, H., Sauvola, J., & Tuoriniemi, S. (2016, July). Consent management architecture for secure data transactions. In *International conference on software engineering and applications* (Vol. 2, pp. 125–132). SCITEPRESS.
13. Alén-Savikko, A., Byström, N., Hirvonsalo, H., Honko, H., Kallonen, A., Kortesniemi, Y., Kuikkaniemi, K., Paaso, T., Pitkänen, O. P., Poikola, A., & Tuoriniemi, S. (2016). *MyData architecture: consent based approach for personal data management.*
14. Riggs, E. R., Azzariti, D. R., Niehaus, A., Goehringer, S. R., Ramos, E. M., Rodriguez, L. L., Knoppers, B., Rehm, H. L., & Martin, C. L. (2019). Development of a consent resource for genomic data sharing in the clinical setting. *Genetics in Medicine, 21*(1), 81–88.
15. Wilbanks, J. (2018). Design issues in e-consent. *The Journal of Law, Medicine & Ethics, 46*(1), 110–118.
16. Jaiman, V., & Urovi, V. (2020). A consent model for blockchain-based distributed data sharing platforms. *arXiv preprint arXiv:*2007.04847.
17. Shah, M., Li, C., Sheng, M., Zhang, Y., & Xing, C. (2020, August). Smarter smart contracts: Efficient consent management in health data sharing. In *Asia-Pacific Web (APWeb) and Web-Age Information Management (WAIM) joint international conference on web and big data* (pp. 141–155). Springer.
18. Thwin, T. T., & Vasupongayya, S. (2018, August). Blockchain based secret-data sharing model for personal health record system. In *2018 5th international conference on advanced informatics: Concept theory and applications* (ICAICTA) (pp. 196–201). IEEE.
19. Dubovitskaya, A., Xu, Z., Ryu, S., Schumacher, M., & Wang, F., 2017. Secure and trustable electronic medical records sharing using blockchain. In *AMIA annual symposium proceedings* (Vol. 2017, p. 650). American Medical Informatics Association.
20. Zheng, X., Mukkamala, R.R., Vatrapu, R., & Ordieres-Mere, J. (2018, September). Blockchain-based personal health data sharing system using cloud storage. In *2018 IEEE 20th international conference on e-health networking, applications and services* (Healthcom) (pp. 1–6). IEEE.
21. Rantos, K., Drosatos, G., Kritsas, A., Ilioudis, C., Papanikolaou, A., & Filippidis, A. P. (2019). A blockchain-based platform for consent management of personal data processing in the IoT ecosystem. *Security and Communication Networks*, 2019.
22. Cha, S. C., Chen, J. F., Su, C., & Yeh, K. H. (2018). A blockchain connected gateway for BLE-based devices in the internet of things. *IEEE Access, 6*, 24639–24649.
23. Laurent, M., Leneutre, J., Chabridon, S., & Laaouane, I. (2019). Authenticated and privacy-preserving consent management in the internet of things. *Procedia Computer Science, 151*, 256–263.
24. Neisse, R., Baldini, G., Steri, G., & Mahieu, V. (2016, May). Informed consent in Internet of Things: The case study of cooperative intelligent transport systems. In *2016 23rd international conference on telecommunications (ICT)* (pp. 1–5). IEEE.
25. Song, T., Li, R., Mei, B., Yu, J., Xing, X., & Cheng, X. (2017). A privacy preserving communication protocol for IoT applications in smart homes. *IEEE Internet of Things Journal, 4*(6), 1844–1852.
26. Fan, K., Jiang, W., Li, H., & Yang, Y. (2018). Lightweight RFID protocol for medical privacy protection in IoT. *IEEE Transactions on Industrial Informatics, 14*(4), 1656–1665.

27. O'Connor, Y., Rowan, W., Lynch, L., & Heavin, C. (2017). Privacy by design: Informed consent and internet of things for smart health. *Procedia computer science, 113*, 653–658.
28. Jahan, M., Seneviratne, S., Chu, B., Seneviratne, A., & Jha, S. (2017, October). Privacy preserving data access scheme for IoT devices. In 2017 *IEEE 16th international symposium on network computing and applications* (NCA) (pp. 1–10). IEEE.
29. Bhaskaran, K., Ilfrich, P., Liffman, D., Vecchiola, C., Jayachandran, P., Kumar, A., Lim, F., Nandakumar, K., Qin, Z., Ramakrishna, V., & Teo, E. G. (2018, April). Double-blind consent-driven data sharing on blockchain. In *2018 IEEE international conference on cloud engineering (IC2E)* (pp. 385–391). IEEE.
30. Ma, S., Guo, C., Wang, H., Xiao, H., Xu, B., Dai, H. N., Cheng, S., Yi, R., & Wang, T. (2018, October). Nudging data privacy management of open banking based on blockchain. In *2018 15th international symposium on pervasive systems, algorithms and networks* (I-SPAN) (pp. 72–79). IEEE.
31. Lim, S. Y., Fotsing, P. T., Almasri, A., Musa, O., Kiah, M. L. M., Ang, T. F., & Ismail, R. (2018). Blockchain technology the identity management and authentication service disruptor: A survey. *International Journal on Advanced Science, Engineering and Information Technology, 8*(4-2), 1735–1745.
32. Treleaven, P., Brown, R. G., & Yang, D. (2017). Blockchain technology in finance. *Computer, 50*(9), 14–17.
33. Niranjanamurthy, M., Nithya, B. N., & Jagannatha, S. (2019). Analysis of Blockchain technology: Pros, cons and SWOT. *Cluster Computing, 22*(6), 14743–14757.
34. Kathrein, A. (2019). Consent receipt specification – Kantara initiative [online]. Kantara Initiative. Available at: https://kantarainitiative.org/download/7902/. Accessed 1 June 2021.

Part III
Applications of Big Data and AI in Digital Finance

Chapter 9
Addressing Risk Assessments in Real-Time for Forex Trading

Georgios Fatouros, Georgios Makridis, John Soldatos, Petra Ristau, and Vittorio Monferrino

1 Introduction

Risk assessment refers to the processes required to quantify the likelihood and the scale of loss on an asset, loan, or investment. In investment banking, risk estimation and monitoring are of great importance as invalid models and assumptions could cause substantial capital losses. This was the case for several financial institutions in the 2008 financial crisis and the COVID-19 pandemic as well.

Despite the recent hype for digitization in the financial industry, many companies lack accurate real-time risk assessment capabilities. This is due to the current practice of updating portfolios' risk estimation only once in a day, usually overnight in a batch mode, which is insufficient for trading in higher frequency markets like Forex (FX) [1]. Surprisingly, this is still an open challenge for various banks

G. Fatouros (✉)
INNOV-ACTS Limited, Nicosia, Cyprus
e-mail: gfatouros@innov-acts.com

G. Makridis
University of Piraeus, Piraeus, Greece
e-mail: gmakridis@unipi.gr

J. Soldatos
INNOV-ACTS LIMITED, Nicosia, Cyprus

University of Glasgow, Glasgow, UK
e-mail: jsoldat@innov-acts.com

P. Ristau
JRC Capital Management Consultancy & Research GmbH, Berlin, Germany
e-mail: pristau@jrconline.com

V. Monferrino
GFT Technologies, Genova, Liguria, Italy
e-mail: Vittorio.Monferrino@gft.com

J. Soldatos, D. Kyriazis (eds.), *Big Data and Artificial Intelligence in Digital Finance*,
https://doi.org/10.1007/978-3-030-94590-9_9

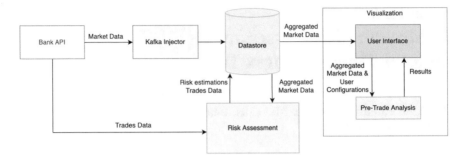

Fig. 9.1 Data pipeline

and asset management firms, where the risk evaluation is outsourced to third party consulting companies, which are in charge of risk monitoring and regulatory compliance with for an agreed investment strategy. Furthermore, in cases where a real-time overview of the trading positions might be provided, the data needed for computing the risk measures and other relevant indicators are updated only once a day. As a result, the risk exposure from intra-day price fluctuations is not monitored properly, which may lead to both trading and investment inefficiencies [2].

When it comes to HFT, the speed of financial risk calculations (derived from algorithm complexity) is the main barrier towards real-time risk monitoring [3]. Traditionally, there is a trade-off between speed and accuracy of the financial risk calculations, as the less computationally intensive risk models, which might be able to yield their assessments near-instantly, are not considered sufficiently reliable in the financial domain [4]. In addition, real-time risk monitoring is also a regulatory requirement [5] that can be met if risk information is updated timely.

This Chapter introduces a new tool, named *AI-Risk-Assessment*, which represents a potential key offering within this space. More specifically it addresses the challenge of real-time risk estimation for Forex Trading, in the scope of the European Union's funded INFINITECH project, under grant agreement no 856632. The primary added-value provided by this tool can be divided into three pillars (components): (1) risk models; (2) real-time management; and (3) pre-trade analysis, which are each described in the following sections along with the tool's core architecture that allows seamless interaction between these components. The proposed tool follows a containerized micro-service design, consisting of different components that communicate with each other in an asynchronous manner, forming a real-time data processing pipeline, illustrated in Fig. 9.1.

2   Portfolio Risk

Portfolio risk is the likelihood that the combination of assets that compromise a portfolio fails to meet financial objectives. Each investment within a portfolio carries its own risk, with higher potential return typically meaning higher risk.

Towards quantifying this risk, each asset's future performance (i.e., returns) should be predicted. To this end, various models and theories regarding the nature of the financial time-series have been proposed. Most of these models assume that the underlying assets' returns follow a known distribution (e.g., Gaussian distribution) and try to estimate its parameters (e.g., mean and variance of asset's returns) based on their historical performance. Moreover, the portfolio risk depends on the weighted combination of the constituent assets and their correlation as well. For instance, according to Modern Portfolio Theory (MPT) [6], which assumes Normal distribution for the financial returns, one could obtain a portfolio's risk performance under a known probability portfolio risk calculating the distribution parameters (μ_p, σ_p^2) from Eqs. 9.1 and 9.2.

$$\mu_p = E(R_p) = \sum_i w_i E(R_i) \tag{9.1}$$

where R_p is the return on the portfolio, R_i is the return on asset i, and w_i is the weighting of component asset i (that is, the proportion of asset "i" in the portfolio).

$$\sigma_p^2 = \sum_i w_i^2 \sigma_i^2 + \sum_i \sum_{i \neq j} w_i w_j \sigma_i \sigma_j \rho_{ij} \tag{9.2}$$

where σ is the (sample) standard deviation of the periodic returns on an asset, and ρ_{ij} is the correlation coefficient between the returns on assets i and j.

Furthermore, the Efficient Market Hypothesis implies that in highly efficient markets, such as the foreign exchange market [7], a financial instrument's value illustrates all the available information about that instrument [8]. As a result, the utilized risk models should be fed with the latest market data to provide valid portfolio risk estimations. The need for continuous updates in the risk estimates is more evident in HFT where traders usually manage numerous portfolios with different strategies (i.e., holding periods, composition, risk profile, and volume) that are mainly based on algorithmic trading [9]. Thus, there is a need for a risk assessment tool that will handle and process a large volume of market and trading data and, based on them to provide risk estimates. The latter should be updated in (near) real-time capturing intra-day volatility in financial markets that could cause both loss of liquidity and capital.

3 Risk Models

The proposed application leverages two standard risk metrics which are broadly adopted not only in risk management but also in financial control, financial reporting and in computing the regulatory capital of financial institutions, namely: Value at Risk (VaR); and Expected Shortfall (ES), which we discuss in more detail below.

3.1 Value at Risk

The first underlying risk metric is the Value at Risk (VaR), which measures the maximum potential loss of an investment under a confidence probability within a specific time period (typically a single day). There are three fundamentally different methods for calculating VaR; parametric, non-parametric, and semi-parametric [10]. *AI-Risk-Assessment* offers four VaR models based on these methods to enhance the risk monitoring process.

Parametric The most popular parametric VaR method (and what is implemented in the tool) is based on the Variance-Covariance (VC) approach [10]. The key assumption here is that the portfolio returns follow the normal distribution, hence the variance of the market data is considered known. Also, a proper history-window (e.g., 250 samples) should be defined, and the variance-covariance matrix of returns should be calculated. The VaR is then obtained by Eq. 9.3.

$$VaR^\alpha = z_{1-\alpha}\sqrt{w^T \Sigma w} \tag{9.3}$$

where z, α, w, Σ are the standard score, the confidence probability of VaR prediction, the portfolio assets weights, and the covariance matrix of the portfolio assets, respectively.

Non-parametric From the non-parametric category, the well-known Historical Simulation (HS) VaR method [11] was selected. In this approach, the historical portfolio returns are taken into account in the VaR calculation along with a selected confidence level. Under the Historical Simulation approach, the first step is to sort the historical portfolio returns for a given time-window. Then the VaR can be derived by selecting the n-worst portfolio performance that corresponds to the required confidence probability ((100-n)%) of the VaR estimation. For instance, to calculate the 1-day 95% VaR for an portfolio using 100 days of data, the 95% VaR corresponds to the best performing of the worst 5% of portfolio returns. Consequently, HS is a computationally effective risk model with acceptable returns when large historical windows are taken into account. However, the latter results in high VaR estimations that restrict the accepted investment strategies.

Semi-parametric As for the semi-parametric VaR method, a Monte Carlo (MC) [12] model was developed. In the context of the proposed solution, the mean and the standard deviation of returns are calculated from the available historical data and then these values are used to produce MC random samples from the Gaussian distribution. To calculate VaR, this process draws the distribution of portfolio returns for the next time step (see Eq. 9.4).

$$VaR^\alpha = q_{1-\alpha} \tag{9.4}$$

where q_α is the α-th percentile of the continuously compounded return.

Semi-parametric with RNNs In addition, this tool introduces a novel semi-parametric VaR model [13] combining deep neural networks with MC simulations. In this approach, the parameters of the returns distribution are initially estimated by a Recurrent Neural Network (RNN) [14] based model, with the network output being used to predict all possible future returns in a MC fashion. VaR based on the input time-series can then be obtained by (Eq. 9.4). The RNN-model in question, named DeepAR, follows the methodology proposed by Salinas et al. [15] and is provided by GluonTS Python library[1] for deep-learning-based time-series modeling [16]. This model is able to capture nonlinear dependencies of the input time-series, such as seasonality, resulting in consistent quantile estimates. Moreover, DeepAR can be fed with several input time-series simultaneously, enabling cross-learning from their historical behavior jointly. As a result, changes in the dynamics of the one time-series may affect the predicted distributions of the other time-series.

3.2   Expected Shortfall

Alternatively, risk assessment can also be performed via Expected Shortfall (ES), also known as Conditional Value at Risk (C-VaR) [17], which is a conservative risk measure calculating portfolio loss with the assumption that this loss is higher than the estimated VaR as illustrated in Fig. 9.2. ES was originally suggested as a practicable and sound alternative to VaR, featuring various useful properties such as sub-additivity that are missing from VaR [18]. This has led many financial institutions to use it as a risk measure internally [19]. Expected Shortfall can be calculated via a parametric equation shown below (see Eq. 9.5) [20].

$$ES_p^\alpha = -\mu_p + \sigma_p \frac{\phi\left(\Phi^{-1}(\alpha)\right)}{a} \tag{9.5}$$

where $\phi(x) = \frac{1}{\sqrt{2\pi}} e^{\frac{-x^2}{2}}$, $\Phi(x)$ is the standard normal c.d.f., so $\Phi^{-1}(a)$ is the standard normal quantile, and μ_p, σ_p the mean and standard deviation of portfolio returns, respectively.

The user interface of the application offers a page for analysis and back-testing of the provided risk assessments (Fig. 9.3). In this way the user is able to compare the true performance of various VAR/ES models in both 95 and 99% confidence level.

---

[1] https://ts.gluon.ai/.

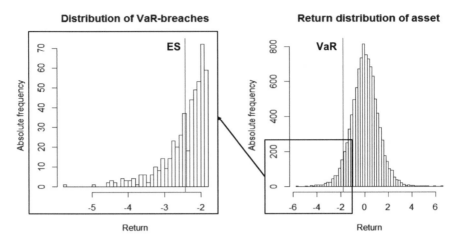

Fig. 9.2 Comparison between value at risk and expected shortfall

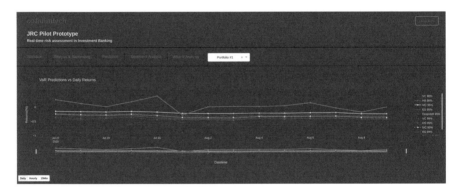

Fig. 9.3 Analysis and Back-Testing application page: The green line is the actual portfolio returns. The dense lines are for 95% confidence probability, while the dash lines for 99%

4 Real-Time Management

Delivering the aforementioned risk assessments while leveraging the latest available data is a challenging task, as FX market prices are updated at inconsistent high frequency time intervals (e.g., 1–8 seconds). Moreover, even small price fluctuations can have a significant impact on portfolios' value, particularly in cases where an high-risk/return investment strategy is being employed. Thus, additional technologies are required that can provide seamless data management with online analytical processing capabilities.

Within Infinitech, this is accomplished via a new data management platform referred to as the InfiniSTORE. The InfiniSTORE extends a state-of-the-art database platform[2] and incorporates two important innovations that enable real-time analytical processing over operational data. Firstly, its hybrid transactional and analytical processing engine allows ingesting data at very high rates and perform analytics on the same dataset in parallel, without having to migrate historical data into a data temporary warehouse (which is both slow and by its nature batch-orientated). Secondly, its *online aggregates* function enables efficient execution of aggregate operations, which improves the response time for queries by an order of magnitude. Additional details can be found in Chap. 2.

The *AI-Risk-Assessment* requires the datastore to firstly store the raw input ticker data along with the risk estimations and secondly to enable online data aggregation and integrated query processing for data in-flight and at rest. These requirements are enabled by the InfiniSTORE and underlying LeanXcale database via its dual interface, allowing it to ingest operational data at any rate and also to perform analytical query processing on the live data that have been added in parallel.

In order for the application to achieve its main objective of measuring intra-day risk in a timely fashion (e.g., updating VaR/ES approximately every 15-minutes), the input tick data (with rate 1–8 second per instrument) should be resampled for the required frequency. LeanXcale provides online aggregates that enables real-time data analysis in a declarative way with standard SQL statements. In this way, only the definition of the required aggregate operations, such as average price per FX instrument per quarter-hour, is required and the result of the execution is pre-calculated on the fly, ensuring consistent transactional semantics. As a result, the typically long-lasting query can be transformed into a very light operation that requires the read-access to a single value, thus removing the need to scan the whole dataset.

Additionally, an interface between the data provider (e.g., a bank's data API) and the datastore is required to handle the real-time data streams ingested. As these microservices need to communicate asynchronously ensuring high availability and fault-tolerance, the most popular intermediate is the use of data queues [21]. Apache Kafka [22] is the most dominant solutions when it comes to data queues [23]. Using Kafka, external components can send data feeds to a specified queue, by subscribing to a specific topic and then sending the data in a common format. Infinitech provides a pre-built container image that contains a Kafka data queue, while the datastore has a built-in Kafka connector which enables interconnection between the LeanXcale datastore and Kafka. The data injection process along with the online data aggregation are illustrated in Fig. 9.4.

---

[2] https://www.leanxcale.com/.

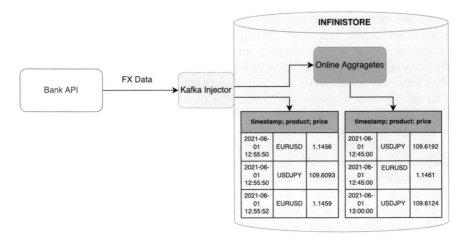

Fig. 9.4 Data injection and online aggregation

5 Pre-trade Analysis

Besides risk monitoring, traders, asset, and risk managers are seeking tools that enable pre-trade analysis. Such a tool could serve as a trader's digital twin, allowing the filtering of potential trading positions based on their risk. In addition, this feature provides direct information to the trader about the change in the risk of his portfolio in case the trader approves the simulated position.

The *AI-Risk-Assessment* incorporates another important feature which is based on what-if-analysis focusing on pre-trade risk assessment. In this way, if a trader sees an opportunity for a trade, they may enter this trade into the provided application (platform) and request a calculation of the risk measures with this new trade added to the portfolio as if it were done already. The pre-trade analysis is a useful tool for traders in order to understand how a potential new trade position may affect the portfolio and its risk Fig. 9.5.

However, in order to enable this feature, the developed risk models should be optimized to yield results instantly using the latest input data (i.e., asset prices). To this end, portfolio risk estimation follows a two-step procedure. Initially, both VaR and ES are calculated for each portfolio instrument separately as a univariate time-series based on the latest market prices available from the real-time management tools described in Sect. 4. It is noted that these calculations are performed under the hood and are updated as new data is made available. The second step is related to the user inputs. For instance, when the user is experimenting with new trading positions, the weight of each asset is obtained, with univariate risk assessments being used to calculate the portfolio VaR/ES. This procedure can be completed very quickly as it requires only simple matrix algebra, as illustrated in Eq. 9.6 [10].

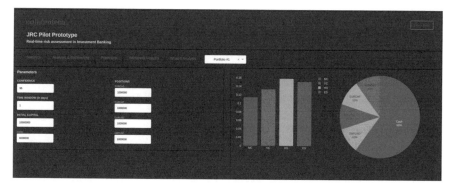

Fig. 9.5 Pre-Trade analysis application page: The user enters (left) the desirable risk parameters (i.e. confidence level and time horizon) along with the positions on the financial instruments of the portfolio to be simulated. The application updates its risk estimations (right) on the fly

$$VaR_p = \sqrt{VRV^T} \qquad (9.6)$$

where V is the vector of VaR estimates per instrument,

$$V = [w_1 VaR_1, w_2 VaR_2, \ldots, w_n VaR_n] \qquad (9.7)$$

and R is the correlation matrix,

$$R = \begin{bmatrix} 1 & \rho_{1,2} & \cdots & \rho_{1,j} \\ \rho_{2,1} & 1 & \cdots & \rho_{2,j} \\ \vdots & \vdots & \ddots & \vdots \\ \rho_{i,1} & \rho_{i,2} & \cdots & 1 \end{bmatrix} \qquad (9.8)$$

where $\rho_{1,2}$ is the correlation estimate between instruments' 1 and 2 returns. This process, described by Algorithm 1 also applies to portfolio ES calculation.

6 Architecture

This section focuses on technical details regarding the deployment of the proposed application in a production environment following the reference architecture described in Chap. 1. Each component is deployed as a docker container in an automated manner, using the Kubernetes container orchestration system. In this way, all software artefacts of the integrated solution are given a common namespace, and they can interact with each other inside the namespace as a sandbox. This deployment method allows for portability of the sandbox, as it can be deployed in different infrastructures.

Algorithm 1 Portfolio risk estimation

1. Parse latest FX prices in 15-minute frequency from datastore
2. Transform prices to 15-minute return
3. **for** *FX instrument i* **do**
       Estimate HS_i, VC_i, MC_i, ES_i
       Calculate weight w_i on each FX instrument i according to user inputs
end
4. Calculate covariance matrix R of returns
5. **for** *risk model R ∈ [HS, VC, MC, ES]* **do**
       Calculate portfolio risk using Eq. 5
end

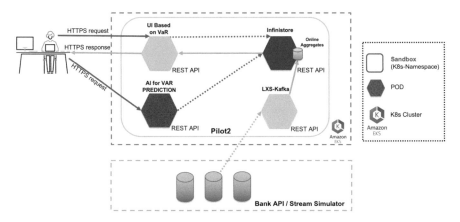

Fig. 9.6 Application deployment scheme

A high-level overview of the Infinitech implementation is illustrated in Fig. 9.6. Each component runs within a Pod, which is the smallest compute unit within Kubernetes. A Pod encapsulates one (or more) containers, has its own storage resources, a unique network IP, access port and options related to how the container should run. This setup enables the auto-scaling of the underlying services according to the volume of external requests. For instance, in case of influx of clients (i.e., traders) requiring risk assessment of their portfolios, the corresponding (red) Pod will automatically scale-out as needed.

It is also noted that as each service is accessed via a REST API, they are able to communicate as required by client requests, while additional features (e.g., financial news sentiment analysis) could be added as a separate component without requiring modification to the existing services.

7 Summary

This chapter addresses one of the major challenges in the financial sector which is real-time risk assessment. In particular, the chapter covers risk models implemented within Infinitech and their integration with associated technological tooling to enable processing of large amounts of input data. The provided Infinitech application leverages several well-established risk models, in addition to introducing a new one based on deep learning techniques. These assessments can be updated in (near) real-time using streaming financial data, and used to drive what-if analysis for portfolio modification. The Infinitech solution can be a valuable asset for practitioners in high frequency trading where consistent risk assessment is required in real-time. It is worth mentioning that the underlying architecture can also be used with different risk models, optimized for different types of financial portfolios, while additional features could be integrated as microservices following Infinitech reference architecture.

Acknowledgments The research leading to the results presented in this Chapter has received funding from the European Union's funded Project INFINITECH under grant agreement no 856632.

References

1. Andersen, T. G., Bollerslev, T., Diebold, F. X., & Vega, C. (2007). *Journal of international Economics, 73*(2), 251.
2. Mengle, D. L., Humphrey, D. B., & Summers, B. J. (1987). *Economic Review 73*, 3.
3. Kearns, M., & Nevmyvaka, Y. (2013). *High frequency trading: New realities for traders, markets, and regulators.*
4. Bredin, D., & Hyde, S. (2004). *Journal of Business Finance & Accounting, 31*(9–10), 1389.
5. Préfontaine, J., Desrochers, J., & Godbout, L. (2010). The Analysis of comments received by the BIS on principles for sound liquidity risk management and supervision. *International Business & Economics Research Journal (IBER), 9*(7). https://doi.org/10.19030/iber.v9i7.598
6. Markowitz, H. (1955). *The optimization of a quadratic function subject to linear constraints.* Technkcal Report. RAND CORP SANTA MONICA CA.
7. Nguyen, J. (2004). The efficient market hypothesis: Is it applicable to the foreign exchange market? Department of Economics, University of Wollongong. https://ro.uow.edu.au/commwkpapers/104
8. Fama, E. F. (2021). *Efficient capital markets II.* University of Chicago Press.
9. Gomber, P., & Haferkorn, M. (2015). *Encyclopedia of information science and technology* (3rd ed., pp. 1–9). IGI Global.
10. Longerstaey, J., & Spencer, M. (1996). *Morgan Guaranty Trust Company of New York, 51*, 54.
11. Chang, Y. P., Hung, M. C., & Wu, Y. F. (2003). *Communications in Statistics-Simulation and Computation, 32*(4), 1041.
12. Hong, L. J., Hu, Z., & Liu, G. (2014). *ACM Transactions on Modeling and Computer Simulation (TOMACS), 24*(4), 1.
13. Bingham, N. H., & Kiesel, R. (2002). *Quantitative Finance, 2*(4), 241.
14. Gers, F. A., Schmidhuber, J., & Cummins, F. (2000). *Neural Computation, 12*(10), 2451.

15. Salinas, D., Flunkert, V., Gasthaus, J., & Januschowski, T. (2020). *International Journal of Forecasting, 36*(3), 1181.
16. Alexandrov, A., Benidis, K., Bohlke-Schneider, M., Flunkert, V., Gasthaus, J., Januschowski, T., Maddix, D. C., Rangapuram, S., Salinas, D., Schulz, J., Stella, L., Türkmen, A. C., & Wang, Y. (2019). Preprint arXiv:1906.05264.
17. Rockafellar, R. T., & Uryasev, S. (2000). Optimization of conditional value-at-risk. *Journal of Risk, 2*, 21–42.
18. Tasche, D. (2002). *Journal of Banking & Finance, 26*(7), 1519.
19. Sarykalin, S., Serraino, G., & Uryasev, S. (2008). *State-of-the-art decision-making tools in the information-intensive age* (pp. 270–294). Informs.
20. Khokhlov, V. (2016). *Evropský časopis ekonomiky a managementu, 2*(6), 70.
21. Dinh-Tuan, H., Beierle, F., & Garzon, S. R. (2019). *2019 IEEE International Conference on Industrial Cyber Physical Systems (ICPS)* (pp. 23–30). IEEE.
22. Kreps, J., Narkhede, N., & Rao, J. (2011). Kafka: A distributed messaging system for log processing. In *Proceedings of the NetDB* (Vol. 11).
23. John, V., & Liu, X. (2017) *Preprint arXiv:1704.00411.*

Chapter 10
Next-Generation Personalized Investment Recommendations

Richard McCreadie, Konstantinos Perakis, Maanasa Srikrishna, Nikolaos Droukas, Stamatis Pitsios, Georgia Prokopaki, Eleni Perdikouri, Craig Macdonald, and Iadh Ounis

1 Introduction to Investment Recommendation

In developed nations, there is growing concern that not enough people are saving for their later life and/or retirement. For instance, 35% (18.4 million) of the UK adult population reported they do not have a pension, and 43% admit they do not know how much they will need for retirement according to a study by Finder.[1] One of the most stable ways to save effectively is through long-term investment in a broad portfolio of financial assets [13]. However, for the average citizen, investing is a difficult, time consuming and risky proposition, leading to few doing so. Hence, there is significant interest in technologies that can help lower the barriers to entry to investment for the average citizen.

Meanwhile, recent advances in Big Data and AI technologies have made the idea of an automated personal assistant that can analyse financial markets and recommend sound investments seem possible. Indeed, there are already a range of online services that will provide financial investment recommendations for a fee.

---

[1] https://www.finder.com/uk/pension-statistics.

R. McCreadie (✉) · M. Srikrishna · C. Macdonald · I. Ounis
University of Glasgow, Glasgow, Scotland
e-mail: richard.mccreadie@glasgow.ac.uk; maanasa.srikrishna@glasgow.ac.uk; craig.macdonald@glasgow.ac.uk; iadh.ounis@glasgow.ac.uk

K. Perakis · S. Pitsios
UBITECH, Chalandri, Greece
e-mail: kperakis@ubitech.eu; spitsios@ubitech.eu

N. Droukas · G. Prokopaki · E. Perdikouri
National Bank of Greece, Athens, Greece
e-mail: nikolaos.droukas@nbg.gr; georgia.prokopaki@nbg.gr; eleni.perdikouri@nbg.gr

© The Author(s) 2022
J. Soldatos, D. Kyriazis (eds.), *Big Data and Artificial Intelligence in Digital Finance*,
https://doi.org/10.1007/978-3-030-94590-9_10

Manual financial investment advice services such as Seeking Alpha[2] and Zacks Investment Research[3] have existed for over a decade, while new automated 'Robo-Advisors' like SoFi Automated Investing[4] and Ellevest[5] have been gaining traction over the last few years. The primary advantages of such automatic systems are that they are more accessible and scalable than other forms of investment advice, while also being flexible with regard to how 'hands-on' the investor wants to be with their portfolio [35]. Hence, such systems are seen as one viable solution to providing personalized financial recommendations for the public.

The European Commission's H2020 INFINITECH project is investigating these technologies within its 'Automated, Personalized, Investment Recommendations System for Retail Customers' pilot, led by the National Bank of Greece. The primary use-case within this pilot is producing automated financial asset recommendations for use by the bank's financial advisors, either during a physical meeting between advisor and customer within a bank branch or remotely. The goals here are threefold: (i) enhanced productivity (i.e. improved productivity of investment consultants of the bank, through faster access to recommendations tailored to their retail customer accounts), (ii) better advice for investments based on a better understanding of customer preferences and behaviour and (iii) increased trading volumes, by widening the pool of investors the bank services.

In the remainder of this chapter, we will discuss the whys and hows of developing an automated financial asset recommender for this use-case. In particular, the chapter is structured as follows. In Sect. 2, we begin by describing the regulatory environment for such systems in Europe, to provide some context of where regulatory boundaries exist that need to be considered. Section 3 then formally introduces the definition of what a financial asset recommender needs to perform in terms of its inputs and outputs, while Sect. 4 discusses how to prepare and curate the financial data used by such systems. In Sect. 5, we provide a review of the scientific literature that is relevant to creating different types of financial asset recommendation systems. Additionally, in Sect. 6, we present experimental results produced within the scope of the INFINITECH project, demonstrating the performance of different recommendation systems using past customers from the National Bank of Greece and over the Greek stock market. A summary of our conclusions is provided in Sect. 7.

---

2   Understanding the Regulatory Environment

Before diving into the practicalities of developing a financial asset recommendation system, it is important to consider the regulatory environment that such a system needs to exist within. Within Europe, there are two main pieces of legislation of interest, which we discuss below:

General Data Protection Regulation (GDPR) GDPR is an EU regulation that was introduced in 2018 and governs the use of personal data regarding EU citizens and those living within the European Economic Area (EEA), and however it has subsequently been used as the basis for similar laws in other countries and so is more widely applicable. GDPR is relevant to the development of financial recommendation systems, as such systems by their nature will need to process the personal data of each customer to function. The key provisions in GDPR to consider therefore are:

- *Data Residency*: All personal data must be stored and remain securely within the EU, and there must be a data protection officer responsible for that data.
- *Pseudonymization*: Personal data being processed must be sufficiently anonymized such that it cannot be subsequently attributed to a specific person without the use of further data.
- *Data Access and Correction*: The customer must be able to access the data stored about them and be able to update that data on request.
- *Transparent Processing*: The customer needs to be made aware of how their data will be used and explicitly accept the processing of their data.

Markets in Financial Instruments Directive (MiFID) MiFID has been applicable across the European Union since November 2007 and forms the cornerstone of the EU's regulation of financial markets, seeking to improve their competitiveness by creating a single market for investment services and activities and to ensure a high degree of harmonized protection for investors in financial instruments. Later in 2011, the European Commission also adopted a legislative proposal for the revision of MiFID, resulting in the Directive on Markets in Financial Instruments repealing Directive 2004/39/EC and the Regulation on Markets in Financial Instruments, commonly referred to as MiFID II and MiFIR, which are currently in effect. These regulations are quite wide-ranging, as they cover the conduct of business and organizational requirements for investment firms, authorization requirements for regulated markets, regulatory reporting to avoid market abuse, trade transparency obligations for shares and rules on the admission of financial instruments to trading.

The MiFID Directive regulates all investment products, including Equities, Bonds, Mutual Funds, Derivatives, Structured Deposits (but not other products such as Loans or Insurance). From the perspective of these regulations, there are three key provisions to account for when developing financial asset recommendation systems (note that if the system is also performing buy/sell orders, there are other provisions in addition to the following to consider):

- *Investors should be classified based on their investment knowledge and experience*: This means that any financial asset recommendation system needs to factor in the investment history of the customer and provide information that is appropriate to their level of expertise.
- *Suitability and appropriateness of investment services need to be assessed*: From a practical perspective, not all financial assets or vehicles will be suitable for all types of investors, this provision means that the types of assets recommended must be filtered to only those that are appropriate for the situation of the customer.
- *Fully inform investors about commissions*: If a commission fee is charged by the underlying services that are being recommended, these must be clearly displayed.

3   Formalizing Financial Asset Recommendation

From a customer's perspective, the goal of financial asset recommendation is to provide a ranked list of assets that would be suitable for the customer to invest in, given the state of the financial market and the constraints and desires of that customer. In this case, 'suitability' is a complex concept that needs to capture both the profitability of the asset with respect to the market and the likelihood that the customer would actually choose to invest in that asset (requiring solutions to model factors such as the perceived risk of an asset in combination with the customer's risk appetite). The notion of suitability also has a temporal component, i.e. how long the customer wants to invest their capital for (known as the investment horizon). With regard to what precisely is recommended, these might be individual financial assets, but more likely are a group of assets (since it is critical to mitigate risk exposure by spreading investments over a broad portfolio). Despite this, from a system perspective, it is often more convenient to initially treat financial asset recommendation as a scoring problem for individual assets, i.e. we identify how suitable an asset is in isolation first, and then tackle aggregation of individual assets into suitable asset groups as a second step. Hence, individual asset scoring can be formulated as follows:

Definition 1 (Individual Asset Scoring) Given a financial asset 'a' (e.g. a stock or currency that can be traded) from a set of assets within a financial market $a \in M$, a customer 'c' with a profile $\mathcal{P}_c$ and an investment time horizon $\mathcal{T}$ produce a suitability score $s(\mathcal{H}_a, \mathcal{P}_c, M, \mathcal{T})$, where $0 \leq s \leq 1$ and a higher score indicates that a is more suitable for investment. Here $\mathcal{H}_a$ represents the historical properties of the asset a, $\mathcal{P}_c$ represents the customer profile, M comprises contextual data about the market as a whole and $\mathcal{T}$ is the investment time horizon (e.g. 3 years).

Asset History $\mathcal{H}_a$ When considering an asset's history $\mathcal{H}_a$, we typically consider the past pricing data for that asset on a daily basis (such as open/close prices, as well as highs/lows for that asset each day). Such time series data can be converted into multiple numerical features describing the asset across different time periods (e.g. the last [1,3,6,12,24,36] months). For instance, we might calculate the Sharpe Ratio over each period to provide an overall measure of the quality of the asset or use volatility to measure how stable the gains or losses from an asset are (representing risk).

Customer Profile $\mathcal{P}_c$ To enable personalization of the recommendations for a particular customer, it is important to factor in the financial position and preferences defined by their profile $\mathcal{P}_c$. The most important factors to consider here are the liquidity of the customer (i.e. how much they can invest safely) and the customer's risk appetite (which should influence the weighting of associated high/low risk assets). We may also have other historical data about a customer in their profile as well, such as past financial investments, although this is rare as most customers will not have invested in the past.

Factoring in the Market $\mathcal{M}$ Markets will behave differently depending on the types of assets that they trade, which is important since we should consider profitability in the context of the market. In particular, there are two ways that we can consider profitability: (1) raw profitability, typically represented by annualized return on investment (where we want returns upward of 7% under normal conditions) and (2) relative profitability, which instead measures to what extent an investment is 'beating' the average of the market (e.g. by comparing asset return to the market average). Relative profitability is important to consider as markets tend to follow boom/bust cycles, and hence a drop in raw profitability might reflect the market as a whole rather than just the asset currently being evaluated.

Investment Time Horizons $\mathcal{T}$ Lastly, we need to consider the amount of time a customer has to invest in the market, as this can strongly impact the suitability of an asset. For example, if working with a short horizon, then it may be advantageous to avoid very volatile assets that could lose the customer significant value in the short term unless they are particularly risk seeking.

By developing an asset scoring function $s(\mathcal{H}_a, \mathcal{P}_c, \mathcal{M}, \mathcal{T})$, we can thereby use the resultant scores to identify particularly good assets to either recommend to the customer directly or include within a broader portfolio of assets. The same scoring function can similarly be used to update asset suitability (and hence portfolio quality) over time as more data on each asset $\mathcal{H}_a$ and the market as a whole $\mathcal{M}$ is collected over time. Having formulated the task and introduced the types of data that can be used as input, in the next section, we discuss how to prepare and curate that data.

4 Data Preparation and Curation

4.1 Why Is Data Quality Important?

Data quality issues can have a significant impact on business operations, especially when it comes to the decision-making processes within organizations [5]. As a matter of fact, efficient, accurate business decisions can only be made with clean, high-quality data. One of the key principles of data analytics is that the quality of the analysis strongly depends upon the quality of the information analysed. However, according to Gartner,[6] it is estimated that more than 25% of critical data in the world's top companies are flawed, at the same time when data scientists worldwide spend around 80% of their time on preparing and managing data for analysis, with approximately 60% of their time being spent on cleaning and organizing data and with an additional 19% of their time being spent on data collection.[7] It is thus obvious that in our big data era, actions that can improve the quality of diverse high volume (financial) datasets are critical and that these actions need to be facilitated by (automated, to the maximum extent possible) tools, optimizing them in terms of (time) efficiency and effectiveness.

The complete set of methodological and technological data management actions for rectifying data quality issues and maximizing the usability of the data are referred to as *data curation* [7]. Data curation is the active and on-going management of data through its lifecycle of interest and usefulness, from creation and initial storage to the time when it is archived for future research and analysis, or becomes obsolete and is deleted [4]. Curation activities enable data discovery and retrieval, maintaining quality, adding value and providing for re-use over time. Data curation has emerged as a key data management activity, as the number of data sources and platforms for data generation has grown. *Data preparation* is a sub-domain of data curation that focuses on data pre-processing steps, such as aggregation, cleaning and often anonymization. Data preparation is the process of cleaning and transforming raw data prior to processing and analysis. It often involves reformatting data, making corrections to that data and combining datasets to add value. The goal of data preparation is the same as other data hygiene processes: to ensure that data is consistent and of high quality. Inconsistent low-quality data can contribute to incorrect or misleading business intelligence. Indeed, it can create errors and make analytics and data mining slow and unreliable. By preparing data for analysis up front, organizations can be sure they are maximizing the intelligence potential of that information. When data is of excellent quality, it can be easily processed and analysed, leading to insights that help the organization make better decisions. High-

---

[6] https://www.reutersevents.com/pharma/uncategorised/gartner-says-more-25-percent-critical-data-large-corporations-flawed.

[7] https://www.forbes.com/sites/gilpress/2016/03/23/data-preparation-most-time-consuming-least-enjoyable-data-science-task-survey-says/#34d3b5ad6f63.

quality data is essential to business intelligence efforts and other types of data analytics, as well as better overall operational efficiency.

4.2   Data Preparation Principles

To make financial data useable, that data must be cleansed, formatted and transformed into information digestible by the analytics tools that follow in a business intelligence pipeline. The actual data preparation process can include a wide range of steps, such as consolidating/separating fields and columns, changing formats, deleting unnecessary or junk data and making corrections to that data.[8] We summarize each of the key data preparation steps below:

Discovering and Accessing Data Data discovery is associated with finding the data that are best-suited for a specific purpose. Data discovery can be a very painful, frustrating and time-consuming exercise. An essential enabler of efficient discovery is the creation and maintenance of a comprehensive, well-documented data catalogue (i.e. a metadata repository). In this context, key data sources include asset pricing data, company and customer profiles, as well as investment transaction data.

Profiling and Assessing Data Profiling data is associated with getting to know the data and understanding what has to be done before the data becomes useful in a particular context and is thus key to unlocking a better understanding of the data. It provides high-level statistics about the data's quality (such as row counts, column data types, min, max and median column values and null counts), and visualization tools are very often exploited by users during this phase, enabling them to profile and browse their data. This is particularly important in the finance context to identify and understand unusual periods in financial time series data, e.g. caused by significant real-world events like COVID-19.

Cleaning and Validating Data Cleaning up the data is traditionally the most time-consuming part of the data preparation process, but it is crucial for removing faulty data and filling in any data gaps found. Some of the key data cleansing activities include:

- Making corrections to the data, e.g. correcting timestamps with a known amount of lag
- Deleting unnecessary data, e.g. deleting extra or unnecessary fields used across different sources in order to cleanly consolidate the discrete datasets
- Removing outliers, e.g. replacing outliers with the nearest 'good' data or with the mean or median, as opposed to truncating them completely to avoid missing data points

---

[8] https://blogs.oracle.com/analytics/what-is-data-preparation-and-why-is-it-important.

- Filling in missing values, i.e. predicting and imputing data missing at random or not at random
- Otherwise, conforming data to a standardized pattern

Once data has been cleansed, it must be validated by testing for errors in the data preparation process up to this point. Often times, an error in the system will become apparent during this step and will need to be resolved before moving forward. Outlier removal is particularly important here, as individual data points such as very high or low prices (e.g. caused by an erroneous trade or a malicious actor) can confuse machine learning systems.

Transforming Data Data transformation is the process of changing the format, structure or values of data. Data transformation may be constructive (adding, copying and replicating data), destructive (deleting fields and records), aesthetic (standardizing salutations or street names) or structural (renaming, moving and combining columns in a database). Proper data transformation facilitates compatibility between applications, systems and types of data.

Anonymizing Data Data anonymization is the process of protecting private or sensitive information by masking, erasing or encrypting identifiers (e.g. names, social security numbers and addresses) that connect an individual to stored data. Several data anonymization techniques exist, including data masking, pseudonymization, generalization, data swapping, data perturbation and more, each one serving different purposes, each coming with its pros and cons and each being better suited than the others depending upon the dataset at hand, but all serving the common purpose of protecting the private or confidential information included within the raw data, as required by law (see Sect. 2).

Enriching Data Data enrichment is a general term that applies to the process of enhancing, refining and improving raw data. It is the process of combining first-party data from internal sources with disparate data from other internal systems or third-party data from external sources. As such, data enrichment could be interpreted both as a sub-step of the data cleansing process, in terms of filling in missing values, and as an independent step of the process. This may require the association (or linking) of the raw data with data from external sources or the enrichment of the raw data with meaningful metadata, thus facilitating their future discovery and association and thus empowering the extraction of deeper insights. A common example here is connecting asset pricing data with profiles for associated companies that influence those assets.

Storing Data Once prepared, the data can be stored or channelled into a third-party application—such as a business intelligence tool—clearing the way for processing and analysis to take place.

4.3 The INFINITECH Way Towards Data Preparation

Within the context of the INFINITECH project, the consortium partners have designed and developed a complete data ingestion/data preparation pipeline, aiming to provide an abstract and holistic mechanism for the data providers, addressing the various connectivity and communication challenges with the variety of data sources that are exploited in the finance and insurance sector, as well as the unique features that need to be considered when utilizing a range of heterogeneous data sources from different sectors. Hence, the scope of the Data Ingestion mechanism is fourfold:

1. To enable the acquisition and retrieval of heterogeneous data from diverse data sources and data providers
2. To facilitate the mapping of the entities included in the data to the corresponding entities of an underlying data model towards its annotation (metadata enrichment)
3. To enable the data cleaning operations that will address the data quality issues of the acquired data
4. To enable the data anonymization operations addressing the constraints imposed by GDPR and other related national and/or European (sensitive) data protection regulations and directives

The INFINITECH complete data ingestion/data preparation pipeline is composed of four main modules, as also illustrated in Fig. 10.1, depicting the high-level architecture of the pipeline. We describe each of these modules in more detail below:

Data Retrieval This module undertakes the responsibility to retrieve or receive the new datasets from a data source or data provider either periodically or on-demand. The scope of the Data Retrieval module is to facilitate the data acquisition from any relational database, HDFS deployments, FTP or HTTP servers, MinIO storage servers, as well as from any API of the data source. The prerequisite is that the appropriate information is provided by the data provider. Additionally, the Data Retrieval module enables the ingestion of new datasets that are pushed from the data provider to its exposed RESTful APIs. Hence, the Data Retrieval module supports all the aforementioned data source types which are considered the most commonly used data sources in current Big Data ecosystems. Nevertheless, the modular architecture of the described solution facilitates the future expansion of the list of supported data source types in an effortless and effective manner.

Data Mapper This module is responsible for the generation of the mapping between the entities of a retrieved dataset and the ones of an underlying data model (derived from the data provider's input). The scope of the Data Mapper module is to enable the mapping of the data entities from a new dataset to the data model that is given by the data provider. In this sense, the data provider is able to create the mappings for each entity of the new dataset to a specific data entity of the data model. To achieve this, at first the Data Mapper module offers the means to

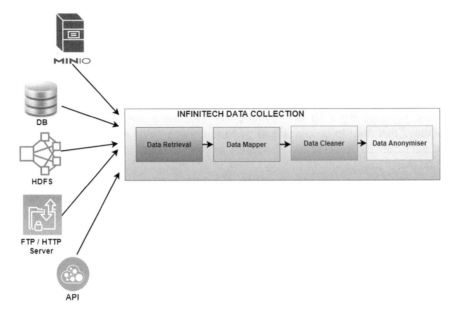

Fig. 10.1 INFINITECH data preparation pipeline

integrate a data model during its initial configuration. Then, during processing, the data entities of the provided dataset are extracted and displayed to the data provider via its user friendly and easy-to-use user interface. Through this user interface, the data provider is able to select the corresponding entities of the integrated data model that will be mapped to the entities of the dataset. The generated mappings are stored for later reuse in a JSON format.

Data Cleaner This module undertakes the responsibility to perform the data cleaning operations on the retrieved dataset, based on the data provider's input. The scope of the Data Cleaner is to provide the data cleaning operations that will ensure that the provided input datasets, which originate from a variety of heterogeneous data sources, are clean and complete to the maximum extent possible. The specific functionalities of this module enable the detection and correction (or removal) of inaccurate, corrupted or incomplete values in the data entities, with the aim of increasing data quality and value. To this end, the data cleaning process is based on a set of data cleansing rules that are defined by the data provider on a data entity level via the Data Cleaner's user friendly and easy-to-use user interface and is a four-step process that includes:

1. The validation of the values of the data entities against a set of constraints
2. The correction of the errors identified based on a set of data correction operations
3. The data completion of the values for the required/mandatory data entities with missing values with a set of data completion operations

4. The maintenance of complete history records containing the history of errors identified and the data cleaning operations that were performed to address them

Data Anonymizer This module undertakes the responsibility of addressing the various privacy concerns and legal limitations imposed on a new dataset as instructed by the data provider. To this end, the anonymizer provides the advanced privacy and anonymization toolset with various data anonymization techniques that can be tailored by the data provider in order to filter or eliminate the sensitive information based on his/her needs. To achieve this, the anonymizer employs a generic data anonymization process which is highly customizable through the definition of anonymization rules which are set by the data provider leveraging his/her expertise.

All four processes can be performed as background processes, provided that the data providers have created in advance the corresponding data retrieval/mapping/cleaning/anonymization profiles for a specific dataset, which will in turn be used in an automated way for the execution of the processes. In the next section, we will discuss approaches to use our curated data to perform financial asset recommendation.

5   Approaches to Investment Recommendation

With the underlying data needed to perform asset scoring prepared, we next need to understand how our asset scoring function can be built. In this section, we will provide a brief background into one class of methods for developing such a scoring function, namely using recommendation models, with a particular focus on recommendation as it applies to financial products and services.

What Is a Recommender? Recommendation systems are a popular class of algorithms that aim to produce personalized item suggestions to a user. Popular examples of such algorithms are movie recommendation (e.g. on Netflix) [3] or product recommendation (e.g. Amazon recommended products) [36]. For our later reported results, we will only be considering supervised machine-learned recommenders, i.e. approaches that leverage machine learning to analyse the past history of items, user interactions with those items and/or the user's relationships to the items, with the goal of producing a recommendation model [15]. However, we will discuss a couple of unsupervised approaches later in this section. The process of creating a supervised machine learned model is known as training. A trained recommendation model can be considered as a function that (at minimum) takes a user and item as input and produces a score representing the strength of relationship between that user and item pair. An effective recommendation model should find a right balance between relevance, diversity and serendipity in the recommended items, using the different types of evidence discussed in Sect. 3. Hence, for financial asset recommendation, we want to train a model to represent the function $s(\mathcal{H}_a, \mathcal{P}_c, \mathcal{M}, \mathcal{T})$ (see Sect. 3), where $\mathcal{H}_a$ represents the item (a financial asset) and $\mathcal{P}_c$ represents the user (a customer).

Types of Recommender Recommendation algorithms can be divided along three main dimensions:

1. The types of evidence they base the recommendations upon.
2. Whether items as scored in isolation or as part of a set or sequence of items.
3. To what extent the recommender understands/accounts for temporal dynamics during the training process.

In terms of evidence, a model might utilize explicit interactions between users and items (e.g. when a user watches a movie) [18], intrinsic information about an item (e.g. what actors star in a movie) [41], time series data about an item (e.g. popularity of a movie over time) [37], intrinsic information about a user (e.g. the user says they like action movies) [24] or time series data about the user (e.g. the user has been watching a lot of dramas recently) [43]. When considering recommendation context, most approaches consider each item in isolation, but some approaches examine items within the context of a set (e.g. recommending an item to add to a customer's basket that already contains items [27]) or when recommending a sequence of items (e.g. recommending a series of places to visit when on holiday [19]). Finally, depending on how the model is trained, simple approaches consider all prior interactions as an unordered set, while more advanced techniques will factor in that recent interactions are likely more relevant than older interactions [40].

Within the context of the financial asset recommendation, explicit interactions between users and items are somewhat rare and difficult to obtain, as they represent past investments made by customers (which is not public data). Hence, approaches that can more effectively leverage available intrinsic and historical data about the customers and assets for recommendation are more desirable here. With regard to recommendation context, we will be focusing on asset recommendations in isolation here, and however we would direct the reader to the literature on basket recommenders if interested in recommendation of items to add to an existing portfolio [1]. Meanwhile, capturing the temporal dynamics of our historical data and interactions is intuitively important within financial markets, as past asset performance is not always reflective of future performance, and hence models that can capture temporal trends will be more effective in this domain.

In the remainder of this section, we will discuss a range of recommendation approaches that can be applied for financial asset recommendation, as well as highlight their advantages and disadvantages. To structure our discussion, we will group models into six broad types. We will start with four supervised approaches that we will experiment with later in Sect. 6, namely *Collaborative filtering*, *User Similarity*, *Key Performance Indicator Predictors*, and *Hybrid*. We then discuss two unsupervised approaches, namely *Knowledge Based* and *Association Rule Mining*.

5.1   *Collaborative Filtering Recommenders*

Collaborative filtering recommenders focus on providing item recommendations to users based on an assessment of what other users, judged to be similar to them, had positively interacted with in the past. These models work with a matrix of user–item interactions, which is typically sparse, owing to the fact that most customers would not have interacted with (invested in) more than a small subset of items (financial assets) previously. The intuition behind these models is that missing ratings for items can be predicted, or imputed, as they are likely correlated across the user and item space. By learning the similarities between users and items, inferences can be made about the missing values. As this approach leverages past interactions between the user and items in order to provide recommendations, it typically works best with large quantities of interactions such that re-occurring patterns can be found. These interactions can be in the form of explicit feedback, wherein a user provides an explicit, quantitative rating for the item, or as implicit feedback, where user interactions are gauged from their behaviour using system, such as clicking a link, or spending a significant amount of time browsing a specific product [1].

Advantages and Disadvantages Collaborative filtering models can be advantageous since they tend to generate diverse recommendations, since they are looking for items interacted with (invested in) by customers that are similar to you, but you have not interacted with. On the other hand, particularly when neighbourhood-based, they suffer from the problem of sparsity, where it is difficult to provide recommendations for a user who has little to no interaction data available, as correlations cannot easily be drawn. More advanced model-based techniques alleviate the sparsity problem, but often still require large amounts of existing interaction data, which presents a particular problem in the financial domain due to the difficulty of curating and obtaining sufficient anonymized financial transactions [14]. Also of note is that many of these models employ dimensionality reduction (to speed up computation and remove noise) and learn *latent* factors between users and items in order to perform recommendation. This makes explaining why certain recommendations were made challenging, as the latent factors learned may not be interpretable by humans.

Applications in Finance Collaborative filtering techniques are often applied for financial recommendation. Some notable approaches include the implementation of a fairness-aware recommender system for microfinance, by Lee et al. [16], which uses item-based regularization and matrix factorization, as well as risk-hedged venture capital investment recommendation using probabilistic matrix factorization, as proposed by Zhao et al. [45]. Luef et al. [17] used user-based collaborative filtering for the recommendation of early stage enterprises to angel investors, as a method of leveraging the investors' circle of trust. Meanwhile, Swezey and Charron [38] use collaborative filtering recommendations for stocks and portfolios, reranked by Modern Portfolio Theory (MPT) scores, which incorporate metrics of risk, returns and user risk aversion. From these approaches, it can be gleaned that while there is value in utilizing user–product interaction data to provide financial recommendation, this is not sufficient as a representation of the suitability of the recommended financial products, and augmentation with asset intrinsic information is often necessary. Furthermore, gathering the required interaction data for collaborative filtering recommendation can present a significant challenge due to the rarity of investment activity among the general populace.

5.2   User Similarity Models

Like collaborative filtering models, user similarity models are based on the intuition that if two users are similar, then similar items should be relevant to both. However, instead of directly learning latent similarity patterns between users and items from past interaction data, user similarity models generate a feature vector for each user and directly calculate similarity between all pairs of users (e.g. via cosine similarity). Once a top-n list of most similar users is found, a ranked list of items can be constructed based on items those similar users have previously interacted with (e.g. invested in), where the more users interact with an item, the higher its rank.

Advantages and Disadvantages The primary advantage of user similarity models is that they are highly explainable (e.g. 'x users like you invested in this asset'). However, in a similar way to collaborative filtering models, they are reliant on sufficient prior interaction data to identify 'good' items for groups of similar users.

Applications in Finance Luef et al. [17] implement a social-based recommender model for investing in early stage enterprises, which asks users to specify an inner circle of trusted investors and then provide recommendations on the basis of what those users invested in. Yujun et al. [44] propose a fuzzy-based stock recommender algorithm, which recommends instruments that were chosen by other similar users. Such models that focus on similar users tend to be seen as more convincing by end users, as the results can be presented in not only an explainable but also a personable manner by alluding to other experts in the field who made similar decisions.

5.3   Key Performance Indicator Predictors

KPI predictor-based recommenders utilize descriptive features of the users and items in order to produce recommendations, rather than relying on the existence of explicit or implicit interactions between those users and items. These features are typically expressed as a single numerical vector representation per user–item pair. Using these vectors as a base, content-based recommenders will typically train a model to predict a rating, score or label (KPI) that can act as a proxy for item suitability to the user within the context of the target domain. For example, in the financial asset recommendation context, such a model might aim to predict the profitability of the asset given the user's investment horizon.

Advantages and Disadvantages The advantage that KPI predictors hold over collaborative filtering is that they do not rely on past interaction data, and hence new items that have not been interacted with in the past can be recommended based on their descriptive features. Such models can also be applied to cold-start users (those with no investment history). The drawback of these models is that they rely on a proxy KPI rather than a true measure of suitability, meaning that performance will be highly dependent on how correlated the KPI is with actual suitability. Additionally, these models tend to only have a limited capacity to personalize the recommendations for each individual user, i.e. the item features tend to dominate the user features in such models.

Applications in Finance Seo et al. [33] produced a multi-agent stock recommender that conducts textual analysis on financial news to recommend stocks. This is one of several models that utilize natural language processing techniques to extract stock predictors from textual data [31, 32]. Musto et al. also develop a case-based portfolio selection mechanism that relies on user metadata for personalization [22], while Ginevicius et al. attempt to characterize the utility score of real estate using complex proportional evaluation of multiple criteria [8].

5.4   Hybrid Recommenders

In practical applications, multiple recommender models are often combined or used in tandem to improve overall performance. Such combined approaches are known as hybrid recommenders. There are two broad strategies for combining recommenders: unification and voting. Unification refers to the combination of recommender models to produce a singular algorithm that returns a result. This might involve feeding the output of multiple existing models into a weighted function that combines the scores from each (where that function may itself be a trained model). Meanwhile, voting approaches consider the output of the different learned models as votes for each item, where the items can be ranked by the number of votes they receive. Recommenders can also be cascaded, wherein multiple recommenders are given strict orders of priority. The absence of relevant information for one recommender will then trigger recommendations from the model of lower priority. This can present a useful solution to the problem of datasets being sparse or the unavailability of consistent feedback from users across various methods.

Advantages and Disadvantages The main advantage of hybrid approaches is that they alleviate the disadvantages of each individual approach and are able to recommend a diverse range of items. On the other hand, they add significant complexity to the recommendation task (particularly when additional model(s) need to be trained to perform the combination) and make outcome explainability more challenging [14].

Applications in Finance Taghavi et al. [39] proposed the development of an agent-based recommender system that combines the results of collaborative filtering together with a content-based model that incorporates investor preferences and socioeconomic conditions. Luef et al. [17] also employed a hybrid recommendation approach that compares the rankings produced by their previously described social user-based recommender and knowledge-based recommender using Kendall's correlation. Matsatsinis and Manarolis [20] conduct the equity fund recommendation task using a combination of collaborative filtering and multi-criteria decision analysis and the associated generation of a utility score. Mitra et al. [21] combine a collaborative filtering approach along with an attribute-based recommendation model in the insurance domain, in order to account for the sparsity of user ratings. Hybrid techniques can provide recommendations that are representative of the different variables that influence the suitability of financial assets for investment, such as investor preferences, textual analysis, expertly defined constraints, prior investment history and asset historical performance and perhaps present the best theoretical solution for the current problem setting. However, discovering the optimal combination of these models can be challenging, especially as their numbers increase.

5.5 Knowledge-Based Recommenders

Having discussed supervised approaches to the problem of financial asset recommendation, to complete our review, we also highlight a couple of unsupervised approaches, starting with knowledge-based recommenders.

Knowledge-based recommendation is based on the idea of an expert encoding a set of 'rules' for selecting items for a user. In this case, the system designer defines a set of filtering and/or scoring rules. At run-time, a user fills in a questionnaire on their constraints and preferences (this information might also be derived from an existing customer profile), needed to evaluate those rules. The rules are then applied to first filter the item set, then score and hence rank the remaining items, forming the recommended set [6].

Advantages and Disadvantages In terms of advantages, they allow for highly customizable and explainable recommendations, as they are based on human-defined rules. They are also usable for cold-start users with no prior history (as long as they correctly fill in the questionnaire). On the other hand, they are reliant on humans defining good generalizable rules, which is both time consuming and error-prone for diverse domains like finance and needs to be continually updated manually [6].

Applications in Finance Gonzalez-Carrasco et al. [9] utilized a fuzzy approach to classify investors and investment portfolios, which they then match along with one another. Set membership for investments is calculated using a set of predefined rules. For instance, a product in prime real estate with normal volatility might be classified as socially moderate and psychologically conservative, but an investment in solar farms with a low long-term return is classified as socially and psychologically aggressive. Luef et al. [17] evaluated the performance of multiple recommender models to recommend early stage enterprises to investors, and one such model was knowledge-based, where investor profiles are used to determine constraints on recommendation, with profile features acting as hard filters and investor preferences acting as soft constraints on enterprise profiles. These recommenders allow for the fine-grained specification of requirements and constraints upon the recommendation and are helpful in providing context to support investment decisions by presenting an integrated picture of how multiple variables influence investment instruments. These recommenders can provide highly customized results that can be tweaked as the user desires alongside their changing preferences; however, they cannot glean latent patterns representative of the userbase and their investment trends, which might prove useful in providing recommendations that leverage a user's circle of trust.

5.6   Association Rule Mining

Association rule mining is an unsupervised, non-personalized method of recommendation which seeks to identify frequent itemsets—which are groups of items that typically co-occur in baskets. In doing so, it is often used as a method of recommending items that users are likely to interact with given their prior interactions with other items identified within those itemsets. These relationships between the items, or variables, in a dataset are termed association rules and may contain two or more items [2]. This technique is typically applied to shopping settings where customers place multi-item orders in baskets and therefore is also termed market basket analysis.

Advantages and Disadvantages Association rules are straightforward to generate in that they do not require significant amounts of data or processing time, and as an unsupervised technique, the input data does not need to be labelled. However, as an exhaustive algorithm, it discovers all possible rules for the itemsets that satisfy the required thresholds of co-occurrence. Many of these rules may not be useful or actionable; however, this requires expert knowledge to evaluate and provide insight upon the meaningfulness of the generated rules.

Applications in Finance One such approach by Nair et al. [23] uses temporal association rule mining to account for the general lack of consideration of time information in typical association rule mining approaches. They leverage a genetic algorithm called Symbolic Aggregate approXimation (SAX) to represent time series data in a symbolic manner and then feed these representations into the model to obtain association rules and provide recommendations on the basis of these rules. Other approaches by Paranjape-Voditel and Deshpande [25], as well as Ho et al. [11], look at the usage of fuzzy association rule mining to investigate hidden rules and relationships that connect one or more stock indices. The former recommends stocks, by using the principle of fuzzy membership in order to calculate relevance of various items in the portfolio, and also introduces a time lag to the inclusion of real-time stock price information, so that any emerging patterns in price movement are aptly captured. This becomes especially applicable in the financial domain, wherein quantitative data about instruments can be classified better using a fuzzy approach as opposed to one using crisp boundaries, so as to identify the ranges of the parameters.

6   Investment Recommendation within INFINITECH

Having summarized how we can generate recommendation scores for assets using different types of recommendation algorithms, in this section, we illustrate how these models perform in practice. In particular, we will first describe the setup for evaluating the quality of financial asset recommendation systems and then report performances on a real dataset of financial transactions data from the Greek stock market.

6.1   *Experimental Setup*

Dataset To evaluate the quality of a financial asset recommendation system, we need a dataset comprising financial assets and customers, as well as their interactions. To this end, we will be using a private dataset provided by the National Bank of Greece, which contains investment transactions spanning 2018–2021, in addition to supplementary information on demographic and non-investment behaviour data from a wider subset of the bank's retail customers, and historical asset pricing data. From this dataset, we can extract the following main types of information:

- **Customer Investments**: This forms the primary type of evidence used by collaborative filtering recommendation models, as it contains the list of invest-ment transactions conducted by all customers of the bank. These timestamped transactions contain anonymized customer IDs, instrument IDs and ISINs, as well information on the instrument names and the volume and direction of the transactions.
- **Customer Intrinsic and Historical Features**: This supplementary data includes aggregated indicators derived from the retail transactions and balances of all the anonymized customers. Such indicators include aggregated credit card

Table 10.1 Greek stock market sample dataset statistics by year

Dataset information		2018	2019	2020	2021
Market data	Unique assets	4087	4328	4883	3928
	Price data points	655,406	796,288	839,637	145,885
	Average annualized return	−6.74%	11.65%	19.22%	6.71%
	Assets involved in investment	1089	983	903	636
Customers	Customers	52,365	52,365	52,365	52,365
	Customers with investment transactions	18,359	15,925	17,060	8126
	Total transactions	72,020	73,246	121,703	47,501

spends across multiple sectors and deposit account balances with aggregations spanning different time windows and demographic features on the customers. These KPIs are recorded in entirety across 2018, 2019 and 2020, with only incomplete aggregates available across 2021. Only a limited subset of users possess investment transactions.

- **Asset Intrinsic and Historical Features**: This data contains the historical prices of the assets spanning 2018–2021, alongside other descriptive information on the assets such as their name, ISIN and financial sector. The price data points are available for almost all days (although there are a few gaps due to data collection failures, see Table 10.1).

Models For our experiments, we will experiment with eight representative supervised recommendation models from those described earlier, starting with two collaborative filtering models:

- **Matrix Factorization (MF) [Collaborative Filtering]** [28]: This is the conventional matrix factorization model, which can be optimized by the Bayesian personalized ranking (BPR [26]) or the BCE loss.
- **LightGCN [Collaborative Filtering]** [10]: Building on NGCF [42], LightGCN is a neural graph-based approach that has fewer redundant neural components compared with the NGCF model, which makes it more efficient and effective.

To represent a user similarity-based approach, we implement a user encoding strategy inspired by the proposed hybrid approach of Taghavi et al. [39], albeit not relying on the use of agents, and using a different content-based model:

- **Customer Profile Similarity (CPS) [User Similarity]**: This strategy represents users in terms of a subset of the non-investment indicators concerning their retail transactions and balances with the bank. These indicators are weighted by age where possible, i.e. customer purchase features from two years ago have a lower weight than those from one year ago. Given these vector representations of each customer, we use these customer vectors to represent the financial assets. Specifically, for each financial asset, we assign it a vector representation that is the mean of the vectors of the customers that invested in that asset previously. The intuition here is that these aggregate vectors will encode the key indicators that make assets suitable for each customer. Finally, to score each asset for a new customer, we can simply calculate the cosine similarity between that customer and the aggregate asset representation, as both exist in the same dimensional space, where a higher similarity indicates the asset is more suitable for the customer.

Meanwhile, since it makes intuitive sense to use profitability as a surrogate for suitability in this context, we also include a range of key performance indicator predictors, which rank financial assets by their predicted profitability:

- **Predicted Profitability (PP) [KPI Predictor]**: This approach instead of attempting to model the customer–asset relationship instead attempts to predict an aspect of each asset alone, namely its profitability. In particular, three regression models are utilized to predict profitability of the assets over the aforementioned test windows. These models utilize features descriptive of the performance of the asset over incremental periods of time prior to testing, namely, [3,6,9] months prior, and include volatility, average closing price and expected returns over these periods.

  - **Linear Regression (LR)** [29]: Linear regression attempts to determine a linear relationship between one or more independent variables and a dependent variable by assigning these independent variables coefficients. It seeks to minimize the sum of squares between the observations in the dataset and the values predicted by the linear model.
  - **Support Vector Regression (SVR)** [34]: Support vector regression is a generalization of support vector machines for continuous values, wherein the aim is to identify the tube which best approximates the continuous-valued function. The tube is represented as a region around the function bounded on either side by support vectors around the identified hyperplane.
  - **Random Forest Regression (RFR)** [30]: Random forest regression is an ensemble method that utilizes a diverse array of decision tree estimators and conducts averaging upon their results in a process that significantly reduces the variance conferred by individual estimators by introducing randomness. Decision trees, in turn, are models that infer decision rules in order to predict a target variable.

In addition, we also implement a range of hybrid models that combine the aforementioned collaborative filtering approaches with customer and item intrinsic and historical features, in a similar manner to the experiments conducted by Taghavi et al. [39], Luef et al. [17] and others who use the intuition of combining intrinsic attributes of customers and items alongside collaborative filtering inputs:

- **MF+CI [Hybrid]**: This model combines the ranked recommendation lists of matrix factorization and the customer similarity model using rank aggregation. The specific rank aggregation method used is score voting.

- **LightGCN+CI [Hybrid]**: This model combines the ranked recommendation lists of LightGCN and the customer similarity model using rank aggregation. The specific rank aggregation method used is score voting.

Training Setup To train our supervised models, we need to divide the dataset into separate training and test subsets. We use a temporal splitting strategy here, where we define a time point to represent the current moment where investment recommendations are being requested. All data prior to that point can be used for training, while the following 9 months of data can be used to evaluate success. We create three test scenarios in this manner, where the three time points are 1st of October 2019 (denoted 2019), 1st of April 2020 (denoted 2020A) and the 1st of July 2020 (denoted 2020B). The collaborative filtering models are each trained for 200 epochs. Due to the large size of the dataset, a batch size of 100000 is chosen for training in order to speed up training time. Each model is run 5 times, and performance numbers are averaged out across the 5 runs.

Metrics To quantitatively evaluate how effective financial asset recommendation is, we want to measure how satisfied a customer would be if they followed the investment recommendations produced by each approach. Of course, this type of evaluation is not possible in practice, since we would need to put each system into production for an extended period of time (e.g. a year). Instead, we rely on surrogate metrics that are more practically available and capture different aspects of how satisfied a user might be if they followed the recommendations produced. There are two main aspects of our recommendations that we examine here:

- *Investment Prediction Capability*: First, we can consider how well the recommendations produced by an approach match what a customer actually invested in. The assumption here is that if the customers are making intelligent investments, then the recommendation system should also be recommending those same financial assets. Hence, given a customer and a point in time, where that customer has invested in one or more assets after that time point, we can evaluate whether the recommendation approach included those assets that the customer invested in within the top recommendations. We use the traditional ranking metric normalized discounted cumulative gain (NDCG) to measure this [12]. NDCG is a top-heavy metric, which means that an approach will receive a higher score the closer to the top of the recommendation list a relevant financial asset is. NDCG performance is averaged across all users in the test period considered. There are two notable limitations with this type of evaluation, however. First, it assumes that the customer is satisfied with what they actually invested in, which is not always the case. Second, only a relatively small number of users have investment history, so the user set that this can evaluate over is limited.
- *Investment Profitability*: An alternative approach for measuring the quality of the recommendations is to evaluate how much money the customer would have made if they followed the recommendations produced by each approach. In this case, for an asset, we can calculate the return on investment when investing in the top few recommended assets after a year, i.e. annualized return. The issue with this

type of metric is it ignores any personalized aspects regarding the customer's situation and can be highly volatile if there are assets who's price experiences significant growth during the test period (as we will show later).

For both of these aspects, we calculate the associated metrics when analysing the top 'k' assets recommended, where 'k' is either 1, 5 or 10.

6.2 Investment Recommendation Suitability

In this section, we will report and analyse the performance of the aforementioned financial asset recommendation models when applied to the Greek Stock market for customers of the National Bank of Greece. The goal of this section is not to provide an absolute view of asset recommendation approach performance, since there are many factors that can affect the quality of recommendations, most notably the characteristics of the market in question for the time period investigated. Instead, this analysis should be used as an illustration of how such systems can be evaluated and challenges that need to be considered when analysing evaluation output.

Investment Prediction Capability We will start by evaluating the extent to which the recommendation approaches recommend the assets that the banking customers actually invested in. Table 10.2 reports the investment prediction capability of each of the recommendation models under NDCG@[1,5,10] for each of the three scenarios (time periods tested). The higher NDCG values indicate that the model

Table 10.2 Financial asset recommendation model performances when evaluating for investment prediction capability. The best performing model per scenario (time period) is highlighted in bold

Scenario	Approach type	Recommender model	NDCG@1	NDCG@5	NDCG@10
2019	Collaborative filtering	MF	0.292	0.219	0.187
		LightGCN	**0.447**	**0.351**	**0.291**
	User similarity	CustomerSimilarity	0.026	0.009	0.007
	Hybrid	MF+CI	0.372	0.189	0.132
		LightGCN+CI	0.427	0.229	0.160
2020A	Collaborative filtering	MF	0.336	0.243	0.199
		LightGCN	**0.461**	**0.349**	**0.283**
	User similarity	CustomerSimilarity	0.011	0.012	0.015
	Hybrid	MF+CI	0.376	0.190	0.139
		LightGCN+CI	0.409	0.222	0.161
2020B	Collaborative filtering	MF	0.359	0.263	0.223
		LightGCN	**0.462**	**0.358**	**0.294**
	User similarity	CustomerSimilarity	0.012	0.012	0.011
	Hybrid	MF+CI	0.355	0.185	0.135
		LightGCN+CI	0.390	0.218	0.158

is returning more of the assets that the customers invested within in the top ranks. The highest performing model per scenario is highlighted in bold. Note that we are omitting the KPI prediction models from this analysis, as they can (and often do) recommend items that no user has invested in, meaning that their NDCG performance is not truly comparable to the other models under this type of evaluation.

As we can see from Table 10.2, of all the models tested, the type of model that produces the most similar assets to what the customers actually choose to invest in is the collaborative filtering type models, where the more advanced graph-based neural model (LightGCN) is better than the classical matrix factorization (MF) model. Given that collaborative filtering models aim to extract patterns regarding past investments across customers, it indicates the customers in our dataset largely invest in similar items over time (hence there are patterns that the collaborative filtering approaches are learning). On the other hand, the user similarity-based model recommends very different assets to what the customers invested within, indicating that the customers do not cluster easily based on the features we extracted about them. Finally, we observe that the hybrid models that integrate more features about the customers and items result less similar assets being recommended than the stock collaborative filtering only approaches. This likely means that the dataset is too small (i.e. has too few examples) to confidently extract generalizable patterns from the large number of new features being given to the model here.

However, just because the collaborative filtering models have the highest scores here does not necessarily mean that they will be the best. Recall that this metric is capturing whether the model is producing the same assets as what the customer actually invested in, not whether those were effective investments. Hence, we will next examine the profitability of the assets recommended.

Investment Profitability We next evaluate the profitability of the recommended assets in order to assess the effectiveness of the recommendation models. Profitability is evaluated as the expected return over each 9-month testing period and then annualized. Table 10.3 reports annualized profitability when considering the top [1,5,10] recommended financial assets by each recommendation model. Profitability is calculated as a percentage return on investment over the year period.

From Table 10.3, we can observe some interesting patterns of behaviour between the different types of recommendation models. Starting with the collaborative filtering-based approaches, we can see that for the 2019 test scenario, these models would have lost the customer a significant amount of value. However, for the 2020A/B scenarios, these models would have returned a good profit of around 30%. This indicates that there was a marked shift in the types of financial asset that was profitable between 2018 and 2020. It also highlights a weakness of these collaborative filtering models, in that they assume that past investments were profitable and hence similar investments now will also be profitable, which was clearly not the case between 2018 (initial training) and the 2019 test scenario. Furthermore, we also see the hybrid models that are based on the collaborative filtering strategies suffering from the same issue, although it is notable that when

Table 10.3 Financial asset recommendation model performances when evaluating for asset profitability. The best performing model per scenario (time period) is highlighted in bold

Scenario	Approach type	Recommender model	Return@1	Return@5	Return@10
2019	Collaborative filtering	MF	−28.76%	−39.51%	−30.71%
		LightGCN	−29.4%	−37.99%	−25.42%
	User similarity	CustomerSimilarity	3.7%	−0.84%	−0.5%
	Hybrid	MF-CI	−42.72%	−42.43%	−42.43%
		LightGCN-CI	−43.65%	−43.07%	−43.07%
	KPI prediction	LR	**114.05%**	**40.37%**	15.42%
		SVR	14.6%	36.61%	29.19%
		RFR	26.23%	39.02%	**31.04%**
2020A	Collaborative filtering	MF	35.7%	34.66%	34.44%
		LightGCN	**36.74%**	22.99%	20.65%
	User similarity	CustomerSimilarity	21.69%	23.59%	23.94%
	Hybrid	MF-CI	43.31%	42.89%	42.89%
		LightGCN-CI	44.91%	39.61%	39.6%
	KPI prediction	LR	0.52%	10.49%	59.98%
		SVR	22.03%	17.02%	**64.05%**
		RFR	0.52%	**59.46%**	31.45%
2020B	Collaborative filtering	MF	56.08%	52.29%	50.34%
		LightGCN	**76.72%**	43.44%	35%
	User similarity	CustomerSimilarity	21.89%	23.83%	23.53%
	Hybrid	MF-CI	75.94%	77.11%	77.11%
		LightGCN-CI	82.82%	74.16%	74.11%
	KPI prediction	LR	−20.15%	**291.04%**	**176.97%**
		SVR	−1.31%	−1.31%	39.21%
		RFR	−3.92%	−6.33%	36.8%

they are working as intended, they recommend more profitable assets than the models that use collaborative filtering alone, showing that the extra features that were added are helping identify profitable asset groups. In terms of the user similarity-based model, we again see poorer performance in the 2019 scenario than the 2020A/B scenarios, which is expected as this type of model is based on analysis of the customers investment history. However, the returns provided by this model are reasonably consistent (around 20–23%) in 2020.

Next, if we compare to the KPI Predictor models that are attempting to explicitly predict assets that will remain profitable, we see that these models can be quite volatile. For instance, for the 2019 test scenario and the LR model, the top recommended asset exhibited a return of 114%, but that same model in 2020B placed an asset in the top rank that lost 20% of its value. This volatility is caused since these models are relying on (recent) past performance of an asset remaining a strong indicator of its future performance. For example, if we consider the 2020B scenario with the 20% loss for the top-ranked asset, that is a case where the asset's price had been recently inflated, but dropped rapidly during the test period. However, if we consider the KPI predictors holistically, as a general method for building a broader portfolio (e.g. if we consider the returns for the top 10 assets), these approaches seem to be consistently recommending highly profitable assets in aggregate and generally recommend more profitable assets than the collaborative filtering-based approaches.

Overall, from these results, we can see some of the challenges when evaluating models automatically without a production deployment. Measuring investment prediction capacity of a model can provide a measure of how similar the recommendations produced are to how the customers choose to invest. However, our analysis of profitability indicates that there is no guarantee that these are good investments from the perspective of making a profit. On the other hand, we can clearly see the value that these automatic recommender systems can bring for constructing portfolios of assets, with some quite impressive returns on investment when averaged over the top-10 recommended assets.

7   Summary and Recommendations

In this chapter, we have provided an introduction and overview for practitioners interested in developing financial asset recommendation systems. In particular, we have summarized how an asset recommendation system functions in terms of its inputs and outputs, discussed how to prepare and curate the data used by such systems, provided an overview of the different types of financial asset recommendation model, as well as their advantages and disadvantages, and finally included an analysis of those models using a recent dataset of assets from the Greek stock market and customers from a large bank. Indeed, we have shown that this type of recommendation system can provide effective lists of assets that are highly profitable, through the analysis of historical transactional and asset pricing data.

For new researchers and practitioners working in this area, we provide some recommendations below for designing such systems based on our experience:

- **Data cleanliness is critical**: Good data preparation allows for efficient analysis, limits errors, eliminates biases and inaccuracies that can occur to data during processing and makes all of the processed data more accessible to users. Without sufficient data preparation, many of the models discussed above will fail due to noisy data. For instance, we needed to spend significant time removing pricing outliers from the asset pricing dataset due to issues with the data provider.
- **Understand what data you can leverage**: There are a wide range of different types of information that you might want to integrate into an asset recommendation model: past investments, asset pricing data, customer profiles and market data, among others. However, often some of this data will be unavailable or too sparse to be usable. Pick a model that is suitable for the data you have available, and avoid trying to train models with too few examples.

- **Analyse your models thoroughly**: Sometimes trained machine learned models hallucinate patterns that are not there, simply because they have not seen counter examples to sufficiently calibrate, particularly in scenarios like this where assets can exhibit short price spikes or troughs. For instance, we have seen one of our KPI predictors suggests that an asset could have an annualized return in the thousands of %, which is obviously impossible. You may wish to apply additional filtering rules to your machine learned models to remove cases where the model is obviously miss-evaluating an asset and use multiple metrics like those presented here to quantify how well those models are functioning before putting them into production.

Acknowledgments The research leading to these results has received funding from the European Community's Horizon 2020 research and innovation programme under grant agreement no: 856632 (INFINITECH).

References

1. Aggarwal, C. C., et al. (2016). *Recommender systems* (Vol. 1). Springer.
2. Agrawal, R., Imieliński, T., & Swami, A. (1993, June). Mining association rules between sets of items in large databases. *SIGMOD Record, 22*(2), 207–216.
3. Bennett, J., Lanning, S., et al. (2007). The Netflix prize. In *Proceedings of KDD cup and workshop* (Vol. 2007, p. 35). Citeseer.
4. Cragin, M. H., Heidorn, P. B., Palmer, C. L., & Smith, L. C. (2007). An educational program on data curation.
5. Curry, E., Freitas, A., & O'Riáin, S. (2010). The role of community-driven data curation for enterprises. In *Linking enterprise data* (pp. 25–47). Springer.
6. Felfernig, A., Isak, K., & Russ, C. (2006). Knowledge-based recommendation: Technologies and experiences from projects. In *Proceedings of the 2006 Conference on ECAI 2006: 17th European Conference on Artificial Intelligence August 29 – September 1, 2006, Riva Del Garda, Italy*, NLD (pp. 632–636). IOS Press.
7. Freitas, A., & Curry, E. (2016). Big data curation. In *New horizons for a data-driven economy* (pp. 87–118). Cham: Springer.
8. Ginevičius, T., Kaklauskas, A., Kazokaitis, P., & Alchimovienė, J. (2011, September). Recommender system for real estate management. *Verslas: teorija ir praktika, 12*, 258–267.
9. Gonzalez-Carrasco, I., Colomo-Palacios, R., Lopez-Cuadrado, J. L., Garciá-Crespo, Á., & Ruiz-Mezcua, B. (2012). Pb-advisor: A private banking multi-investment portfolio advisor. *Information Sciences, 206*, 63–82.
10. He, X., Deng, K., Wang, X., Li, Y., Zhang, Y., & Wang, M. (2020). LightGCN: Simplifying and powering graph convolution network for recommendation. In *Proc. of SIGIR* (pp. 639–648).
11. Ho, G., Ip, W., Wu, C.-H., & Tse, M. (2012, August). Using a fuzzy association rule mining approach to identify the financial data association. *Expert Systems with Applications, 39*, 9054–9063.
12. Järvelin, K., & Kekäläinen, J. (2017). IR evaluation methods for retrieving highly relevant documents. In *ACM SIGIR Forum* (Vol. 51, pp. 243–250). New York, NY: ACM.
13. Kay, J. (2012). The Kay review of UK equity markets and long-term decision making. *Final Report, 9*, 112.
14. Khusro, S., Ali, Z., & Ullah, I. (2016). Recommender systems: Issues, challenges, and research opportunities. In K. J. Kim & N. Joukov (Eds.), *Information Science and Applications (ICISA*

2016, Singapore (pp. 1179–1189). Singapore: Springer.
15. Kumar, P., & Thakur, R. S. (2018). Recommendation system techniques and related issues: a survey. *International Journal of Information Technology, 10*(4), 495–501.
16. Lee, E. L., Lou, J.-K., Chen, W.-M., Chen, Y.-C., Lin, S.-D., Chiang, Y.-S., & Chen, K.-T. (2014). Fairness-aware loan recommendation for microfinance services. In *Proceedings of the 2014 International Conference on Social Computing (SocialCom '14)* (pp. 1–4). New York, NY: Association for Computing Machinery.
17. Luef, J., Ohrfandl, C., Sacharidis, D., & Werthner, H. (2020). A recommender system for investing in early-stage enterprises. In *Proceedings of the 35th Annual ACM Symposium on Applied Computing, SAC '20* (pp. 1453–1460). New York, NY: Association for Computing Machinery.
18. Ma, H., Yang, H., Lyu, M. R., & King, I. (2008). SoRec: Social recommendation using probabilistic matrix factorization. In *Proceedings of the 17th ACM Conference on Information and Knowledge Management* (pp. 931–940).
19. Manotumruksa, J., Macdonald, C., & Ounis, I. (2017). A deep recurrent collaborative filtering framework for venue recommendation. In *Proceedings of the 2017 ACM on Conference on Information and Knowledge Management* (pp. 1429–1438).
20. Matsatsinis, N. F., & Manarolis, E. A. (2009). New hybrid recommender approaches: An application to equity funds selection. In F. Rossi & A. Tsoukias (Eds.), *Algorithmic Decision Theory* (pp. 156–167). Berlin: Springer.
21. Mitra, S., Chaudhari, N., & Patwardhan, B. (2014). Leveraging hybrid recommendation system in insurance domain. *International Journal of Engineering and Computer Science, 3*(10), 8988–8992.
22. Musto, C., Semeraro, G., Lops, P., de Gemmis, M., & Lekkas, G. (2014, January). Financial product recommendation through case-based reasoning and diversification techniques. *CEUR Workshop Proceedings* (Vol. 1247).
23. Nair, B. B., Mohandas, V. P., Nayanar, N., Teja, E. S. R., Vigneshwari, S., & Teja, K. V. N. S. (2015). A stock trading recommender system based on temporal association rule mining. *SAGE Open 5*(2). https://doi.org/10.1177/2158244015579941
24. Niu, J., Wang, L., Liu, X., & Yu, S. (2016). FUIR: Fusing user and item information to deal with data sparsity by using side information in recommendation systems. *Journal of Network and Computer Applications, 70*, 41–50.
25. Paranjape-Voditel, P., & Deshpande, U. (2013). A stock market portfolio recommender system based on association rule mining. *Applied Soft Computing, 13*(2), 1055–1063.
26. Rendle, S., Freudenthaler, C., Gantner, Z., & Schmidt-Thieme, L. (2009). BPR: Bayesian personalized ranking from implicit feedback. In *Proc. of UAI* (pp. 452–461).
27. Rendle, S., Freudenthaler, C., & Schmidt-Thieme, L. (2010). Factorizing personalized Markov chains for next-basket recommendation. In *Proceedings of the 19th International Conference on World Wide Web* (pp. 811–820).
28. Rendle, S., Krichene, W., Zhang, L., & Anderson, J. (2020). Neural collaborative filtering vs. matrix factorization revisited. In *Proc. of RecSys* (pp. 240–248).
29. Schneider, A., Hommel, G., & Blettner, M. (2010, November). Linear regression analysis part 14 of a series on evaluation of scientific publications. *Deutsches Ärzteblatt International, 107*, 776–782.
30. Schonlau, M., & Zou, R. Y. (2020). The random forest algorithm for statistical learning. *The Stata Journal, 20*(1), 3–29.
31. Schumaker, R. P., & Chen, H. (2009, March). Textual analysis of stock market prediction using breaking financial news: The AZFin text system. *ACM Transactions on Information Systems, 27*(2), 1–19.
32. Sehgal, V., & Song, C. (2007). Sops: Stock prediction using web sentiment. In *Seventh IEEE International Conference on Data Mining Workshops (ICDMW 2007)* (pp. 21–26).
33. Seo, Y.-W., Giampapa, J., & Sycara, K. (2004). Financial news analysis for intelligent portfolio management.
34. Shmilovici, A. (2005). *Support vector machines* (pp. 257–276). Boston, MA: Springer US.

35. Singh, I., & Kaur, N. (2017). Wealth management through Robo advisory. *International Journal of Research-Granthaalayah, 5*(6), 33–43 (2017)
36. Smith, B., & Linden, G. (2017). Two decades of recommender systems at amazon.com. *IEEE Internet Computing, 21*(3), 12–18.
37. Steck, H. (2011). Item popularity and recommendation accuracy. In *Proceedings of the Fifth ACM Conference on Recommender Systems* (pp. 125–132).
38. Swezey, R. M. E., & Charron, B. (2018, September). Large-scale recommendation for portfolio optimization. In *Proceedings of the 12th ACM Conference on Recommender Systems*.
39. Taghavi, M., Bakhtiyari, K., & Scavino, E. Agent-based computational investing recommender system. In *Proceedings of the 7th ACM Conference on Recommender Systems* (pp. 455–458).
40. Tian, G., Wang, J., He, K., Sun, C., & Tian, Y. (2017). Integrating implicit feedbacks for time-aware web service recommendations. *Information Systems Frontiers, 19*(1), 75–89.
41. Vasile, F., Smirnova, E., & Conneau, A. (2016). Meta-prod2vec: Product embeddings using side-information for recommendation. In *Proceedings of the 10th ACM Conference on Recommender Systems* (pp. 225–232).
42. Wang, X., He, X., Wang, M., Feng, F., & Chua, T.-S. (2019). Neural graph collaborative filtering. In *Proc. of SIGIR* (pp. 165–174).
43. Wu, C., & Yan, M. (2017). Session-aware information embedding for e-commerce product recommendation. In *Proceedings of the 2017 ACM on Conference on Information and Knowledge Management* (pp. 2379–2382).
44. Yujun, Y., Jianping, L., & Yimei, Y. (2016, January). An efficient stock recommendation model based on big order net inflow. *Mathematical Problems in Engineering, 2016*, 1–15.
45. Zhao, X., Zhang, W., & Wang, J. (2015). Risk-hedged venture capital investment recommendation. In *Proceedings of the 9th ACM Conference on Recommender Systems, RecSys '15* (pp. 75–82). New York, NY: Association for Computing Machinery.

Chapter 11
Personalized Portfolio Optimization Using Genetic (AI) Algorithms

Roland Meier and René Danzinger

1 Introduction to Robo-Advisory and Algorithm-Based Asset Management for the General Public

The takeover of the financial advisory service using robots (so-called "robo-advisors") in the classic field of wealth management, asset management and personalized investments started as a new emerging trend after the 2008 global financial crisis. New and tighter regulations on banks and insurance companies led technology-based financial companies (i.e. "fintechs") to offer "robo-advisory" services. The first robo-advisors were still considered to be real exotics. However, this situation changed significantly around 2015. This was quickly recognized as a "booming" market, also driven by the impressive development of the US market for robo-advisor offerings. Typical of the first generation of online asset managers was the exclusive use of the so-called passive investment strategies and products. The main selling proposition of robo-offerings lies in their low overall cost level, which was combined with a fast and seamless digital sales and on-boarding process. The development of this selling proposition was propelled by the use of low-cost ETFs. The use of AI and mathematical algorithms was not a top priority at that time. Rather, the "robo-advisory" consisted of digital on-boarding and risk profiling processes, which enabled the classification of customers and their mapping to appropriate risk levels. Users were typically divided in up to ten risk levels, and one out of ten standard portfolios appropriate to the customer's personal risk level was selected. All these new "robo-advisors" offered an easy entry point for new customers for their investing. Secondly, the offer represented an innovative opportunity for banks to address new consumer segments.

R. Meier · R. Danzinger (✉)
Privé Technologies, Vienna, Austria
e-mail: roland.meier@privetechnologies.com; rene.danzinger@privetechnologies.com

© The Author(s) 2022
J. Soldatos, D. Kyriazis (eds.), *Big Data and Artificial Intelligence in Digital Finance*,
https://doi.org/10.1007/978-3-030-94590-9_11

Robo-advisors served as a starting point for shifting financial advisory processes to digital channels. According to the Boston Consulting Group [1], which surveyed 42,000 adults in 16 countries, customers worldwide are becoming increasingly digital. As of 2017, only 15% called for exclusively face-to-face advice, while hybrid advice offers and purely digital offers were preferred by approximately 42% of the customers. Two years earlier, in 2015, approximately 37% of the customers still preferred face-to-face (F2F) advice, and only 25% preferred digital offers and channels.

The COVID-19 pandemic has accelerated the adoption of digital channels across industries. According to a McKinsey survey [2], industries across various regions experienced an average of 20% growth in "fully digital" between November 2020 and April 2021, building on previous gains earlier in the pandemic. Still, some 44% of digital consumers say that they don't fully trust digital services. The ways companies can build and establish trustworthy relationships with consumers differ by industry. Consumer trust in banking is overall the highest among all industries. This rapidly advancing digitization asks for the employment of new methods and processes to meet the needs of new customer groups. The latter include, above all, personalized solutions that respond to individual needs and provide customers with tailored offers at the push of a button.

In recent years, virtual assistants and robo-advisory supported the advisory process. Specifically, artificial intelligence, supported by genetic algorithms, results in a paradigm shift that is significantly advancing the client relationship processes of the financial services industry. Processes that were previously considered impossible to digitize, such as the generation of portfolio proposals, can now be automated and digitized.

Likewise, personal interaction is digitally supported. Personal interaction between (bank) customers and their advisors can also be digitally supported, e.g. by online advisory solutions. In a (digital) advisory meeting, digital solutions make it possible to combine the competencies and skills of the human advisors with the bank's research and business intelligence in real time for the benefit of the retail customer. Digital tools and processes are increasingly becoming an integral part of hybrid sale processes and are proven as particularly valuable in supporting advisors. Digitization does only simplify and improve the customer experience. It also enables financial institutions to improve and optimize the quality of their customer advisory offerings. At the same time, it enables advisors to offer highly personalized products and services, i.e. optimized and personalized portfolios.

The remainder of this chapter is structured as follows: Section 2 following this introductory section describes traditional portfolio optimization methods along with their limitations. Section 3 introduces genetic algorithms and their merits for portfolio optimization. It also presents our novel, flexible approach to portfolio optimization, which leverages genetic algorithms, along with relevant validation results. Section 4 summarizes and papers and draws main conclusions.

2 Traditional Portfolio Optimization Methods

The primary goal in creating optimized investment portfolios is to generate robust and well-diversified portfolios. In contrast to genetic algorithms, traditional methods are one-step optimization models. The proposed portfolio is the direct result of the calculation process based mainly on the historical price data of the considered assets or asset classes. Similarly, the objective function is one-dimensional with potential parameters the maximum volatility, Sharpe ratio targets or a predefined minimum return that meets certain investment risk and performance parameters. The following two subsections present the two most common concepts that have been employed for several years to perform portfolio optimization

2.1 The Modern Portfolio Theory

The most prominent traditional method is the modern portfolio theory (MPT) which was published by Harry Markowitz in 1952 and for which he has received the Nobel Prize for economics in 1990. The goal of MPT is to create a diversified portfolio consisting of a variety of assets to be prepared for future uncertainties. The risk/return ratio of a broadly diversified portfolio is superior to any investment in just one single asset – no matter how well chosen. The MPT is a one-step method that simply maximizes the portfolio return subject to a given risk constraint. Portfolios compiled with the modern portfolio theory have some limitations. The major one is the high sensitivity to the given input data and the lack of not being able to handle multiple input factors.

A major issue of MPT is the estimation of the input parameters' "expected returns" and "variances" which are required for the calculation of optimal portfolios. The assumption of the future values of these parameters is based on data from the past, which are used for predictions of future developments. Consequently, in the MPT, effective values deviate from the estimated values, which leads to misallocation and thus to only optimal portfolios in theory, but not in reality. Many studies have shown that the application of the MPT leads to serious problems in practical usage. Especially the estimation/forecasting errors in the determination of the expected return have a large impact on the portfolio structure. The implications of an estimation error in determining the correlation are rather small.

Portfolios generated based on the Markowitz theory are mathematically correct but deliver no feasible results in practice. The method generally overweighs assets with high estimated returns, negative correlations and small variances. Experience shows that portfolios often generate higher losses than expected because the MPT does not cover extreme risk scenarios which happen more frequently than expected. It also systematically underestimates the risk of loss and overestimates the opportunity for gain. Furthermore, it fails to protect against risk stemming from diversification, especially in times of turbulent markets. The assumption of a normal distribution of returns does not correspond to reality. Extreme price movements occur much more frequently than the normal distribution suggests. Therefore, the risks of loss tend to be systematically underestimated, and profit opportunities are rather overestimated as in reality volatility and correlations strongly deviate from theoretical assumptions. Thus, despite its scientific foundation, the MPT is not very widely used in practical operations within financial services, and its acceptance within asset management is rather limited.

2.2 *Value at Risk (VaR)*

As stated above, one of the main parameters required to generate optimized portfolios that need to be quantified is portfolio risk. In the 1990s, value at risk (VaR) was established among banks and asset managers to measure the risk of trading portfolios. VaR has been widely implemented in the regulatory frameworks of the financial sector and is used to quantify requirements, within the internal risk management of banks, for ongoing back-testing and for stress-testing. VaR is a key risk measure for determining the highest expected loss. It is defined as the highest expected loss of a security that will not be exceeded with a predefined probability within a fixed period of time. Today, VaR is a standard risk measure in the financial sector.

For the calculation of the potential loss and to determine the VaR, the following three parameters are required: the holding period of the positions in days, weeks or months; the reference period – for example, 1 year; and the so-called confidence level in per cent, which is referred to as confidence probability. The VaR is usually recalculated daily by financial institutions. VaR has established itself as a valuable tool for risk managers. In portfolio management, however, the method is highly controversial. The value-at-risk concept assumes that events in the future will behave as events have behaved in the past. This assumption is especially wrong if a crisis phase arises after a longer period without severe market turbulences. To put it bluntly, VaR can also be interpreted as a fair weather risk measure. The highest possible loss of a portfolio is ignored. VaR is therefore not "black swan proof". Rare events that strike cruelly break the back of VaR.

3 Portfolio Optimization Based on Genetic Algorithms

3.1 *The Concept of Evolutionary Theory*

The theory of evolution and the concept of natural selection have their origin in biology and have found their way into various fields such as mathematics, physics and computer science. These concepts inspired new powerful optimization methods for solving complex mathematical problems that depend on a large number of parameters. Many optimization problems comprise tasks and models that resemble evolutionary processes, which are very similar to biological processes. Specifically, the word evolution originates from Latin and its meaning is "slow development". Moreover, the theory of evolution, founded by Charles Darwin in 1859, states that all animal and plant species we know today evolved from other species. Also, relevant research had recognized that living things, even if they belong to the same species, have different characteristics. Some are larger; others are better camouflaged or faster than their peers. Creatures whose traits are particularly favourable for survival have the most offspring. They pass on their genes to the next generation, ensuring that their species become better and better adapted to changing living conditions. In millions of years of evolutionary history, nature has produced organisms perfectly adapted to the surrounding environment. Darwin referred to this form of development as natural selection: Only the offspring of a species that have the best adaptation to their environment survive.

In this context, genetic algorithms are commonly considered and employed to solve problems that are associated with evolutionary processes. This class of algorithms represent one of the most modern and best available methods for solving highly complex tasks.

3.2 Artificial Replication Using Genetic Algorithms

Genetic algorithms are primarily suitable for optimizing and finding solutions to complex tasks [3]. In the evolutionary process of biology, genes of organisms are subject to natural mutation, which results in species diversity or genetic variability. The effects of mutated genes on new generations can be positive, negative or neutral. Through reproduction (recombination) of individuals, species adapt to changing conditions over a long period of time. This process is artificially reproduced by software solutions, where the quality of different parameters and solutions is evaluated with fitness functions to allow comparability and selection of solutions. The repetition of this process is continued until a defined termination criterion is reached. The result of the natural selection is the solution with the highest fitness score.

3.3 Genetic Algorithms for Portfolio Optimization

To enable the next generation of portfolio construction, advanced algorithms need to be developed. The genetic theory enables the development of algorithms that build robust and dynamic portfolios based on multiple input parameters. A concrete set of such algorithms are introduced and described in the following paragraphs.

3.3.1 Multiple Input Parameters

Genetic algorithms are capable of considering multiple input parameters. This ability is the foundation for meeting clients' diversified needs in terms of their investments in assets and securities. Modern wealth management faces a diverse group of clients with varying (investment or risk) goals, personal needs and many different focus points. Hence, there is a need to develop optimization algorithms and build an artificial intelligence engine to aid investment propositions for various client groups. The latter groups range from wealth management and private banking to retail clients.

Possible input parameters to these optimization algorithms include asset prices, dividends, volatility, correlation parameters as well as diverse asset breakdowns such as currencies, regions, asset classes and credit ratings. Furthermore, it is possible to consider additional parameters such as economic data, different types of ratings (e.g. sustainability ratings), individually defined investment themes (e.g. investments in renewable energies, healthcare, technology, megatrends, gold, aging society, infrastructure, impact investing) and more. Overall, there is a need for algorithms that consider and use a wide array of datasets.

3.3.2 Data Requirements

To meet the challenges of a modern optimization algorithm based on multiple input parameters, a vast number of data points from usually different data sources are required. Some of the most prominent examples of data sources that are used for portfolio optimization include:

- Financial market price data fetched from several market data providers
- Financial asset master data fetched from several data providers
- Customer risk profile data fetched directly from the financial institution
- Mutual fund, ETF and structured product allocation and breakdown data fetched from several market data providers
- Individual data points received from financial institutions and maintained as customer-specific data points
- Customer economic outlook fetched directly from financial institutions based on questionnaires and customer (risk) profiles

3.3.3 A Novel and Flexible Optimization Approach Based on Genetic Algorithms

Our approach to developing a genetic optimizer for portfolio optimization seeks to improve the quality of asset allocation of a portfolio of financial assets (e.g. stocks, bonds, funds or structured products) with respect to quantitative and qualitative financial criteria. It utilizes concepts of genetic algorithm, including an evolutionary iterative process of portfolio modification and a subsequent evaluation of the quality of the financial portfolio, via a scoring assessment method (i.e. "fitness" assessment). Specifically, in the scope of our approach, a "fitness factor" evaluates a specific quality of a portfolio. Furthermore, a "fitness score" represents an overall metric to describe the quality of a portfolio, i.e. it evaluates the quality of the portfolio across multiple factors.

We offer flexibility in the use of different fitness factors, which helps in defining the objectives of the optimization process based on different customer "optimization preferences". Hence, users can design and assign weights to the factors that are important to the individual investor. It enables the users to integrate their market views using a functionality that is conveniently characterized as "frozen". This functionality fixes the weights of holdings an investor wants to have untouched, i.e. the holdings that he/she does not want to be considered in the scope of the optimization process. Given this flexibility of the methodology, the optimization provides highly customized portfolios based on clients' needs.

The genetic algorithm is also designed to reach a recommended portfolio within a very short period of time, as the speed of execution is one of the key drivers behind the potential adoption of this methodology. Given that there are multiple stages of iterations that need to be optimized based on a large underlying product universe with hundreds or thousands of assets to select from, the optimizer shall provide a solution within a response time of less than 8–10 s.

An additional goal for the optimizer is to ensure the compatibility of genetic optimization with other optimization methods and potential strategies. For example, a traditional mean-variance optimization can be defined by a Sharpe ratio factor, which will enable the genetic optimization process to generate results for portfolios with the highest Sharpe ratio. As another example, a strategic approach based on a certain asset allocation strategy can be defined by an asset class allocation factor, which will enable the genetic optimization process to recommend a portfolio according to the strategic asset allocation of a financial institution or asset owner.

Note also that our methodology provides multi-asset support: It can work with all different asset classes and product types, including stocks, bonds, funds, ETFs and structured products, as soon as the needed price and master data are available from relevant data sources.

3.3.4   Fitness Factors and Fitness Score

A "fitness score", as we designed it, is calculated as the weighted aggregate of a number of factors. These factors represent mathematically determined formulas based on certain criteria of the given financial assets. The corresponding weightings of the fitness factors can be defined individually for each optimization request or can be predefined as constants based on standard definitions. In general, the sum of the weighted fitness factors gives the total fitness score. Each of these fitness factors is a parameter of the total fitness score, covering a certain aspect or financial criteria for the given financial assets. One fitness factor may have an offsetting or amplifying character with another fitness factor. Nevertheless, the factors are generally designed based on independent and uncorrelated attributes of the financial assets.

The fitness functions are applied to the complete portfolio, which is getting optimized. They can take into account all financial holdings within the investment portfolio, including equities, bonds, mutual funds, structured notes and cash. Each fitness factor is assigned with a given weight which determines the magnitude of the impact of that factor on the optimization process. Fitness factors considered for the fitness score can cover various parameters, including but not being limited to the examples presented. Hence, a flexible framework allows the creation of basically any fitness factor related to the investment assets of the asset universe, subject to the condition of data availability about the asset.

Fitness score will be calculated as:

$$FS = \sum_{i=0}^{n} w_i \, F_i$$

$i = 1,2,...,n.$
F_i is the corresponding fitness factor
w_i is the corresponding weight for each fitness factor i.

All fitness factors are scaled from 0.0 to 1.0. The weight of the fitness factors can be any number from 0.0 to 1.0.

The weights work in a relative way meaning the fitness factors with high weight will affect the resulting portfolio more than the fitness factors with low weight. For example, if the weight of Factor A = 0.9 and Factor B = 0.1, the resulting portfolio will be impacted by Factor A more than by Factor B. The genetic algorithm framework enables the development of fitness factors considering various financial key figures, portfolio-related constraints as well as characteristics of the underlying assets.

Fitness factors can be quite flexible created with different targets, which can be grouped in general as follows:

Risk/Return-Related Fitness Factors
- *Volatility-capped performance factor* minimizes the difference between the return of the recommended portfolio and the target return of the client under the condition that the recommended portfolio has a volatility that is not higher than the risk limit. The risk limit (in %) is in most cases linked to the risk level.
- *Portfolio volatility factor* measures the risk of the recommended portfolio; it minimizes the volatility of the recommended portfolio compared to the current or defined model portfolio.
- *Portfolio performance factor* measures the performance of the recommended portfolio; it maximizes the return of the recommended portfolio.
- *Sharpe ratio factor* measures the Sharpe ratio of the target goal allocation portfolio.

Portfolio Constraint-Related Fitness Factors

- *Product diversification factor* measures is the diversification of different financial products within a portfolio.
- *Reuse of existing asset factor* measures reuse of the investor's existing holdings for the new portfolio. The objective of that factor would be to reduce turnover of holdings in the customer's portfolio and primarily define whether an asset should be kept within a customer's portfolio. The so-called frozen assets can be completely excluded from the optimization calculations.
- *Preferred asset factor* measures the use of preferred assets (premier funds or other defined preferences) in the recommended portfolio.
- *FX minimization factor* minimizes the foreign exchange (FX) exposure in implementing the recommended portfolio by reducing FX transactions. It measures for each currency (cash and securities denominated in these currencies) in the recommended portfolio what percentage of such currency is already in the existing portfolio.
- *Preferred currency factor* measures the extent of the recommended portfolio that complies with the investor's preference for the investment currency.

Asset Characteristics' Related Fitness Factors

- *Asset class allocation factor* measures how well the recommended portfolio follows a given target portfolio/target allocation with respect to selected or predefined asset classes.
- *Sustainability/ESG allocation factor* measures how well the recommended portfolio follows a given target portfolio/target allocation with respect to ESG/sustainability criteria/rating.
- *Region allocation factor* measures how well the recommended portfolio follows a given target portfolio/target allocation with respect to specified regional preferences based on asset breakdowns.
- *ETF allocation factor* measures how well the recommended portfolio follows a given target portfolio/target allocation with respect to ETFs as a "low-cost" investment product selection.
- *Core/satellite allocation factor* measures how well the recommended portfolio follows a given target portfolio/target allocation with respect to assets tagged as core investments or satellite investment products.
- *Sentiment factor* measures how well the recommended portfolio follows a given target portfolio/target allocation with respect to a sentiment dataset or criteria/rating.

3.3.5   Phases of the Optimization Process Utilizing Genetic Algorithms

Genetic algorithms are utilized to discover the best portfolio out of all possible portfolio construction options considering the user-defined constraints defined and described above as fitness factors. Genetic algorithms are built based on the evolutionary processes of our nature. Guided random search is used for finding the

Table 11.1 Sample portfolio ranking table

	Weighted fitness factor 1	Weighted fitness factor 2	Weighted fitness factor 3	...	Fitness score (weighted sum of all factors)
Portfolio 1	0.13	0.29	0.14		**0.88**
Portfolio 2	0.10	0.07	0.03		**0.24**
Portfolio 3	0.34	0.16	0.21		**0.90**
...					
Portfolio 254	0.01	0.01	0.00		**0.05**
Portfolio 255	0.03	0.06	0.05		**0.20**
Portfolio 256	0.29	0.18	0.31		**0.83**

optimal portfolio, starting with a random initial portfolio. The following iteration process and the concept of natural selection filter portfolios in each generation based on its fitness score. The optimal portfolio results as the one with the highest fitness score after a predefined number of iterations.

Optimization Process

After receiving the inputs, the first generation of portfolios will be generated. There will be [256] portfolios in every generation. The number in brackets ([ ]) is an adjustable parameter which is determined through a qualitative and quantitative assessment. A random generation process is used for this calculation. The first portfolio of the first generation will be the input portfolio (from the data input). The financial assets and their weights within the portfolio are randomly assigned. The fitness score, which is the weighted aggregate of different factors, is calculated for all portfolios as shown in Table 11.1. The portfolios are then ranked according to their scores.

Within the further process, portfolios with fitness score in the top 50% range will be selected, while the bottom 50% will be discarded as illustrated in Fig. 11.1.

The "surviving" 50% [128] portfolios will be treated as parents to generate [128] offsprings. Each child is generated by two adjacent parents: For example, the first portfolio will be combined with the second one, the second one with the third one and so on. The [128]th portfolio will pair with a randomly created new portfolio. Mutation is performed by introducing new financial assets, rather than from existing portfolios. These steps will be repeated for [100] times. After this automated genetic calculation, the portfolio with the highest fitness score is selected as the "fittest portfolio" and the final output.

3.3.6 Algorithm Verification

The described and developed methodology has undergone an extensive verification process with a wide range of combinations of different inputs. The input variables include 5 sets of portfolios, 5 sets of preferred currencies, 11 sets of fitness factor

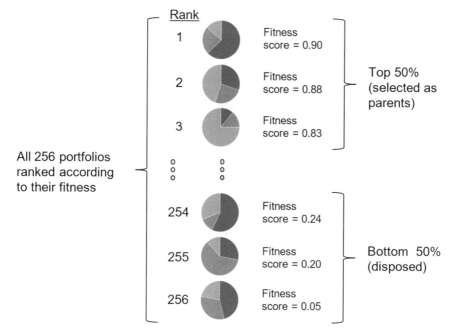

Fig. 11.1 Sample portfolio selection table

weights, 6 sets of frozen amounts and 5 different product universes. It has already been tested with over 8000 different predefined test cases. The resulting response time, fitness score and other related metrics have been examined and compared.

Fitness Improvement

By changing the selected inputs of the 8000+ cases, the improvement of the fitness score of these optimized portfolios achieved a level of 30% on average as shown in Fig. 11.2 and Table 11.2. This result is consistent with the traditional portfolio theory, while providing more customization for a client such as the currency of the assets, maximum risk acceptance, etc. would lead to "better results".

Convergence Test

Additional analysis has been done to study the behaviour of the algorithm by plotting the fitness score of the portfolios in each generation. Figure 11.3 illustrates an example of how the fitness of the portfolios improves over the different generations. It can be observed that there is a greater dispersion in "fitness quality" in portfolios of earlier generations, while portfolios of later generations are yielding

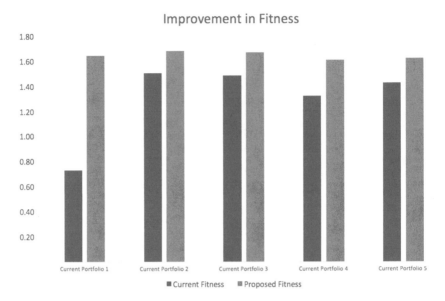

Fig. 11.2 Portfolio fitness improvement

Table 11.2 Sample portfolio ranking table

(Average results)	Fitness	Return (%)	Volatility (%)	Sharpe
Before optimization	1.26	4.77	14.04	0.36
After optimization	1.64	8.22	13.11	0.69

generally higher fitness scores. It can also be observed that through the optimization process, the fitness score of each individual portfolio gradually converges to a level where there is no or only a marginal room to increase the fitness score further.

3.3.7 Sample Use Case "Sustainability"

In the past years, "Green" or ESG (environmental, social and governance) compliant investments are in growing demand among institutional and retail investors. Assets under management in ESG funds and corresponding ETFs have recently grown disproportionately strongly. Among other things, this has to do with the fact that many new sustainability products have been launched and regulators are increasingly demanding compliance with and disclosure of sustainability criteria, thus also creating drivers for change. New regulations, for example, the EU Disclosure Regulation 2019/2088, introduce disclosure requirements regarding sustainability in the financial services sector.

The proposed optimizer framework based on genetic algorithms enables the development and consideration of fitness factors for various criteria, including sustainability. It will also be possible to develop a fitness factor for each of the

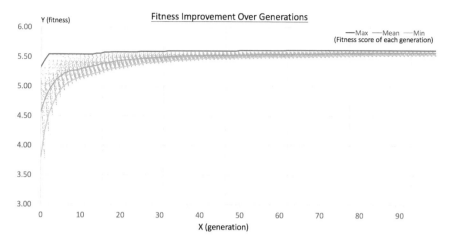

Fig. 11.3 Fitness improvement over generations

17 Sustainable Development Goals (SDGs) proposed by the United Nations to generate specific portfolios that meet certain ESG/sustainability requirements. As a first step, an additional sustainability fitness factor was developed to optimize assets classified as sustainable according to qualitative criteria. It enables users to select the percentage weighting of the desired ESG compliance or sustainability grade of a portfolio proposal. The fitness factor ensures that the desired weighting or even exclusivity for ESG compatible investments in the portfolio is targeted.

Sustainability Fitness Factor Test Series

The created sustainability fitness factor was verified in an extensive test series. The underlying investment universe of this test series consisted of 127 mutual funds, of which 31 were classified as sustainable, and for the second batch of tests, a restricted investment universe of 30 mutual funds, including 9 sustainable funds, marked as sustainable selection. The optimization was performed over a 5-year period using live market data. In total, 3100 portfolios were generated with different preferences for volatility targets, different regions, ETF allocation and sustainability targets.
    The sustainability preference was set as follows (Fig. 11.4):

(a) No preference for sustainability, i.e. 0% fixed weight for the sustainability fitness factor
(b) Sustainable assets preferred, i.e. soft target for sustainability weight of approximately 60% allocation in the portfolio
(c) Exclusive allocation with sustainable assets 100%

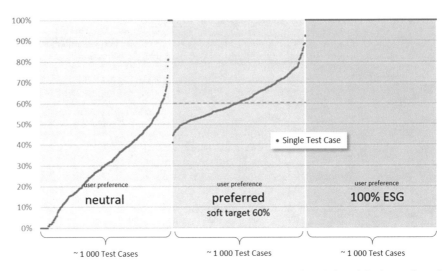

Fig. 11.4 Result of sustainability fitness factor test series, where from left to right the number of test cases and on y-scale the result of each single case is displayed within the blue line

Test Series Results

The preference (a) results show an intensive allocation of sustainable funds, with 93% of all portfolios containing sustainable funds. The median weighting was 30.5%, the first quartile (25th percentile) was 16.1%, and the third quartile (75th percentile) achieved an allocation of 44.1%. This "neutral" result can be taken as reference since the sustainable fitness factor had – as no preference was selected – no influence on the allocation of sustainable funds. The allocation of those funds is the result of risk/return factors as well as other selected preferences.

The preference (b) results confirm the proper function of the algorithm since the sustainable fitness factor was in use as a soft target. Nevertheless, it was not the only criteria the genetic algorithm had to consider. The fitness factor with the highest weighting was the volatility-capped performance factor to ensure the given volatility target is met. Also, the soft target for the sustainable fitness factor of 60% allocation was very well achieved. The median weighting was 60.4%, and first quartile (25th percentile) amounted to 54.3%, and the third quartile (75th percentile) achieved an allocation of 68.7%. Overall, 788 or 72% of all portfolios had an allocation of sustainable funds in the targeted range between 50% and 70%.

The preference (c) results were 100% achieved and the optimizer allocated only sustainable funds to the portfolios. The results of this test series confirmed the functioning of the genetic algorithms as well as the chosen configuration of the test setup.

4 Summary and Conclusions

This chapter presented a personalized portfolio optimization system, which was developed in the scope of the H2020 INFINITECH project. The system aims to define, structure and develop an end-to-end process that digitizes a "private ranking" like advisory journey to enable personalized and professional advisory services to the general public. This includes the personal risk assessment of the user resulting in a risk score or SRRI (synthetic risk and reward indicator), based on a new designed UI/UX setup, the definition and selection of personal investment preferences such as sustainability/ESG criteria, regional preferences or sentiment preferences. The solution aims to have further "customized" preferences as potential enhancements, including but not exclusively based also on additional "third-party content" providers.

The core of the portfolio construction and optimization process utilizes genetic algorithms based on the concept of evolution theory. It applies an iterative approach and natural selection process to generate portfolio proposals based on a wide range of individual and customized preferences with different investment goals. Preferences are defined and set up as fitness factors which can cover risk-/return-related key figures, various portfolio constraints (e.g. diversification and asset characteristic variables such as sustainability, asset class breakdowns or sentiment data). The goal of the overall approach is to enable advisors and potentially end customers to directly manage smaller portfolios (i.e. <25k €) in a professional investment advice manner. The solution is designed and implemented as a SaaS (software as a service) tool. Therefore, it can be flexibly integrated with existing advisors, banks, insurance or brokerage online systems. The proof of concept of this open framework solution has been validated based on a substantial test series, including several self-calculated fitness factors. As part of ongoing developments, the solution will be enhanced by adding a "sentiment fitness factor". The latter will be provided by a sentiment analysis service provider, which will act as an independent third-party data source.

The final goal for the system is to provide a set of predefined fitness and optimization factors, which shall be offered as a SaaS solution. This SaaS solution will become available for integration with existing backend solutions from financial institutions or alternatively provided via APIs which can be approached from other system providers that need to service their clients based on the presented approach to portfolio construction and optimization. Hence, such system providers will have the opportunity to leverage the presented approach and algorithms while offering them to customers based on their own UI/UX settings.

Acknowledgements The research leading to the results presented in this chapter has received funding from the European Union's funded Project INFINITECH under Grant Agreement No. 856632.

References

1. Boston Consulting. (2017). *Retail Customer Survey.*
2. Hajro, N., Hjartar, K., Jenkins, P., & Vieira, B. (2021). *McKinsey digital, what's next for digital consumers*https://www.mckinsey.com/business-functions/mckinsey-digital/our-insights/whats-next-for-digital-consumers
3. Schillinger, J., & Musset, J. (2019). Overview of artificial intelligence genetic optimization.

Chapter 12
Personalized Finance Management for SMEs

Dimitrios Kotios, Georgios Makridis, Silvio Walser, Dimosthenis Kyriazis, and Vittorio Monferrino

1 Introduction

Small Medium Enterprises are vital to all economies and societies worldwide. According to OECD [1], SMEs in its member countries account for 99% of all businesses and almost 60% of value added. Respectively, in Europe SMEs hold a vital role as they account for 99.8% of all enterprises in the EU-28 non-financial business sector (NFBS), generating 56% of added value and driving employment with 66% in the NFBS [2]. Despite the global importance of SMEs, a lot of them struggle to keep up with the pace of change in their digital transformation journey. The complex and challenging environment, with various technological disruptions and radical business changes, holds as the main barrier to them. Additionally, the ongoing COVID-19 pandemic, altering customer's behavior and business trends, made it also necessary to push SMEs' e-commerce activities and the utilization of online channels. On the other hand, it provoked significant liquidity concerns to businesses unable to utilize new digital tools [3].

The introduction of digital technologies and state-of-the-art analytics tools can empower SMEs by helping to reduce operating costs, saving time, and valuable resources, especially for firms that illustrate reduced economic activity and smaller

D. Kotios (✉) · G. Makridis · D. Kyriazis
University of Piraeus, Piraeus, Greece
e-mail: dimkotios@unipi.gr; gmakridis@unipi.gr; dimos@unipi.gr

S. Walser
Bank of Cyprus, Nicosia, Cyprus
e-mail: silvio.walser@bankofcyprus.com

V. Monferrino
GFT Technologies, Genova, Liguria, Italy
e-mail: Vittorio.Monferrino@gft.com

J. Soldatos, D. Kyriazis (eds.), *Big Data and Artificial Intelligence in Digital Finance*,
https://doi.org/10.1007/978-3-030-94590-9_12

volumes of production, which also tend to have limited market reach and lower negotiation power with stakeholders [4].

Introduction of modern predictive and descriptive analytics can affect almost every aspect of an SME's operation, leading to data-driven strategic decision-making processes. The digital transformation of an SME could offer a new perspective to its business and financial management, leading to a competitive advantage by increased productivity and quality control, introducing new marketing techniques and the ability to identify new markets and foresee business opportunities.

However, the digital transformation journey of an SME poses various risks and challenges. Based on the 2019 OECD SME Outlook [1], the currently limited digital skills found in most SMEs management, the inability to identify, attract, and retain ideal employees, and the lack of required resources or financing options and strict regulations regarding data protection pose the main barriers toward SME digitalization.

Of course, not all sectors face the same challenges and barriers regarding their digital transformation. The generation of data and utilization of data analytics appears to be the highest in the financial sector [5], with some SMEs competing large financial software providers who are offering commercial data analytics applications for SMEs through cloud computing services, allowing SMEs to access tailored AI services even when they lack the required resources to develop them internally [6]. Other commercialized applications utilizing Big Data analytics are included in ERP or Accounting Software packages, with stand-alone Business Financial Management (BFM) software and tools also being available for commercial use. Most of the offered solutions are geared toward analyzing historical transactions of data residing within the ERP system. Banks, retaining a variety of data of SME customers as required for their core activities, could offer a solution by utilizing all available data and provide a variety of data analytics tools aiming at increased business financial management efficiency for their customers, while offering value-added services on top of their core business. Toward this direction, banks can harness all available operational and customer data to provide accurate business insights and analytical services to SMEs resulting, as noted by Winig [7], in increased customer base and engagement.

However, developing personalized segmented services, especially when considering the enormous variations found in the SME market, from business model and goal to business scale, is not an easy task for a bank as it poses a variety of business and technical challenges.

This chapter introduces a data-driven approach to facilitate the development of personalized value-adding services for SME customers of the Bank of Cyprus, under the scope of Europeans Union's funded INFINITECH project, grant agreement no 856632.

As a detailed presentation of all microservices developed and how those are interconnected would far exceed the constraints of this chapter, we showcase the development process and the new possibilities unlocked by the foundation of the proposed mechanism, namely the BFM Hybrid Transaction Categorization

Engine. The interconnection of the various microservices is provided with a brief presentation of the underlying DL model used for categorical time-series forecasting needs, namely the BFM Cash Flow Prediction Engine. The categorization of all SME transactions is required in order to label all historical data and unlock most features of an innovative BFM toolkit. Based on the classified data, the developed cash flow prediction model is one of the key BFM tools which adds value to SMEs by providing a holistic approach to income and expenses analysis. For the transaction categorization model, a hybrid approach was followed, applying initially a rule-based step approach based on transaction, account, and customer data aggregation. Then, at an operational phase, a tree-based ML algorithm is also implemented, creating a smart classification model with high degree of automation and the ability to take users' re-categorization into account. Given that lack of prominent research, the provided dataset sourcing from a real-world banking scenario was labeled based on various rules and the input of banking experts, incorporating various internal categorical values present in the dataset. In this direction, 20 Master Categories with 80 respective Sub-categories tailor-made for SMEs were created. As expected in a real-world banking scenario, the generated categories of the transactions were highly imbalanced, and thus, a CatBoost [8] model was preferred based on the findings of [9], where the most well-established boosting models were reviewed on multi-class imbalanced datasets, concluding that CatBoost algorithms are superior to other boosting algorithms on multi-class imbalanced conventional datasets. The second microservice illustrated, the Cash Flow Prediction Engine, aims at developing an accurate and highly scalable time-series model which can predict inflows and outflows of SMEs per given category. To achieve this, after exploring traditional time-series forecasting models and newly introduced ML/DL approaches, deep learning techniques were utilized to provide information regarding the future cash flow of SMEs. The analysis focuses on time-series probabilistic forecasting, utilizing a Recurrent Neural Network model.

2 Conceptual Architecture of the Proposed Approach

The Conceptual Architecture and a workflow of the various components included in the Business Financial Management platform developed are presented in Fig. 12.1.

Bank of Cyprus (BoC) is developing a testbed based on the AWS cloud computing services. The pilot's cloud infrastructure is being developed as a blueprint testbed, with other pilots of the INFINITCH project utilizing similar cloud solutions as the one being established. For the data collection process, tokenized data from designated BoC databases, as well as data from open sources and SME ERP/Accounting software will be migrated to the data repository of the BoC testbed. Upon returning to BoC datastore, a reverse psedonymization (i.e., mapping tokenized ID to user) will be performed in order for the respective analytic output to reach their designated SME clients. The pilot utilizes both historical and real-time data for its various Business Financial Management tools, as the need to

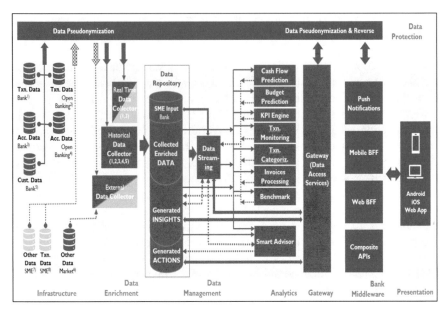

Fig. 12.1 BFM platform conceptual architecture

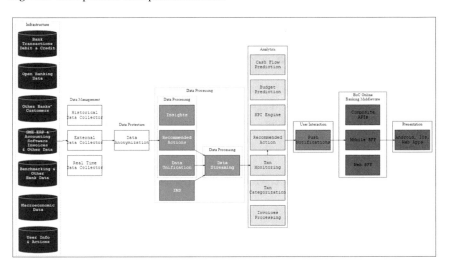

Fig. 12.2 BFM platform within the INFINITECH Reference Architecture

provide real-time business intelligence that relies on live data is crucial for the pilot's development. Based on the INFINITECH Reference Architecture, which is based on BDVA's RA, the pilot's workflow is translated as Fig. 12.2.

Also following the respective INFINITECH CI/CD process and also taking into consideration the EBA 2017 guidelines regarding outsourcing banking data to cloud, the first two components were deployed as depicted in Fig. 12.3.

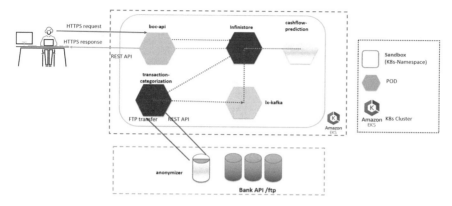

Fig. 12.3 Initial component development in INFINITECH testbed

All components are packaged following a microservices approach as docker containers and then deployed in a Kubernetes cluster (i.e., AWS EKS). More specifically, every component is deployed in a Pod within the cluster with specific features (e.g., the capability of auto-scaling when needed in terms of the demand). Thus, as a sandbox is defined, all the components deployed under the same namespace allow the interaction and connection between them.

3   Datasets Used and Data Enrichment

3.1   Data Utilized for the Project

For the development of the data analytics components presented, the utilized dataset had been provided by Bank of Cyprus as a real-world use case, in the scope of INFINITECH Project. All data received by the testbed, except the various internal codes kept by the bank for operational purposes, have already been tokenized, with no actual way of retrieving indicating customer-related personal and/or sensitive information. This anonymized dataset includes transaction, customer, and account data of over a thousand SMEs for the years 2017–2020, exceeding 3.5 million data entries. Despite the existence of the various internal codes, all transaction data were unlabeled with no initial indication of their underlying category. The key data source, both for the transaction categorization and for the cash flow prediction model, was the available transaction data which contained more than 40 variables indicating from root information like Date, Amount, Credit/Debit indicators, and Description, to more sophisticated information like Touchless Payment indicators, Merchant Category Codes, Standing Order indicators, etc. The SME Accounts dataset contained tokenized information of the individual accounts of the customers, available balances, and the respective NACE rev2 code, which is a statistical

Table 12.1 Datasets used within the BFM toolkit

Dataset name	Dataset description
Transaction data from the bank	SME customers transactions Dataset from BOC
Transaction data from open banking (PSD2)	SME customers transactions dataset from financial institutions other than BOC. BATCH input (e.g., every 6 hours or every night)
Accounts data from bank	Account data regarding SME customers of BOC. E.g., balances, available amount, account type
Accounts data from open banking	Account ID to be anonymized. E.g., balances, available amount, account type
Customer data from bank	Customer demographics
Other data (Market)	Macroeconomic SME-related data from public/private resources (e.g., https://www.data.gov.cy/ and https://ec.europa.eu/eurostat/data/database)
Other data from SME	Other data that is provided by the SME ERP/accounting system (e.g., number of customers, suppliers, stock). Non-transactions-related data
Transaction data from SME	Data that is provided by the SME ERP/accounting system and relates directly or indirectly to account debit/credit transactions. For instance, invoice data; non-PSD2 data (i.e., payment account related), e.g., saving accounts in financial institutions other than BOC
SME Input	Data obtained through BFM feedback loop transaction re-categorization which accounts should be used for the analysis, etc.

classification of economic activities used across the European Community [10] and is used by the bank for the identification of the SMEs operating sector. The availability of SMEs economic activity is significant for this chapter and our future work as well, since external national and European data utilizing horizontally the NACE code system can be utilized in our models to provide more personalized sector information. The datasets used for the BFM tools operation and for future intended data enrichment are summarized in Table 12.1.

3.1.1 Data Enrichment

Besides the data originating from the bank, various external sources as presented in Table 12.1 will be utilized in order to provide accurate business insights and additional information to the SMEs. The data used by external sources are retrieved by the various (historical, external, or real-time) data collectors, based on developed REST APIs within the designed microservices. Data enrichment has three main objectives:

1. Offers more holistic financial management potentials to the SMEs via Open-banking data integration.
2. Provides sector specific information and personalized insights by utilizing open-source market data.
3. Offers account reconciliation and additional innovative services by ingesting ERP/Accounting data to the BFM platforms underlying ML/DL models.

However, data enrichment will not be implemented until future work on upcoming data analytics components is completed, with the details on how the external data are retrieved or the required data streams designed being out of this chapter's context.

4   Business Financial Management Tools for SMEs

The proposed solution offers a variety of data analytics microservices, all aiming to assist SMEs in monitoring their financial health, get a deeper understanding of their operating costs, allocate resources with supported budget predictions, and retrieve useful information relevant to their underlying business sector.

Since a detailed presentation of all microservices developed and underlying interconnection would far exceed the constraints of this chapter, the focus is given on the development process of the BFM toolkit foundation, namely the Hybrid Transaction Categorization Engine. Besides the development of a smart personalized classification engine that takes the user's re-categorization into account and offers a high degree of automation, this task can assist in the categorization of open banking data and offers fertile ground for the implementation of explainable AI frameworks to better comprehend the outcomes of our classification ML model.

The interconnection of the various microservices is showcased with a brief presentation of the DeepAR RNN model utilized [11], where the produced categories are taken into account to serve the time-series forecasting needs of the model, offering highly personalized and accurate probabilistic predictions to SMEs per Master Category.

The rest of the analytics components are briefly described just to offer a glimpse of the BFM tools developed and how they can empower the SMEs, utilizing a set of descriptive and predictive analytics services based on their personalized needs.

4.1   Hybrid Transaction Categorization Engine

As mentioned above, the classification of SME transactions is vital for the additional development of financial management microservices. The absence of labeled data is the main challenge when developing a transaction categorization model. Two prevalent approaches arise when creating a classification model. The first one is utilizing unsupervised machine learning techniques to create clusters with no prior knowledge of the expected outcomes. However, this cannot be applied in our transaction categorization scenario as labels are fixed and of distinct nature in the finance sector, so this approach suffers from difficulty interpretating the outcomes and leads to a less robust and interpretable model. The second and proposed approach is initially hand labeling a representative subset based on expert knowledge creating a rule-based model, which can then be integrated with a supervised machine learning model, offering a high degree of update automation and transaction re-classification.

4.1.1 Rule-Based Model

A step approach was followed for the rule-based model, incorporating various internal codes of the bank, some of them being case specific interpreted only by the banking experts (i.e., Transaction Type Code) and others being used universally in the business world (i.e., Merchant Category Code and NACE codes). Before mapping the variables above to given categories, it was vital to capture all transactions between accounts belonging to the same SME as those would be classified as "Transfers between own accounts." The exact flowchart of the rule-based model and the swift to a hybrid transaction engine is illustrated in Fig. 12.4.

4.1.2 CatBoost Classification Model

A key challenge in developing the hybrid classification model is alleviating the bias inserted by the rule-based model. Toward this direction, it was crucial to enable the user to re-categorize a given transaction category. This process of updating the existing knowledge and adapting the model to re-categorizations is fundamental for continuous model optimization, i.e., increased accuracy and personalization.

To this end, a CatBoost model was periodically retrained in order to adopt to changes made by the end users (i.e., SMEs). CatBoost is a novel algorithm for gradient boosting on decision trees, which has the ability to handle the categorical features in the training phase. It is developed by Yandex researchers and is used for search, recommendation systems, personal assistant, self-driving cars, weather prediction, and many other tasks at Yandex and in some other companies. CatBoost makes use of CPU as well as GPU which accelerates the training process. Like all gradient-based boosting approaches, CatBoost consists of two phases in building trees. The first is choosing the tree structure and the second is setting the value of leaves for the fixed tree. One of the important improvements of the CatBoost is the unbiased gradient estimation in order to control the overfit. To this aim, in each iteration of the boosting, to estimate the gradient of each sample, it excludes that sample from the training set of the current ensemble model. The other improvement is the automatic transformation of categorical features to numerical features without

Fig. 12.4 Hybrid
classification model flowchart

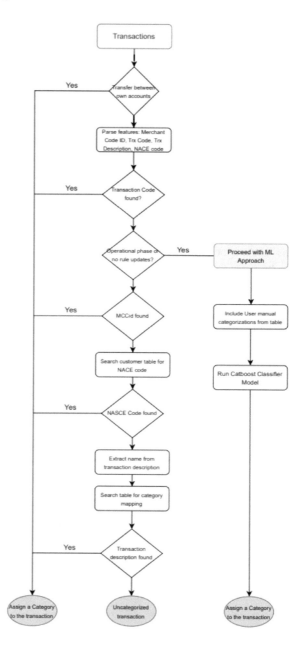

any preprocessing phase. CatBoost is applicable for both binary and multi-class
problems.

Given the nature of the different steps utilized in the rule-based model and aiming
at high efficiency and categorization accuracy, a hybrid model was preferred instead
of a fully AI one, since the two first steps of the rule-based process provide accurate

categorization results, producing root categories like cash withdrawals, deposits, and banking fees.

In more detail, as some root categories produced by the first two steps (i.e., transfer between accounts and Transaction Type Code mapping) of the rule-based model are predefined, the aim of this model is to learn and mimic the last three steps, while also taking into account the changes done by the user at an operational phase. The number of the remaining Master Categories that can be produced in these steps is 16, so the evaluation on multi-class tasks in terms of various metrics was applied. Given that the dataset was highly imbalanced, finding a proper normalization factor was another challenge which has been overcome by hyper-parameter optimization.

The main outcome that can be derived from the results is that the model can learn the rules utilized incorporating key merchant features (i.e., NACE code and Merchant Code) and correctly categorize the results with 98% accuracy. Furthermore, it is worth mentioning that some of the transactions were categorized as "Uncategorized Expense." The transactions falling into this category are expected to be categorized by the respective SME. Consequently, when the model will be retrained, the additional knowledge gained from the SME performed categorization will be incorporated into the model.

4.1.3   Explainable AI in Transaction Categorization

Although statistics, with its various hypothesis testing and the systematic study of the variable importance, is well established and studied, the same does not apply for the explainability of various techniques in machine and deep learning. Even though multiple measures and metrics of performance have been extensively applied, some of them can be rather misleading as they do not convey the justification behind the decision made. Stronger forms of interpretability offer several advantages, from trust in model predictions and error analysis to model refinement.

In the context of our research, the interpretation of the results is as significant as the results themselves, as it can lead to significant technical insights regarding the transaction categorization engine's evaluation. Explainable methods can be categorized into two general classes:

1. Machine learning model build-in feature importance
2. Post hoc feature importance utilizing models such as LIME [12] and SHAP [13]. In our classification scenario, both LIME and SHAP techniques were leveraged as a qualitative evaluation of the results.

The SHAP values denoting the importance of each feature included in the CatBoost model are depicted in Fig. 12.5. It is evident in the figure that the model learned the rules that we based on Merchant Code ID (i.e., MCCCodeID) and transaction beneficiary NACE code as these two are the most important features. Additionally, the significance of the Account Key (skAcctKey) as the third most important feature in the model strengthens the proposed approach of a user-oriented updating approach.

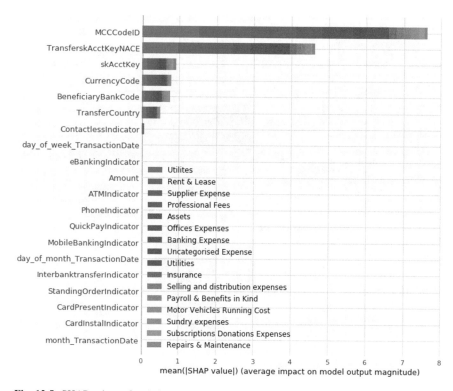

Fig. 12.5 SHAP values of each feature included in the CatBoost model

Apart from the feature importance based on SHAP analysis, as far as the CatBoost evaluation is concerned, it is of high importance to qualitatively check some of the outcomes based on the Local Interpretable Model-agnostic Explanation (LIME), which is a recent technique capable of explaining the outcomes of any classifier or regressor using local approximations (sample-based) of models with other interpretable models.

Figure 12.6 offers five examples of how LIME framework can assist to interpret the predictions of our transaction categorization model. For instance, as illustrated in the figure, the first transaction is categorized as "Banking Expense" with probability of 39% and features contributing toward this outcome are the MCCCode, the NACE Code, and the specific Account. Likewise, with a probability of 23%, it can be categorized as "Uncategorized Expense," and with a probability of 16%, it can be classified as "Selling and Distribution Expense," with the features contributing toward this decisions also being depicted. In the second example observed, the given category is "Selling and Distribution Expense" with a confidence level of 96% (Fig. 12.6).

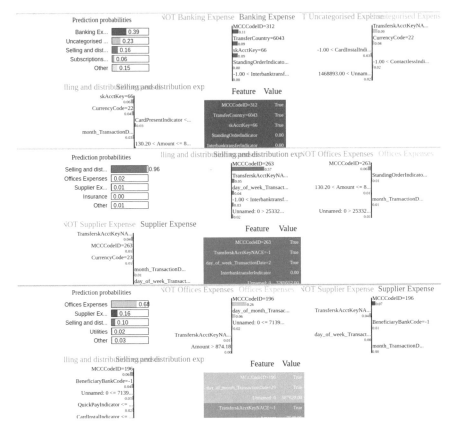

Fig. 12.6 LIME analysis of three specific transaction categorization examples, explaining the features and the value contribution of the model outcomes with their respective probabilities

4.1.4 Paving the Way for Open Data Enrichment: Word Embeddings in Transaction Descriptions

The Transaction Categorization Engine is enriched with another innovative contribution of creating word embeddings from the transaction descriptions. These embeddings are used in the transaction categorization model and serve as common ground between integrating open banking data and the proprietary (internal) data that is utilized. This approach not only increases the categorization accuracy but also paves the way of classifying transactions provided by other institutes as part of PSD2 or can be used as features in other complementary downstream machine learning processes such as fraud detection, which is also implemented partly in our Transaction Monitoring microservice. This effort however raises another challenge as information in short texts is often insufficient which makes them hard to classify. Recently, continuous word representations in high-dimensional spaces brought a

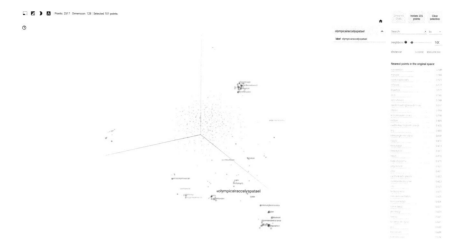

Fig. 12.7 Tensorboard Illustration of word embeddings example created through transaction descriptions

great impact in Natural Language Processing (NLP) community by their ability to unsupervisedly capture syntactic and semantic relations between words, phrases, and even complete documents. Employment of these representations produced very promising results with the help of available large text bases in the fields of language modeling and language translation. Motivated from the success of continuous word representations in the NLP world, this work proposes to represent the financial transaction data in a continuous embedding space to take advantage of the large unlabeled financial data. The resulting vector representations of the transactions are similar for the semantically similar financial concepts. We argue that, by employing these vector representations, one can automatically extract information from the raw financial data. We performed experiments to show the benefits of these representations. In Fig. 12.7, a tensorboard example of a Word2Vec Skip-Gram [14] model having to do with transaction descriptions related to "olympicair" is illustrated, presenting grouping indications of similar transaction categories.

4.2 Cash Flow Prediction Engine

The second microservice showcased is the Cash Flow Prediction Engine, which aims to accurately predict the cash inflows and outflows of the given categories produced by our hybrid transaction categorization model for each SME. The engine's objective and the nature of the task implied the necessary data transformation in time-series representation, enabling the experimentation with various general forecasting models, or prevalent DL models. Thus, both resampling and aggregating the amount of the transactions based on each specific account and date have been

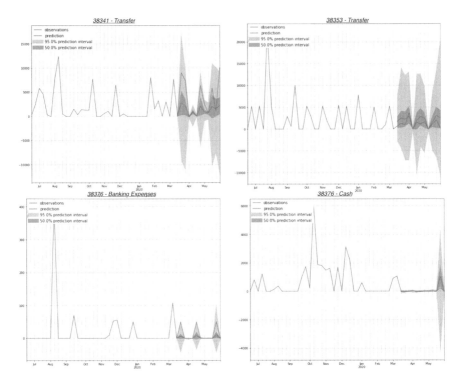

Fig. 12.8 Examples of DeepAR application for Cash Flow Prediction

applied to our data sample. The forecasting models considered consisted of a batch of SARIMA variations, which were however unable to cover the complex needs of our Cash Flow Predictions engine. Facebook's Prophet, a modular regression forecasting model with interpretable parameters that can be intuitively adjusted by analysts with domain knowledge about the time-series, as presented in [15], was also examined. However, the results were not as satisfying when trying to predict transaction inflows and outflows in our scenario, opposed to [16], where the model is compared with DeepAR model in order to forecast food demand, showing promising results. The relevant plots are depicted in Fig. 12.8, showing the estimators on specific time series. Specifically in each plot the predicted mean value (green line) and the actual values (blue line) are depicted along with the green gradient area denoting 2 confidence intervals (i.e., 50%, 95%). DeepAR, a DL approach implementing an RNN-based model, close to the one described in [11], was chosen as the most suitable one, originating from the open-source GluonTS toolkit [16]. More specifically, the chosen model applies a methodology for producing accurate probabilistic forecasts, based on training an autoregressive Recurrent Neural Network model on many related time-series. RNNs have the concept of "memory" that helps them store the states or information of previous inputs to generate the next output of the sequence. The RNN that predicts the mean and the variance of the underlying time-series is coupled with a Monte Carlo simulation yielding results represented as a distribution. Moreover, the chosen model learns seasonal behavior patterns from the given covariates that strengthens its forecasting capabilities. As expected, while configuring the model to our forecasting needs, various long-established challenges regarding time-series forecasting arose. These challenges can be summarized in (a) the "cold start" problem, which refers to the time-series that have a small number of transactions, (b) the stationarity–seasonality trade-off, where it is assumed that in order to have predictable time-series they have to be stationary without trend and seasonality factors present, (c) the existence of noisy data and outliers observed, and (d) the adequacy of the length of the dataset in order to apply ML/DL techniques. The aforementioned challenges were dealt with the use of surrogate data, DeepAR model optimization, injected transactions thresholds, and hyper parameters configuration, which however exceed the showcase purposes of the Cash Flow Prediction Engine included in this chapter. As for the evaluation scheme of the model presented, since cross-validation methods reflect a pitfall in a time-series forecasting scenarios as they may result in a significant overlap between train and test data, the optimum approach is to simulate models in a "walk-forward" sequence, periodically retraining the model to incorporate specific chunks of transaction data available at that point in time.

4.3   Budget Prediction Engine

Having not only a good budget in place but also an effective real time budget monitoring and adjustment capability is essential for the business success of an SME. The Budget Prediction engine takes into consideration cash flow, benchmark, macroeconomic, and other available SME data which is key to come up with smart budget targets. The derived smart budgets dynamically consider a changing environment and provide actionable insights on potentially required budget adjustments.

A microservice allows the user to set budgets per category and to allocate available resources. The set budgets will not only be evaluated based on already scheduled invoices (inflows and outflows) but also on predictions derived from historical incomes and spending. The budget prediction engine is closely connected to the cash flow prediction model presented above, as it utilizes the same DeepAR model in its core.

4.4   Transaction Monitoring

A major objective of the BFM smart advisor is to reduce the administrative burden for the SME. The transaction monitoring engine aims to support this purpose by acting as a kind of transaction guard that identifies abnormal transactions.

Abnormal transactions refer to the following transaction categories. Those that show irregularly high transaction amount for the specific merchant, originate from a new merchant, signal double charging notification or represent potential fraudulent transactions. The transaction guard would also "watch out" for transactions that could be of significant interest to the business such as transactions relating to refunds or insurance claims.

4.5 KPI Engine

The KPI engine delivers key metrics that allow the SME in an easy way to understand the state of their financial health and performance in real time. Besides the actual diagnosis, the engine not only comes with smart alerts that immediately point out anomalies but also with a comparison of actual versus best practice target values accompanied with a strong indication on how best practice figures can be potentially achieved and/or current values be improved. Altogether, the KPI engine effectively guides the SME in their decision-making process and ultimately contributes toward a stable financial environment.

4.6 Benchmarking Engine

Benchmarking has been underestimated for a long time within the SME sector and many times being avoided due to its cost and time impact. The benchmarking engine focuses on bringing a valuable comparison insight to the SME in a cost-/time-effective way, doing so by comparing the respective SME with other SMEs operating in the same/similar environment and under the same attributes. As a result the SME can locate key areas (e.g., operations, functions, or products) for improvement and take actions accordingly to potentially increase its customer base, sales, and profit.

4.7 Invoice Processing Invoices (Payments and Receivables)

Invoice Processing Invoices represent a vital input to other engines like the Cash Flow Prediction or KPI engine. Furthermore, the retrieved invoice data is also utilized to come up with VAT or other provisioning insights. Today, SMEs invest significant effort into the invoice monitoring, collection, and reconciliation process. The Invoice engine supports these processes and over and above can assist in the liquidity management by integrating with factoring services.

5 Conclusion

This chapter illustrates how the utilization of state-of-the-art data analytics tools and technologies, combined with the integration of available banking data and external data, can offer a new perspective. The proposed mechanism offers automation and personalization, increasing the productivity of both SMEs and financial institutions. The provided BFM tools empower SMEs through a deeper understanding of their operation and their financial status, leading to an increased data-driven decision-making model. Respectively, financial institutions can harness all available data and offer personalized value-added services to SMEs on top of their core business. The generated data of the BFM tools assist banks to better understand their SME customers and their transaction behaviors, identifying their financial needs and supporting the design of tailor-made financial products for the SMEs. Moreover, the conceptual architecture presented, which is based on the INFINITECH RA, enables new perspectives in the fields of data management, analytics, and testbed development, enabling the effortless introduction of new SME microservices and refinement of existing ones, all aiming at increased business financial management capabilities.

Acknowledgments The research leading to the results presented in this chapter has received funding from the European Union's funded Project INFINITECH under grant agreement no: 856632.

References

1. OECD SME and Entrepreneurship Outlook 2019, Policy highlights. Available to http://www. oecd.org/industry/smes/SME-Outlook-Highlights-FINAL.pdf. (Accessed 04.10.2019).
2. Muller, P., Robin, N., Jessie, W., Schroder, J., Braun, H., Becker, L. S., Farrenkopf, J., Ruiz, F., Caboz, S., Ivanova, M., Lange, A., Lonkeu, O. K., Muhlshlegel, T. S., Pedersen, B., Privitera, M., Bomans, J., Bogen, E., & Cooney, T. (2019). *Annual Report on European SMEs 2018/2019 - Research & Development and Innovation by SMEs*. European Commission.
3. Casalino, N., Żuchowski, I., Labrinos, N., Munoz Nieto, Á. L., & Martín, J. A. (2019). *Queen Mary School of Law Legal Studies Research Paper Forthcoming*
4. Kergroach, S. (2020). *Journal of the International Council for Small Business, 1*(1), 28.
5. Manyika, J., Chui, M., Brown, B., Bughin, J., Dobbs, R., Roxburgh, C., & Hung Byers, A. (2011). *Big data: The next frontier for innovation, competition, and productivity*. McKinsey Global Institute.
6. M. Bianchini, V. Michalkova, Data analytics in SMEs: Trends and policies (2019).
7. Winig, L. (2017). *MIT Sloan Management Review, 58*(2), 57.
8. Prokhorenkova, L., Gusev, G., Vorobev, A., Dorogush, A. V., & Gulin, A. (2017). *Preprint arXiv:1706.09516*.
9. Tanha, J., Abdi, Y., Samadi, N., Razzaghi, N., & Asadpour, M. (2020). *Journal of Big Data, 7*(1), 1.
10. Rev, N. (2008). *Statistical classification of economic activities in the European community*. Technical Report, Methodologies and Working papers. European Communities, Luxembourg (2).

11. Salinas, D., Flunkert, V., Gasthaus, J., & Januschowski, T. (2020). DeepAR: Probabilistic forecasting with autoregressive recurrent networks. *International Journal of Forecasting, 36*(3), 1181–1191.
12. Ribeiro, M. T., Singh, S., & Guestrin, C. (2016). *Proceedings of the 22nd ACM SIGKDD International Conference on Knowledge Discovery and Data Mining* (pp. 1135–1144).
13. Lundberg, S., & Lee, S. I. (2017). *Preprint arXiv:1705.07874.*
14. Mikolov, T., Chen, K., Corrado, G., & Dean, J. (2013). *Preprint arXiv:1301.3781.*
15. Taylor, S. J., & Letham, B. (2018). *The American Statistician, 72*(1), 37.
16. Alexandrov, A., Benidis, K., Bohlke-Schneider, M., Flunkert, V., Gasthaus, J., Januschowski, T., Maddix, D. C., Rangapuram, S., Salinas, D., Schulz, J., Stella, L., Caner Türkmen, A., & Wang, Y. (2019). *Preprint arXiv:1906.05264.*

Chapter 13
Screening Tool for Anti-money Laundering Supervision

Filip Koprivec, Gregor Kržmanc, Maja Škrjanc, Klemen Kenda, and Erik Novak

1 Introduction

Activities in the anti-money laundering and counter-terrorist financing (AML/CTF) sphere are an important part of the provision of a stable financial and economic sector. As an autonomous countrywide regulatory authority, Bank of Slovenia (BOS) supervises the compliance of individual banks, conducts inspections and serves as a consulting body when drawing regulations [1]. As the world is more and more interconnected, and country and geographical borders are getting blurred, money transfers between countries and entities with high volume are getting more ubiquitous than ever. With the interconnection and easy use of the system for financial transfers across the world and within the EU area, efficient and timely supervision is more needed than ever.

With malicious entities moving faster and outpacing the current detection and prevention mechanisms, it is becoming clear that in the digital age of data expansion, manual and human-based detection methods are unable to cope with the sheer amount of transactions and volume of additional data needed for processing. This can be exemplified by not-so-recent events [2] that clearly show that additional data and automatic detection of specific early warning systems for risky transaction patterns should be put in place both on the level of banks and on the level of the supervisory authority. Such a system will be beneficial to both supervisory authorities by enabling greater degree and more efficient supervision as to individual banks by providing a tool for early discovery and reporting of problematic scenarios.

F. Koprivec (✉)
JSI, FMF, IMFM, Ljubljana, Slovenia
e-mail: filip.koprivec@ijs.si

G. Kržmanc · M. Škrjanc · K. Kenda · E. Novak
JSI, Ljubljana, Slovenia
e-mail: gregor.krzmanc@ijs.si; maja.skrjanc@ijs.si; klemen.kenda@ijs.si; erik.novak@ijs.si

© The Author(s) 2022
J. Soldatos, D. Kyriazis (eds.), *Big Data and Artificial Intelligence in Digital Finance*,
https://doi.org/10.1007/978-3-030-94590-9_13

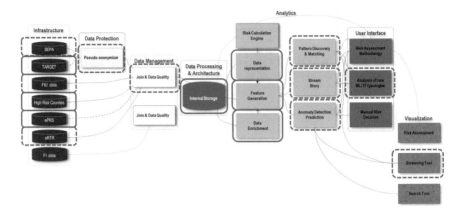

Fig. 13.1 Screening tool pipeline overview

Artificial intelligence and machine learning methods are naturally suited for the problems facing supervisory authorities and banks. Screening tool developed as part of the Infinitech project naturally augments existing tools and provides new information. The main goal of the screening tool is to process and analyse transaction data from different sources, enrich it with additional external info specific to anti-money laundering and counter-terrorist financing needs and efficiently combine different levels of granularity of data. The screening tool tries to recognize unusual transaction patterns and relationships among combined data that indicates typologies and risks of money laundering (ML) or terrorist financing (TF) at a level of specific financial institutions. Results of automated screening and flagged patterns are automatically presented to domain experts for further exploration and consideration. Automation in the process increases the amount of transactions that can be analysed, removes possible human mistakes and enables processing and ingestion of data from additional sources that were untraceable for a human actor. Patterns reported by the screening tool are weighted according to their relative measure of *riskiness* and put into context, where domain experts can further explore and analyse the patterns in enriched context before making a more informed decision.

The screening tool pipeline is presented in Fig. 13.1. The pipeline is composed of several individual and mostly independent components neatly wrapped in containers following the Infinitech way of implementation. This way, the whole tool suite can easily be deployed on-site and enable flexibility and seamless integration. Following the Infinitech philosophy, additional tools and detectors can be easily included and integrated with the screening tool suite without bespoke modifications. The tool suite is built for performance with huge volumes of data and can be used in almost real time while also providing a batch mode for the exploration of historical data and scenario modelling.

2 Related Work

The use of machine learning in graph datasets has seen enormous progress in the last few years and has been actively incorporated in many previously thought to be hard problems. Transaction data is naturally modelled as a graph structure (nodes as actors, transactions as edges, with additional nodes and attributes for metadata). Both supervised and unsupervised learning approaches have been proposed to deal with the task of detecting suspicious patterns in financial transaction data.

2.1 Anti-money Laundering with Well-Defined Explanatory Variables

In [3] the authors propose unsupervised methods that can be used on high-dimensional client profiles. A method for visualization is proposed by using dimensionality reduction. The high-dimensional feature vectors describing customers are projected into 2D space for visualization using a dimensionality reduction procedure such as PCA. That makes it possible for supervision personnel to visually identify outlier groups (clusters) of potentially risky subjects. Additionally, a peeling algorithm is proposed for anomaly detection. Here in each step, the most extreme outlier based on any distance function (the authors propose Mahalanobis distance) is removed and marked as an anomaly and the process is then repeated for any number of steps.

Such methods can be used for any type of anomaly detection problem in n-dimensional Euclidean space. It is, however, crucial to apply them according to the end goal of identifying suspicious transactions. It is important to use features that expose activities that are known to be common among money laundering groups.

A client profiling approach based on k-means clustering of customer profiles was described in [4]. The explanatory variables (feature vectors) have been naively constructed through interactions with domain experts. The optimal number of clusters has been estimated using Silhouette coefficient and a sum of squared errors. After clustering, rules for classification have been generated and tested with multiple rule generation algorithms. The relevance of generated rules for targeting high-risk customers was estimated manually by domain experts. The authors show that it is essential to include features that are relevant to the given problem. Such features include account type, account age, the volume of transactions etc. The end result of such goal-driven development are distinct clusters that are named according to the goal (examples include "Group of risk", "Standard customer" etc.) and can be clearly described using a set of rules.

Additionally, Mastercard's Brighterion claims to be using unsupervised learning for their AML products [5].

In [6] a supervised learning technique that operates directly on transactions (not entities) is presented. The Gradient Boosting model is trained to predict suspicious transactions based on past Suspicious Activity Reports (SARs). The feature vectors used in training incorporate information both about the entity as a whole and the individual transaction. Such features include an indication of any previous bankruptcies related to the entity, company sector type, activity level and amount of transactions in the last 2 months grouped by transaction type. The authors show the model is accurate and efficient, outperforming the bank's rule-based approach.

2.2   Machine Learning on Graph Structured Data

While using manually generated client feature vectors offers greater model explainability, there may still be information hidden in transaction networks that is not captured this way. Recent progress in Graph Machine Learning has it made possible to capture deeper structural information of node neighbourhoods.

Nodes in a graph can roughly be described in two distinct ways: what communities they belong to (homophily) or what roles they have in the network (structural equivalence).

The DeepWalk algorithm [7] generates node representations in continuous vector space by simulating random walks on the network and treating these walks as sentences. The generated embeddings can then be used for any prediction or classification task. The random walk idea has been further extended by node2vec [8]. DeepWalk and node2vec have been successfully used in domains such as biomedical networks [9] and recommendation systems [10].

Another algorithm struc2vec proposed by [11] works in a similar manner to DeepWalk, only that it performs random walks on a modified version of the original network, which better encodes structural similarities between nodes in the original network. Embeddings generated by struc2vec may perform better on structural equivalence-based tasks than node2vec. Struc2vec embeddings have been shown to generally perform better than node2vec or DeepWalk on link prediction tasks in biomedical networks [9].

DeepWalk, node2vec and struc2vec are inherently transductive, i.e. require the whole graph to be able to learn embeddings of individual nodes. This might present a challenge for real-world applications on large evolving networks. Therefore, other representation learning techniques have been proposed recently that are inductive, i.e. can be trained only on parts of the network and can be directly applied to new, previously unseen networks and nodes.

Graph Convolutional Networks (GCNs) [12] are models that leverage node features to capture dependence between nodes via message passing. The original GCN idea is further extended to inductive representation learning by Hamilton et al. [13].

A (Variational) Graph Autoencoder—(V)GAE—model has been proposed by [14] to generate node representations. The authors demonstrate that an autoencoder using a GCN encoder and a simple scalar product decoder generates meaningful node representations of the Cora citation dataset. It is, however, unclear how such a model would perform on graphs where structural equivalence plays a more significant role in descriptions of nodes.

A temporal variation of GCN called T-GCN has been proposed [15]. The authors propose a model that first aggregates spatial information with a GCN layer separately at each time step and then connects time steps with Gated Recurrent Units (GRUs) to finally yield predictions. Such a model outperforms other state-of-the-art techniques (incl. SVR and GRU) on a traffic prediction task on real-world urban traffic datasets, aggregating both spatial and temporal information. Additionally, a temporal graph attention (TGAT) layer has been recently proposed as an alternative for inductive representation learning on dynamic graphs [16].

2.3 Towards Machine Learning on Graph Structured Data for AML

Transaction networks, on which money laundering detection can be performed, usually contain only a small fraction of known suspicious subjects (if any). Such class imbalances need to be dealt with in order to produce robust supervised models.

In [17] the proposed method for detecting money laundering patterns is based on node representation learning on graphs. A transaction graph is constructed from the real-world transaction data from a financial institution. The undirected, unweighted graph is constructed from data within a fixed time period. A small number of subjects is known to be suspicious regarding money laundering. Accounts outside bank's country are aggregated by country. Node representations are learned using DeepWalk and are then classified with three binary classifiers: Support Vector Machine (SVM), Multi-Layer Perceptron (MLP) and Naive Bayes. The extreme class imbalances are overcome using the widely used SMOTE algorithm for synthetic oversampling of the minority class. Another strategy tested is the undersampling of the majority class and random duplication of the minority class. The model is then evaluated on a part of ground truth (not oversampled) entities not included in the training. Best results are achieved using the MLP and random duplication (although differences between results may be insignificant), while it is suggested that SMOTE produces slightly more stable models.

An adaptation of SMOTE for graphs, named GraphSMOTE, is proposed in [18]. GraphSMOTE generates synthetic nodes of the minority class (and not just synthetic embeddings as described in [17]) that are inserted into the graph, on which training can then be performed directly. The authors show that variants of GraphSMOTE out-perform traditional training techniques for imbalanced datasets including weighted loss and variations of SMOTE and generalize well across different imbalance ratios.

In [19] the authors experiment with predicting suspicious entities in four separate directed networks published in ICIJ Offshore Leaks Database. Some nodes are marked as blacklisted by matching with international sanction lists. The datasets are highly imbalanced, containing less than 0.05% of blacklisted nodes. In the first part, embedding algorithms are used in a similar way as in [17]. One-class SVM (O-SVM) is used here to predict suspicious entities in contrast to oversampling. The model is then evaluated on a proportion of ground truth data only. Struc2vec mostly outperforms node2vec in terms of the AUC score, although the difference varies significantly across the four datasets in the database. Additionally, degree centrality measures (PageRank, Eigenvector Centrality, Local Clustering Coefficient and Degree Centrality) are used as features to describe the importance of nodes in the networks. Among these PageRank alone mostly performs best, outperforming struc2vec as well, highlighting the role of node centrality in such tasks. Additionally, all experiments were conducted on undirected, directed and reversely directed versions of graphs. Best results were generally achieved using reversed networks, although the results vary significantly across datasets.

A novel method called Suspiciousness Rank Back and Forth (SRBF) inspired by the PageRank algorithm is additionally introduced in [19]. The authors show that it generally performs better in detecting suspicious subjects compared to both degree centrality measures and the mentioned graph embedding algorithms, achieving the overall best score in 3 out of 4 datasets.

The mentioned methods in [19] are evaluated against the list of known blacklisted entities. However, it remains a challenge to validate entities that appear high risk but are not on the original ground truth blacklist. The use of Open-Source Intelligence (OSINT) is proposed in [20] for verification of these predicted high-risk entities. Such methods, although taking much manual work, have proven to be successful in some cases when enough information was found online to uncover potential hidden links.

In [21] an experiment is conducted using the labelled Elliptic Dataset of Bitcoin transactions. Around 2% of nodes in the dataset are marked as "illicit" (e.g. known to have belonged to dark markets), some (21%) are labelled as "licit" (e.g. belonging to well-established currency exchanges). The data is spread across 49 time steps. Each node is accompanied with approximately 150 features that are constructed from transaction information and aggregated neighbourhood features. They describe an inductive approach to predicting the suspiciousness of nodes using GCNs. Finally, such an approach is compared with different classification methods against the objective to predict whether a node is licit or illicit. The classifiers tested are Multi-Layer Perceptron (MLP), Random Forest and Logistic Regression trained with weighted cross-entropy loss to prioritize illicit nodes as they are the minority class.

Superior classification results are shown using Random Forests using node features only. Additionally, a performance improvement is achieved by making the Random Forest model "graph-aware" by concatenating GCN embeddings to the mentioned features. Using GCNs alone in a supervised setting did not yield as good results; however, using EvolveGCN to capture temporal dynamics yielded slightly better results compared to pure GCN. An intriguing fact arises when looking at the accuracy of the proposed models over time. As a large dark market is closed at some time step, the accuracy of all models drops significantly and is not recovered after following time steps. The robustness of models to such events is a major challenge to address.

In [21] a future work idea is pointed out to combine the power of Random Forests and GCNs using a differentiable version of decision trees as proposed by [22] in the last layer of GCN instead of Logistic Regression.

Anomaly detection on graphs could also be used for detecting anomalous activities. In [23] a fully unsupervised approach for detecting anomalous edges in a graph is presented. A classifier is trained on the same number of existing and non-existing edges to predict whether an edge exists between two nodes. The nodes that have then their edges classified as non-existing can be viewed as anomalous. The approach has been tested on real-world datasets such as online social networks.

3   Data Ingestion and Structure

3.1   Data Enrichment and Pseudo-Anonymization

At the core of the screening tool is an efficient data ingestion, store and representation of large multigraphs of transaction data from two transactional data sources. A pure transaction data graph is valuable and provides a backbone for further analysis, but data is also enriched to enable deeper exploration and an easier understanding of acquired results. Additional information, tailored specifically for anti-money laundering scenarios, is ingested (and automatically updated) alongside. Due to the high sensitivity of transaction data, privacy concerns and legal considerations, the ingestion pipeline follows a very specific and specially tailored pseudo-anonymization and enrichment pipeline. The general dataflow is presented in Fig. 13.2.

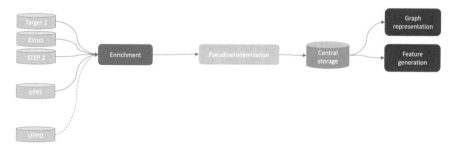

Fig. 13.2 Data ingestion and preparation overview

Transaction data from different sources is enriched with company-specific information provided by the public business registry. Additional company-related information especially targets money laundering and terrorist financing information as well as general features to discover anomalous behaviour patterns e.g. company size, capitalization type, ownership structure, company registration date, possible company closure date, capital origin etc. Account information data is provided by eRTR (public accounts' registers) and information required for AML scenarios is added: account registration, possible account closure, account owner etc. Important piece of information is the type of the account; an account can either be used by a private person or associated with a specific company and in some cases even both. Meta parameters pertaining to more efficient identification of suspicious patterns are periodically ingested: EU list of high-risk third countries, classifying the level of high risk to those countries on the list.

Subject to legal and privacy issues, the future version of the screening tool is expected to include additional transactions into highly risky foreign countries. The list is curated by the office for the prevention of money laundering as mandated by applicable AML and CTF law. The inclusion of this list will provide more broad information for specific entities and improve tracking of suspicious money paths.

After ingestion and data enrichment all directly identifiable company and account data is pseudo-anonymized. Initial exploration showed that great care must be taken during the pseudo-anonymization process. Some information is inevitably irrecoverably lost due to pseudo-anonymization as a way to satisfy privacy concerns (multiple accounts of the same private person, company name, account numbers), but other information might be lost unintentionally. Original transaction data and data from transactions to risky countries must be paired before the pseudo-anonymization process as reporting standards differ and thus make post-pseudo-anonymization pairing impossible. Another negative side effect of pseudo-anonymization is the fact that it can easily conceal data quality issues. For example: badly structured account numbers, manually imputed data and other small inconsistencies invisible when processing data at transaction level before pseudo-anonymization, alongside with spurious white space, letter and number similarities, string collations etc., produce totally different pseudo-anonymized result that is hard to detect. Multiple cross-checks and data validation procedures were used during the developmental phase to uncover a few such pitfalls and adjust the pseudo-anonymization process accordingly. The cross-checks for data quality control are also in place in the production environment to monitor full data flow.

Pseudo-anonymizer developed during the project is able to successfully anonymize incoming stream of data, reusing salt info from the previous invocation. This enables easier ingestion of data at later stages, combined with new data and ingestion from different platforms while still keeping as much of information about graph structure and connectivity. Due to the high amount of data and the need for automatic ingestion, developed pseudo-anonymization service is provided as part of Infinitech components as a standalone Kubernetes service that is able to mask data in real time.

3.2   Data Storage

Ingested and pseudo-anonymized data is transformed and stored to enable fast and efficient queries. Due to large and fast data ingestion, data is stored in three separate databases. The raw database is a simple PostgreSQL container storing pseudo-anonymized unnormalized raw data. The data ingestion engine firstly automatically normalizes raw data and merges it with existing data stored in the master PostgreSQL database specifically configured for high-performance reading and time-based indexing. Master database serves as a core data source for Fourier transformation-based feature calculation Sect. 5.2, stream story and feature generation. After normalization, a battery of sanity checks is performed to confirm the schema validity of ingested data and check for inconsistencies and anonymization errors. Normalized data is imported into the open-source version of the neo4j database, which serves as a backbone for graph-related searches, pattern recognition and anomaly detection.

The graph database schema is specifically tailored to enable easy exploration of transaction neighbourhoods and execution of parameterized queries. Great care was taken to provide proper indexing support for time-based analysis for neighbourhood root system evolution and obtaining financial institution-level data. This enables efficient exploration of transaction structure evolution in relative time and specific risky typology detection. In conjunction with the vanilla neo4j database, neo4j Bloom is also configured and included in the platform for more in-depth analysis and further exploration.

Both database part and normalization and sanity check services are provided as self-contained docker images in line with the Infinitech way and fully configurable for ease of use.

4   Scenario-Based Filtering

BOS acts as a supervisory authority and as such oversees the risk management and risk detection, both at the level of financial institution and also on the level of the financial sector. The screening tool combines data from multiple transactional data sources, merges and normalizes data into a single data source and enables efficient data exploration on multiple levels of granularity. As a supervisory authority one of the main goals is that the financial institutions develops proper control environment for suspicious transaction detection and reporting. Great care was taken to provide suitable explanations and interpretations of different risky scenarios flagged as unrecognized by individual financial institutions and deemed risky.

Domain expert knowledge and initial exploration on historical data combined with high-risk third countries' transaction data showed that the surest way to get highly interpretable and quickly actionable insights is to first apply scenario-based filtering. Typical scenarios used in simple detection come in simple, human-readable rule-based forms:

> Newly established companies or companies closed soon after the establishment, receiving large sums of money from foreign accounts and paying similar amount to private account pose high risk.
> Company risk increases with the level of the risk the payer/receiver country.

The main advantage of rule-based scenarios is straightforward explainability, composability and easy correspondence with existing anti-money laundering regulations. As a supervisory authority, providing a clear explanation for flagging and further processing increases the ability to pursue claims and also improves the actionability of findings. Furthermore, explainable rule-based filters can easily be implemented down the road in on-site checks. A direct contribution of collaboration with experts regarding existing rules for evaluation and flagging of risky behaviour was the development of special *parameterized generalized rules*. They automatically encompass common money laundering and terrorist financing transaction patterns and map them to database queries for further exploration as presented in Sect. 6.

The rule-based approach also translates nicely into the language of *typologies*. Specific transaction patterns can be interpreted as time-based parts of the money laundering process (preparation, diffusion...) or graph-based parts (the part where money is rewired to individuals previously associated with the company, part where money is crossing the EU border...). Since not all transactions are included, detecting even parts of scenarios can provide good insight and lead to further exploration. Specific parts in this are referred to as *typologies* and usually named after historical cases when they were discovered and/or prominent. Detecting and correctly identifying such parts on subgraphs (specific bank transactions, a combination of banks where bank accounts typically mix...) and specific time frames is easy with generalized parameterized rule-based queries and directly translates to language and context already spoken by experts and is actionable.

Rule-based filtering proved to be flexible enough to easily detect specific patterns in historical data that was immediately interpreted as a phase of a potential money laundering scenario. Although flexible, even generalized rules were are still easily interpretable and can be put in the context of anti-money laundering during analysis.

5 Automatic Detection and Warning Systems

5.1 Pattern Detection

The screening tool will provide automatic detection of suspicious typologies and flagging of risky scenarios. Automatic detection and early real-time warning systems give the supervisory authority additional data, additional risk-based analysis, thus easing and supporting the supervisory process if decided so by a domain expert.

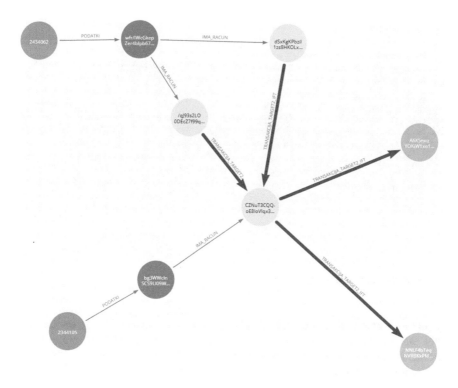

Fig. 13.3 Large transaction forwarding

Figure 13.3 shows a specific flagged scenario detected and further explored in the screening tool. Nodes in the graph correspond to different scenario entities: yellow nodes correspond to accounts owned by companies, blue nodes correspond to accounts owned by private people, green nodes correspond to companies and orange nodes correspond to *flagged and monitored* objects in for chosen scenario. In this specific scenario, orange nodes present to companies closed less than 6 months after the registration, corresponding to the scenario where companies closed soon after the establishment process the anomalous amount of transactions. Edges between nodes correspond to specific dependencies. Edges (in red) between bank account nodes correspond to transactions with edge thickness indicating the amount being transferred (in log scale). Edges between account nodes and purple nodes correspond to account ownership. A company might own one or more accounts and the combination of transactions from the same company might tell a completely different story than an isolated bank account. Lastly edges between the flagged orange nodes and company nodes indicate which company node was flagged, with specific flagging information and results stored as attributes of the orange node. Labels on nodes correspond to anonymized company and bank account indicators.

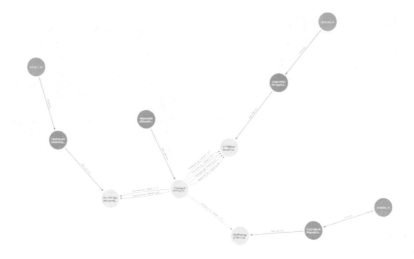

Fig. 13.4 Transaction crumbling

In this scenario, a company received two large transitions (1.2M €), on the same day, and made two payments with practically the same amount to two different individual bank accounts. The proxy company existed for less than 4 months.

Figure 13.4 shows another discovered pattern. This time the company nodes are coloured in purple.

In the second detected example, the same entity performs several identical transactions to two entities opened and closed in a short period of time and to a third entity open one year later.

Both scenarios exhibit signs of a potential money laundering phase called *layering*. The main goal of layering is to make the source of illegal money difficult to detect through various means: shell companies, receiving and processing payments through foreign and non-EU countries, specifically through countries with lax AML standards. The second example shows the possible use of *smurfing* where payments are divided into smaller amounts to conceal and evade ordinary controls and transfers through multiple accounts to disperse the ownership information.

The third anomaly is presented in Fig. 13.5, which consists of a transaction, where a single short-lived company disperses multiple transactions to private person accounts in a short time span. To make it even more anomalous, all transactions cross the country border, thus making efficient money tracing more difficult.

Future directions for automatic detection of specific scenarios is to use relative time scale to compare local relationship evolution of similar company types. This will enable better modelling of graph representation pertaining to specific types of transaction history related to various company types (transaction structure of utility companies and retail companies is widely different from companies offering financial services to institutional clients).

Fig. 13.5 Dispersing transactions to multiple private accounts

5.2   Data Exploration Using Fourier Transforms

Financial data is inherently temporal. It is reasonable to assume that there exist some recurring patterns in the trade activity of subjects. Let ϕ_j be a time series of a daily number of outgoing transactions made by entity j. (The same analysis as demonstrated here can be performed with any other time series.) We use the efficient Fast Fourier Transform (FFT) algorithm [24] to compute frequency-domain representations $A_j(\nu)$ of each ϕ_j.

Baselines of the resulting spectra are corrected using the I-ModPoly algorithm as described in [25] and implemented in the BaselineRemoval Python library (https://pypi.org/project/BaselineRemoval/) using polynomial degree 1. The resulting frequency domain spectrum of each client is resampled by linear interpolation at $n = 1000$ equally spaced points throughout the frequency range to yield n-dimensional

Fig. 13.6 Barycentre spectra of FFT clusters. Approximate number of members of each cluster shown in brackets

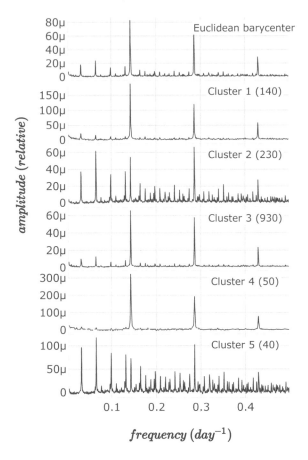

vector representations of entities χ_j. Finally each χ_j is normalized by dividing with $\sum_{t=1}^{|\phi_j|} \phi_j(t)$.

K-means clustering using $k = 5$ and Euclidean distance is performed on the given feature vectors. Entities that do not have enough data points recorded for quality spectra are filtered out with a simple rule; accounts with <365 non-zero data points in ϕ_j are not considered for this analysis. Our dataset is sparse; <1% of all entities in the dataset remains available for frequency-domain analysis after filtering.

It is seen from Fig. 13.6 that all clusters have three distinct peaks. These correspond to frequencies that are multiples of $\frac{1}{7}$. There are inherent weekly dynamics in the data, as there are no transactions processed during weekends. There are distinct differences between clusters. Clusters 1 and 3 are highly similar to the Euclidean barycentre, containing mostly weekly, but also some other (monthly) frequencies. Entities in clusters 2 and 5 exhibit strong monthly dynamics (for example, make payments once a month), while subjects in cluster 4 exhibit weekly dynamics (e.g., make a single transaction on each working day and no transactions on weekends).

6 Use Cases

The full working platform for anti-money laundering and supervision encompasses all previously described segments joined in a simple and intuitive platform. The platform takes care of data ingestion, pseudo-anonymization, data enrichment, automatic anomaly detection and pattern flagging. Neo4j Bloom tool combined with currently developed web-based graphical user interface for graph exploration allows experts in supervisory and compliance to manually inspect flagged scenarios in a broader context.

Automatic scenario detection provides the user with suggested parameterized queries that cover both scenario-based analysis with computed risk measures.

Figure 13.7a shows an example of parameterized query from available scenarios suggested by automatic detection. In this case, the user is able to manually configure what relative time he or she is interested in. Figure 13.7b shows an example result produced by such a query. Depending on the granularity and selected timescale, certain features are more pronounced. Example scenario nicely shows that transaction neighbourhoods form a highly partitioned graph with a large component controlling more than half of the graph and a lot of smaller components. Small components are of particular interest as they usually exhibit irregular behaviour.

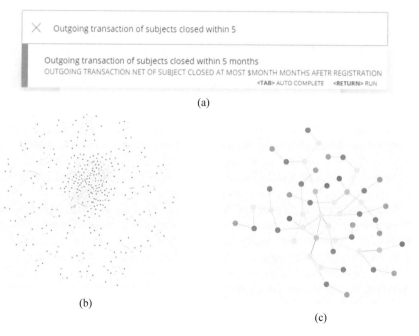

(a)

(b)

(c)

Fig. 13.7 Parameterized query with resulting graph structure. (**a**) Parameterized query example. (**b**) Transaction graph corresponding to parameterized query. (**c**) Medium level cluster close up

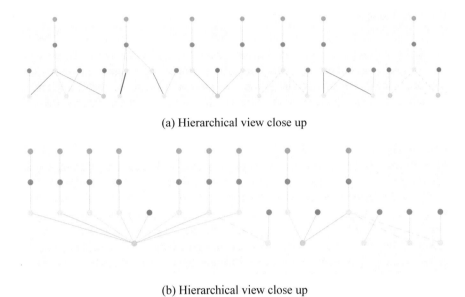

(a) Hierarchical view close up

(b) Hierarchical view close up

Fig. 13.8 (**a**) and (**b**) Hierarchical views close up

The screening tool enables easy exploration and further processing and application of ML algorithms to specific clusters, for example one seen in close up in Fig. 13.7c.

Exploring different graph clusters usually raises additional questions about specific cluster data. As the data is preprocessed, additional transaction and company-related data is readily included and can be uncovered by clicking on specific nodes or edges as seen in Fig. 13.9a. Presentation of additional data depends on the data status. Confidential data such as account number is presented in pseudo-anonymized version due to privacy and legal concerns. Data that does not represent personal, identifying information or confidential data (transaction date) can be presented in full. Depending on anonymizer settings, different levels of authorization can be presented with different pseudo-anonymization levels (bank account numbers, BIC numbers, full dates, specific company type).

Additional close up examples shown in Fig. 13.9b and c show additional transaction topologies with different features. Nodes can also be enriched with additional metrics calculated during anomaly detection and presented with this information and similar nodes. As before, each can be separately explored and escalated if needed.

Close up view can further be specialized to a hierarchical view, which exposes additional structural data dependency. Two close up of such views are presented in Fig. 13.8a and b. The hierarchical view is especially useful to quickly assess cluster structure and separate nodes with a high degree of connectivity connecting multiple clusters as seen in Fig. 13.9.

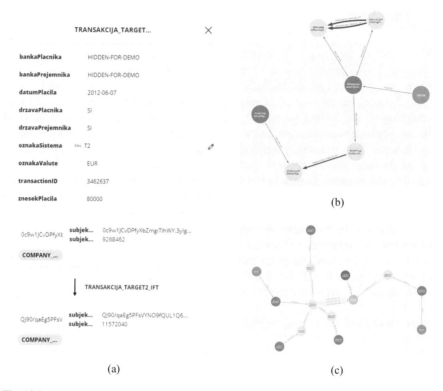

Fig. 13.9 Information and close up examples. (**a**) Additional transaction info. (**b**) Sending money between company accounts. (**c**) Start type typology with transactions to private person

Due to the highly specific nature of some companies and strange transaction patterns of others (utility companies, municipalities, large companies with foreign subsidiaries), a special whitelist is being implemented. Since all data is pseudo-anonymized, it is hard to almost impossible to reason about specific corner cases manually. Externally provided whitelist can be fully anonymized and used in two different scenarios. One can fully include whitelisted companies from calculations in anomaly detection algorithms or perform all the calculations on the whole dataset and only filter them out during presentation and manual exploration. The second option is additionally useful, as whitelisting can be elaborated and excluded but still viewed. An important example is whitelisting a utility company for having a large number of transactions from private accounts but still calculating and presenting risk scores for overseas transactions.

7 Summary

The screening tool is part of a comprehensive PAMLS platform, which enables automated data capture and automatic data quality controls. The screening tool accomplishes three specific goals, firstly it enables data acquisition and enrichment from different transactions' data sources. Secondly, the tool provides an automatic screening of pre-known transaction patterns that are potentially suspicious and finally allows a supervisory authority to investigate the enriched transaction space and discover new potentially suspicious patterns in order to detect the inherent risk of the financial institutions and apply the commensurate level of the supervision. The PAMLS platform is intended for big data processing and detection of potentially risky transaction patterns. The screening tool is one of the tools in PAMLS and enables the mapping of enriched transactional data sources into the space of graphs, where the use of parameterized queries easily maps risk typology into graph topologies and thus human-friendly investigation and pattern recognition. The tool is still in the development phase and will need to be further validated by domain experts. The need for similar solutions is evident both at the level of individual financial institutions and supervisory authorities.

Acknowledgments The research leading to the results presented in this chapter has received funding from the European Union's funded Project INFINITECH under grant agreement no. 856632.

References

1. Yearly report 2020 (2021), Bank of Slovenia.
2. Expert tells NLB Irangate commission good system was not implemented (2017). https://english.sta.si/2441615/expert-tells-nlb-irangate-commission-good-system-was-not-implemented
3. Sudjianto, A., Yuan, M., Kern, D., Nair, S., Zhang, A., & Cela-Díaz, F. (2010). *Technometrics, 52*(1), 5.
4. Alexandre, C., & Balsa, J. (2016). https://arxiv.org/abs/1510.00878.
5. *Next generation anti-money laundering and compliance powered by artificial intelligence and machine learning* (2017). https://brighterion.com/anti-money-laundering-compliance-requires-unsupervised-machine-learning/
6. Jullum, M., Løland, A., Huseby, R. B., Ånonsen, G., & Lorentzen, J. (2020). *Journal of Money Laundering Control* **23**(1). https://doi.org/10.1108/JMLC-07-2019-0055. https://www.emerald.com/insight/1368-5201.htm
7. Perozzi, B., Al-Rfou, R., & Skiena, S. (2014). *Proceedings of the ACM SIGKDD International Conference on Knowledge Discovery and Data Mining* (pp. 701–710). Association for Computing Machinery. https://doi.org/10.1145/2623330.2623732
8. Grover, A., & Leskovec, J. (2016). *Proceedings of the 22nd ACM SIGKDD International Conference on Knowledge Discovery and Data Mining* (pp. 855–864). http://arxiv.org/abs/1607.00653

9. Yue, X., Wang, Z., Huang, J., Parthasarathy, S., Moosavinasab, S., Huang, Y., Lin, S. M., Zhang, W., Zhang, P., & Sun, H. (2020). *Bioinformatics, 36*(4), 1241. https://doi.org/10. 1093/bioinformatics/btz718

10. Chen, J., Wu, Y., Fan, L., Lin, X., Zheng, H., Yu, S., & Xuan, Q. (2019). *arXiv:1904.12605.* http://arxiv.org/abs/1904.12605

11. Ribeiro, L. F. R., Savarese, P. H. P., & Figueiredo, D. R. (2017). *KDD '17: Proceedings of the 23rd ACM SIGKDD International Conference on Knowledge Discovery and Data Mining.* https://doi.org/10.1145/3097983.3098061. https://arxiv.org/abs/1704.03165

12. Kipf, T. N., & Welling, M. (2017). *arXiv:1609.02907* [cs, stat]. http://arxiv.org/abs/1609.02907

13. Hamilton, W. L., Ying, R., & Leskovec, J. (2017). *arXiv:1706.02216* [cs, stat]. http://arxiv.org/ abs/1706.02216

14. Kipf, T. N., & Welling, M. (2016). *Variational Graph Auto-Encoders*. Technical Report. https:// arxiv.org/abs/1611.07308

15. Zhao, L., Song, Y., Zhang, C., Liu, Y., Wang, P., Lin, T., Deng, M., & Li, H. (2020). *IEEE Transactions on Intelligent Transportation Systems, 21*(9), 3848. https://doi.org/10.1109/TITS. 2019.2935152. http://arxiv.org/abs/1811.05320

16. Xu, D., Ruan, C., Korpeoglu, E., Kumar, S., & Achan, K. (2020). *arXiv:2002.07962* [cs, stat]. http://arxiv.org/abs/2002.07962

17. D. Wagner, (2019). In M. Becke (Ed.), *SKILL 2019 – Studierendenkonferenz Informatik* (pp. 143–154). Gesellschaft für Informatik e.V.

18. Zhao, T., Zhang, X., & Wang, S. (2021). *Proceedings of the 14th ACM International Conference on Web Search and Data Mining.* https://doi.org/10.1145/3437963.3441720

19. Joaristi, M., Serra, E., & Spezzano, F. (2019). *Social Network Analysis and Mining, 9*(1), 1. Springer Vienna. https://doi.org/10.1007/s13278-019-0607-5

20. Winiecki, D., Kappelman, K., Hay, B., Joaristi, M., Serra, E., & Spezzano, F. (2020). *Proceedings of the 2020 IEEE/ACM International Conference on Advances in Social Networks Analysis and Mining, ASONAM 2020 pp. 752–759.* https://doi.org/10.1109/ASONAM49781. 2020.9381389

21. Weber, M., Domeniconi, G., Chen, J., Karl Weidele, D. I., Bellei, C., Robinson, T., & Leiserson, C. E. (2019). *Anti-Money Laundering in Bitcoin: Experimenting with Graph Convolutional Networks for Financial Forensics.* Technical Report.

22. Kontschieder, P., Fiterau, M., Criminisi, A., & Bulò, S. R. (2015). *2015 IEEE International Conference on Computer Vision (ICCV)* (pp. 1467–1475). ISSN: 2380-7504. https://doi.org/ 10.1109/ICCV.2015.172.

23. Kagan, D., Elovichi, Y., & Fire, M. (2018). *Social Network Analysis and Mining, 8*(1), 27. https://doi.org/10.1007/s13278-018-0503-4.

24. Cooley, J. W., & Tukey, J. W. (1965). *Mathematics of Computation, 19*, 297.

25. Zhao, J., Lui, H., McLean, D., & Zeng, H. (2007). *Applied Spectroscopy, 61*, 1225. https://doi. org/10.1366/000370207782597003

Chapter 14
Analyzing Large-Scale Blockchain Transaction Graphs for Fraudulent Activities

Baran Kılıç, Can Özturan, and Alper Şen

1 Introduction

Blockchain technologies are having a disruptive effect in finance as well as many other fields. Building on cryptographic technologies, blockchains can provide programmable transaction services directly to the masses in a trustless, speedy, and low-cost manner by removing middleman organizations that operate in a classical centralized fashion. Blockchains can keep ownership records of various assets such as cryptocurrencies and tokens that represent various assets such as company shares, stablecoins (which are tokenized forms of fiat currencies), media, and works of art. Blockchains make it easy to transfer and trade crypto-assets all over the world which creates big opportunities for value creation as well as introduce challenges since crypto-assets can move freely among different jurisdictions bypassing regulatory controls.

Figure 14.1 illustrates the global movements of crypto-assets on decentralized autonomous unregulated blockchains and entering regulated financial systems in different jurisdictions. These global crypto-asset movements can make anti-money laundering laws ineffective and crypto-assets obtained from various illegal activities such as ransomware, scam-related initial coin offerings [1], and stolen crypto-assets difficult to catch. In particular, exchange companies provide trading services that allow crypto-assets to be bought and sold with fiat currencies which can then be deposited into or withdrawn from bank accounts. Such mechanisms may allow fraudulently obtained funds to enter the regulated environments. Hence, there is a need to trace such movements of tainted crypto-assets on blockchains so that necessary actions can be taken in order to block them.

B. Kılıç · C. Özturan (✉) · A. Şen
Bogazici University, Istanbul, Turkey
e-mail: baran.kilic@boun.edu.tr; ozturaca@boun.edu.tr; alper.sen@boun.edu.tr

© The Author(s) 2022
J. Soldatos, D. Kyriazis (eds.), *Big Data and Artificial Intelligence in Digital Finance*,
https://doi.org/10.1007/978-3-030-94590-9_14

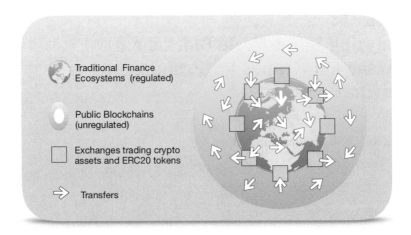

Fig. 14.1 Global movements of crypto-assets on unregulated blockchains and entering regulated financial systems in different jurisdictions

Table 14.1 Blockchain transaction graph analysis companies

	Company	Web site
(a)	Chainalysis	https://www.chainalysis.com
(b)	Elliptic	https://www.elliptic.co
(c)	Ciphertrace	https://ciphertrace.com
(d)	Scorechain	https://www.scorechain.com
(e)	Crystal	https://crystalblockchain.com
(f)	Blockchain Intelligence Group	https://blockchaingroup.io
	Bitrank	https://bitrank.com
(g)	Etherscan	https://info.etherscan.com/ethprotect
(h)	Dune Analytics	https://duneanalytics.com
(i)	Alethio	https://reports.aleth.io
(j)	The Graph	https://thegraph.com

When the Internet first appeared, there was a big need for web page search services. This has led to the development of search engine companies. There is a similar need in the case of blockchain networks and that is the need to get information on addresses, tokens, and transactions on the blockchain. This scenario brought the development of block explorer services being offered by companies like Etherscan which is currently the most popular explorer service for Ethereum. Block explorer services provide basic information on individual addresses, tokens, and transactions such as amounts, balances, and times of transactions. Blockchain transactions, however, form a directed graph, and it is possible to analyze this graph in order to trace fraudulent activities, get provenance information about tokens representing products, and in general use it for business intelligence. Table 14.1 shows a list of well-known companies that provide services for blockchain transaction analysis.

The first six companies, (a)–(f), in the table focus on providing blockchain intelligence services on detection and prevention of financial crime in crypto-assets. These services can be used by financial institutions and government agencies. The seventh company, (g) Etherscan, is the well-known block explorer service company which has recently added a service called ETHProtect that provides transaction tracing information about tainted funds down to their origin. The last three companies in the table, (h), (i), and (j), do not focus on crime detection but rather provide information that can be used for general purpose business analytics.

The importance of analyzing large-scale blockchain graphs is further evidenced by the fact that global money laundering and terrorist financing watchdog Financial Action Task Force (FATF) has published a guidance [2] for Virtual Asset Service Providers (VASPs). In this guidance, it is stated that *"VASPs be regulated for anti-money laundering and combating the financing of terrorism (AML/CFT) purposes, licensed or registered, and subject to effective systems for monitoring or supervision."* The recent Colonial Pipeline ransomware incident [3, 4] in which hackers invaded the company's systems and demanded nearly 5 million USD also led to a serious response from the U.S. authorities. This can also be taken as evidence for the importance of tracking fraudulent transactions on blockchains. Finally, we note that the valuation of newly emerging companies that offer blockchain graph analysis services is quite high. For example, Chainalysis raised 100 USD million venture capital at a 1 billion USD valuation [5]. The Graph [6], which as of July 13, 2021 is valued at 750 million USD (as reported in [7]), provides an indexing protocol for querying networks like Ethereum and lets anyone build and publish subgraphs. Several academic papers have also addressed the issue of analyzing the blockchain transaction graphs [8–14]. All of these facts point to the importance of building a sustainable system for analyzing large-scale blockchain graphs whose sizes are growing and will grow even more rapidly when new blockchain technologies with higher transaction throughput rates will be deployed in the near future.

In the rest of the chapter, we first cover scalability issues in blockchain transaction graph analytics in Sect. 2. In Sect. 3, we cover data structures of transaction graphs. In Sect. 4, we present Big Data Value Association (BDVA) reference model [15] of our blockchain graph analysis system and in Sect. 5 present parallel blacklisted address transaction tracing. In Sect. 6, we present tests of our graph analysis system. Finally, we close the chapter with a discussion and our conclusions in Sect. 7.

2 Scalability Issues in Blockchain Transaction Graph Analytics

Blockchain transaction graph analysis systems have to be scalable. That is, they should continue to perform analysis with slow growing time costs when the transaction numbers increase. As of June 16, 2021, Bitcoin blockchain has 649.5 million

Table 14.2 Transaction throughputs

Blockchain	Type	Transaction throughput (tps)
Bitcoin	Non-permissioned, PoW	7
Ethereum	Non-permissioned, PoW	14–30
Hyperledger	Permissioned	3.5K [16]
Ethereum2	Non-permissioned, PoS	first 2–3K, and then 100K [17]
Cardano	Non-permissioned, PoS	1K per stake pool with Hydra [18]
Avalanche	Non-permissioned, PoS	>4500 [19]

and Ethereum has 1.2 billion transactions. The current proof-of-work (PoW)-based Bitcoin and Ethereum blockchains have very low transaction throughputs as shown in Table 14.2. Newer proof-of-stake-based systems like Ethereum2, Avalanche, and Cardano will be able to achieve thousands of transactions per second (tps). Permissioned Hyperledger that is designed for enterprises can also achieve thousands of tps. Such high transaction throughputs mean that the transaction graphs are expected to grow to billions and billions in size in the forthcoming years. For example, if full 4000 tps is performed, it roughly equals 345 million transactions in a day and 2.4 billion transactions in a week.

In order to handle a massive number of transactions in a graph whose size is growing by billions in a month, the graph analytics software developed must work in parallel and should employ distributed data structures that keep the transaction graph partitioned among multiple processors. If efficient parallel algorithms are employed, such a system can scale simply by increasing the number of processors used. This is the adopted approach in this work, i.e., design a system that employs parallel algorithms and works on distributed memory systems. In Sect. 3, we present an overview of transaction graphs and then in Sect. 4 present the architecture of our system using the BDVA reference model [15, p. 37].

3   Distributed Data Structures of Transaction Graphs

The ledger maintained on the blockchain can be (i) account based or (ii) unspent transaction output (UTXO) based. In account-based blockchains, the coin balance of an address is kept in a single field. In UTXO based systems, there can be several records in the ledger owned by an address, each one keeping track of an amount of coins available for spending. When a payment is performed, a number of records whose total value equals the payment's amount are taken as input to a transaction and new unspent amount records are generated representing the new amounts available for spending. The payment recipients are the owners of these records. The transaction graphs of account- and UTXO-based systems are shown in Fig. 14.2. Ethereum blockchain is account based, whereas Bitcoin is UTXO based. Account-based transactions can be represented as a directed graph where

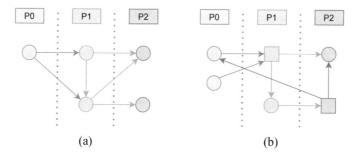

Fig. 14.2 Distributed data structures of (**a**) account-based and (**b**) UTXO-based blockchain graphs

each vertex represents an address and each directed edge represents a transaction (i.e., asset transfer from one address to another address). A UTXO transaction graph can be represented as an AND/OR graph $G(V_a, V_t, E)$ with V_a as the set of addresses, V_t as the set of transactions, and E as the set of edges, defined as $E = \{< a, t >: a \in V_a \text{ and } t \in V_t\} \bigcup \{< t, a >: t \in V_t \text{ and } a \in V_a\}$. Here, each transaction, t, can be thought of as an AND vertex, and an address can be thought of as an OR vertex. In Fig. 14.2b, circle nodes represent the addresses and the square nodes represent the transactions. The amount of the payment can be stored on each edge.

Since our graph system keeps transaction graph in a partitioned fashion, this means the data structures are distributed over the processors of a High Performance Computing (HPC) cluster. Figure 14.2 also illustrates the partitioning of a simple example transaction graph by coloring with blue, green, and purple colors the addresses and the transactions that are stored in a distributed fashion on each of three processors, P0, P1, and P2.

4 BDVA Reference Model of the Blockchain Graph Analysis System

Big Data Value Association has come up with a reference model [15, p. 37] to position Big Data technologies on the overall IT stack. Figure 14.3 shows where the components of the graph analysis system are located on the BDVA reference system.

At the lowest level of the stack, data about crypto-asset movements (cryptocurrencies Bitcoin and Ethereum and ERC20 token transfers) are obtained from blockchains. Note that in this chapter, we only report results for the Bitcoin and Ethereum datasets. We do not currently have real-life token transfer datasets for Hyperledger. We use a cloud-based HPC cluster computer that is managed by StarCluster [20] and programmed using MPI [21]. The transactions are parsed

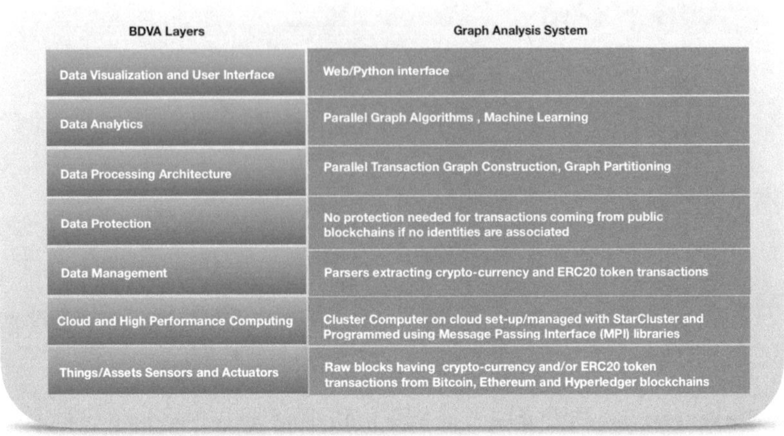

Fig. 14.3 Components of the blockchain graph analysis system located on the BDVA reference model

and cryptocurrency and token transactions are extracted and put in files. Since our datasets are public and no identities are attached to the addresses, the data protection layer is empty. The data processing architecture layer constructs in-memory distributed data structures for the transaction graph. This is done in a parallel fashion. Graph partitioning software like Metis [22] can be used to produce load-balanced partitions with small partition boundaries so that we have low communication costs during parallel running of the various graph algorithms. The Data Analytics layer contains parallel graph algorithms that are covered in this chapter and presented in the results, Sect. 6. In the future, this layer will also contain machine learning algorithms which are currently not implemented. Graph algorithms part, however, enables us to compute various features (node degrees, pagerank, shortest paths from blacklisted addresses, etc.), which in the future can provide data to machine algorithms. The topmost layer provides an interface, currently, through a message queue to Python programs which can implement visualization of subgraphs that contain tracing information to blacklisted addresses. In the next section, we present parallel fraudulent transaction tracing algorithm that is used.

5 Parallel Fraudulent Transaction Activity Tracing

Given a blockchain address of a customer and a set of blacklisted addresses, we want to identify a subgraph that traces transaction activities between the blacklisted addresses and the queried customer address. Figure 14.4 shows two example trace subgraphs: (a) for the DragonEx Hacker [23] on the Ethereum blockchain and (b)

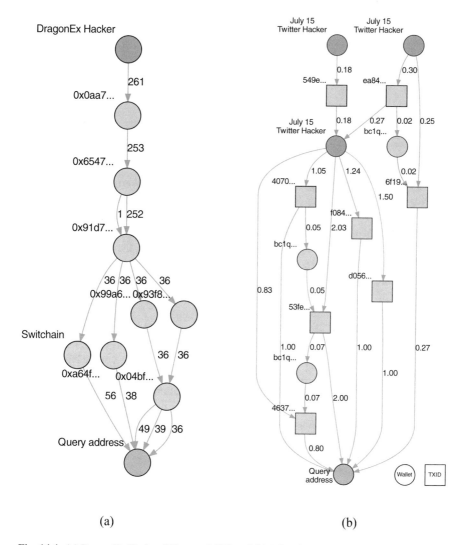

Fig. 14.4 (**a**) DragonEx Hacker (Ethereum) [23] and (**b**) July 15 Twitter Hack (Bitcoin) [24] trace subgraphs returned by the parallel tracing algorithm

for the July 15 Twitter Hack [24] on the Bitcoin blockchain. Such trace subgraphs are returned by the parallel tracing algorithm that is presented in this section.

The parallel blacklisted address transaction tracing algorithm is given in Algorithms 1 and 2. The algorithm finds the subgraph between two sets of nodes. We use it to trace the set of queried addresses Q back to the set of blacklisted addresses B. The algorithm first finds the nodes reachable from B denoted with RB by depth-first traversal (line 2). Then, it finds the nodes that can reach Q denoted with RQ by traversing the edges in the reverse direction (line 3). The intersection of RB and

Algorithm 1 Parallel blacklisted address transaction tracing: part 1

1: **procedure** TRACE($p, P, G^p(V^p, E^p), B, Q, T_s, T_e$)
2:     $RB^p \leftarrow$ DFS($p, P, G^p(V^p, E^p), B, T_s, T_e, false$)   ▷ nodes reachable from B
3:     $RQ^p \leftarrow$ DFS($p, P, G^p(V^p, E^p), Q, T_s, T_e, true$)   ▷ nodes that can reach Q
4:     $V'^p \leftarrow RB^p \cap RQ^p$ ▷ nodes of subgraph
5:     $V' \leftarrow \bigcup_{i=0}^{P-1} V'^p$ ▷ MPI_Allgatherv
6:     $E'^p \leftarrow \{e \mid e \in E^p \wedge e = (s, t, b) \wedge s, t \in V' \wedge b \in [T_s .. T_e]\}$ ▷ edges of subgraph
7:     **return** $G'^p(V'^p, E'^p)$
8: **end procedure**

RQ gives us the set of nodes in the subgraph V'^p (line 4). We use distributed data structures, and the superscript p denotes the part of the data on processor p. Each local set of nodes of the subgraph is exchanged between the processors, and we get the global set of nodes V' in all processors (line 5). Each process finds the local edges whose source and target are in the V' and is between the given block range from T_s to T_e (which also represents the time since all of the transactions in a block have the same timestamp) (line 6). T_s is the block range start and denotes the start time range of trace. T_e is the block range end and denotes the end time range of trace. The algorithm returns the subgraph (line 7).

The depth-first search function takes current processor ID p, total processor count P, the graph G^p, set of starting addresses F, block start T_s, block end T_s, and traversal direction Rev. It returns the set of nodes R^p that are reachable from the set of starting nodes F. C^p is the set of nodes that needs to be visited and is located on remote processors. M^p is the set of remote nodes that are already visited. S^p is the stack of nodes to be visited. At the start, S^p contains the local nodes that are in F (line 13). The algorithm first traverses the local reachable nodes (lines 16–34) and then sends the remote nodes to be visited C^p to corresponding processors (lines 36–48). This loop continues until there is no node left in the stack of any processors. TCN is the total number of nodes in C^p in all processors. It is initialized with 1 to run the first loop (line 14). The nodes in the stack S^p are traversed until the stack is empty. At each step, a node cur is taken from the stack, marked as visited (lines 17–18). The nodes at the tail (head) of the outcoming (incoming) edges of the node cur are the traversed (if Rev is true) (lines 19–33). If the edge is between the given block time range, it needs to be visited (line 25). If it is a local node and is not already visited, it is added to the stack (line 26). If it is a remote node and not already sent to other processors to be visited, it is added to the set of remote nodes to be visited C^p (line 29). The total number of nodes in C^p in all processors is calculated (line 35). If there are any nodes that need to be sent to remote processors and visited in any of the processors, the nodes C^p are exchanged between processors. The C^p is added to M^p not to send them again (line 45). The C^p is cleared. The received nodes are added to the stack S^p.

Algorithm 2 Parallel blacklisted address transaction tracing: part 2

9: **procedure** DFS(p, P, $G^p(V^p, E^p)$, F, T_s, T_e, Rev)
10:   $R^p \leftarrow \varnothing$ ▷ nodes reachable from F
11:   $C^p \leftarrow \varnothing$ ▷ remote nodes to be visited
12:   $M^p \leftarrow \varnothing$ ▷ remote nodes that are already visited
13:   $S^p \leftarrow F \cap V^p$ ▷ stack of nodes to visit
14:   $TCN \leftarrow 1$ ▷ total number of remote nodes to be visited in all processors
15:   **while** $TCN > 0$ **do**
16:    **while** $S^p \neq \varnothing$ **do**   ▷ while there are nodes to visit
17:     $S^p \leftarrow S^p \setminus \{cur\}$ $\exists cur \in S^p$ ▷ pop a node from stack
18:     $R^p \leftarrow R^p \cup \{cur\}$ ▷ mark the node as visited
19:     **for all** $e \in E^p$ **do**
20:      **if** Rev **then**
21:       $(t, s, b) \leftarrow e$
22:      **else**
23:       $(s, t, b) \leftarrow e$
24:      **end if**
25:      **if** $s = cur \wedge b \in [T_s \mathrel{.\,.} T_e]$ **then**
26:       **if** $t \in V^p \wedge t \notin R^p$ **then**   ▷ if node is local and not already visited
27:        $R^p \leftarrow R^p \cup \{t\}$ ▷ mark node as reached
28:        $S^p \leftarrow S^p \cup \{t\}$ ▷ push node to stack
29:       **else if** $t \notin V^p \wedge t \notin M^p$ **then**   ▷ if node is remote and not already visited
30:        $C^p \leftarrow C^p \cup \{t\}$ ▷ save the node send it later
31:       **end if**
32:      **end if**
33:     **end for**
34:    **end while**
35:    $TCN \leftarrow \sum_{i=0}^{P-1} |C^p|$ ▷ MPI_Allreduce
36:    **if** $TCN > 0$ **then**
37:     **for all** $i \in ([0 \mathrel{.\,.} P - 1] \setminus \{p\})$ **do**
38:      Send $\{t \mid t \in C^p \wedge t \in V^i\}$ to processor i
39:     **end for**
40:     $CR^p \leftarrow \varnothing$
41:     **for all** $i \in ([0 \mathrel{.\,.} P - 1] \setminus \{p\})$ **do**
42:      Receive CR_i from processor i
43:      $CR^p \leftarrow CR^p \cup CR_i$
44:     **end for**
45:     $M^p \leftarrow M^p \cup C^p$ ▷ save sent nodes as visited not to send them again
46:     $C^p \leftarrow \varnothing$ ▷ clear the send nodes
47:     $S^p \leftarrow S^p \cup (CR^p \setminus R^p)$ ▷ push the not visited received nodes to the stack
48:    **end if**
49:   **end while**
50:   **return** R^p
51: **end procedure**

---

6 Tests and Results

In order to test our blockchain graph analysis system, we have set up an Amazon cloud-based cluster using the StarCluster cluster management tools. Our cluster had 16 nodes (machine instances). Ethereum tests used c5.4xlarge machine instances,

Table 14.3 Bitcoin and Ethereum blockchain data statistics used in tests

	Statistic	Bitcoin	Ethereum
(a)	Blocks	0–674,999	0–10,199,999
(b)	Time coverage of blocks	3.1.2009–17.3.2021	30.7.2015–4.6.2020
(c)	No. of transactions	625,570,924	766,899,042
(d)	No. of addresses	800,017,678	78,945,214
(e)	No. of 40 major ERC20 token transfer transactions	N/A	43,371,941
(f)	Number of blacklisted addresses (blacklist size)	21,028	5830

Table 14.4 Description of tests and best timings obtained for Bitcoin (btc) and Ethereum (eth) blockchain data

		Best Timing, secs (nodes,processes)	
Tests	Description	btc	eth
T1	Transaction graph construction	1910 (16,64)	219 (16,128)
T2	Graph partitioning using ParMetis [22]	out of memory	5718 (16,16)
T3	Page ranking on transaction graph	445 (16,256)	150 (12,96)
T4	Degree distributions of transaction graph nodes	60 (16,16)	7 (16,32)
T5	No. of transfer transactions of 40 major ERC20 tokens	N/A	19 (16,128)
T6	Connected component count	1984 (16,256)	177 (12,144)
T7	Shortest path-based blacklisted node trace forest	240 (16,192)	32 (16,128)
T8	Extracting node features	4 (16,256)	13 (16,128)
T9	Example of trace subgraph of a blacklisted address	0.7 (16,128)	0.5 (16,128)

each of which had 16 virtual CPUs and 32 GiB memory. Bitcoin tests used r5b.4xlarge machine instances, each of which had 16 virtual CPUs and 128 GiB memory. As datasets, we have used roughly nine years of Bitcoin and five years of Ethereum blockchain data. The full details of transaction data used are given in Table 14.3. Note that our Ethereum dataset also contains major ERC20 token transfer transactions. The Ethereum dataset is publicly available and can be downloaded from the Zenodo site [25]. We have also collected from the Internet various blacklisted blockchain addresses that have been involved in ransomware, scam-related initial coin offerings, and hacked funds. For example, the source [26, 27] provides a list of Bitcoin addresses involved ransomware and sextortion scams.

Table 14.4 shows various computations carried on the Bitcoin and Ethereum datasets. Detailed timings of Ethereum tests were previously reported in an earlier publication [13] on a slightly smaller dataset. Computations on the Bitcoin transaction graph are first reported in this book chapter. The tests were run with node sizes 4, 8, 12, and 16 and with 1, 2, 4, 8, 12, and 16 MPI processes per node. In addition, Table 14.4 shows the best timing obtained and the corresponding number of nodes and the total number of MPI processes for the run.

It is important to note here that if the graph analysis system that we developed was not distributed memory model based, we would have problems fitting the whole graph on a node. In fact, even the distributed memory system can also run into out-of-memory problems when using a small number of nodes (for example, Bitcoin graph did not fit on 4 nodes). The good thing about distributed memory system is that one can simply increase the number of nodes, and then since the graph is partitioned, the partitions will get smaller and fit on the nodes of the cluster. This is how we have been able to run our analyses on large transaction graphs. This is also how we will continue to be able to run our analyses in the future when blockchain graph sizes grow drastically due to changes in the throughputs of the blockchains as given in Table 14.3. Distributed memory allows us to achieve scalability as far as graph storage is concerned. There is, however, another scalability issue that we should be concerned about and that is the scalability of our computational processes. When the number of processors is increased, communication among processors increases. Since communication is expensive, it may lead to execution time increase. The overall execution time should decrease or if growing, it should grow slowly. In general, once the transactions are loaded from disk and graph is constructed (i.e., Test T1), the computational tests, (T2...T9), execution times decrease, levelling off after increasing the number of processors beyond some point. This is as expected because even if the work per processor decreases, the communication cost increases.

Since graph construction (Fig. 14.5) involves reading transactions from disk and then uses a ring communication (see the algorithm in [13]) to implement an all-to-all communication in order to construct the graph, we can expect its timing to level off and even increase after the number of nodes is increased. Figure 14.5 shows this happening. In the case of Ethereum, the timing levelled off. For Bitcoin, since the number of addresses is much higher, we see the times decrease and then increase due to increased communication cost.

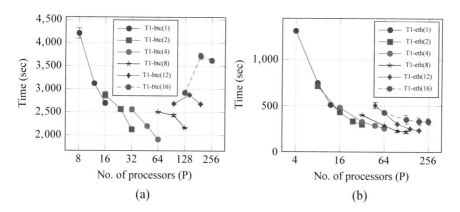

Fig. 14.5 Test T1: Graph construction times for Bitcoin (**a**) and Ethereum (**b**) on the HPC cluster. The label T1-btc(i) and T1-etc(i) means i virtual processors (MPI processes) per node of the cluster

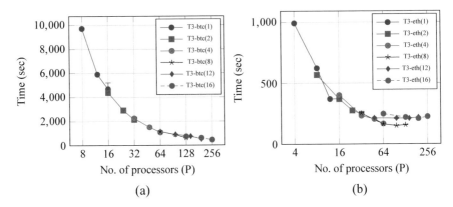

Fig. 14.6 Test T3: Page ranking times for Bitcoin (**a**) and Ethereum (**b**) on the HPC cluster. The label T1-btc(i) and T1-etc(i) means i virtual processors (MPI processes) per node of the cluster

Figure 14.6 shows the execution times for the pagerank computation. Note that after graph construction, the graph resides in memory and the pagerank computations are run on the distributed data structures. We see faster processing times when the number of processors is increasing.

In Table 14.5, we show analysis results obtained from pagerank computation. The first most important addresses are shown in this table. The most important addresses belong to exchange companies and popular gambling and betting sites. Note that pagerank computation helps us to find important addresses such as those of exchange companies or other popular companies. In particular, exchange companies are required to do identity checks on customers. Customers who deposit or withdraw from such sites are very likely to have their identities verified. Hence, even though we do not know the actual identities of people owning addresses, we can infer that their identities have been verified. The pagerank information can be used as a feature in machine learning algorithms for scoring purposes in the future. This is one of the motivations for computing pagerank in our graph analysis system.

In Table 14.6, we have also given statistics on the total number of distinct addresses at the tail/head of incoming/outcoming transactions to/from k most important addresses for the datasets. Given the 800 and 79 million addresses, respectively, in the Bitcoin and Ethereum datasets we have used, what fraction of these addresses directly transacted with the k most important addresses? Since Bitcoin is UTXO based, it is expected that address reuse is seldom, and hence we have lower percentages. On the other hand, since Ethereum is account based, address reuse is more frequent, and the fractions are higher. The fact that about half (48%) of the addresses directly transacted with the most important 1000 addresses, efforts can then be concentrated on this small set of important addresses to see if their identity verification procedures are strong. If they are strong, then lower risk scores can be assigned to the addresses who transacted with them.

Table 14.5 Top 10 ranked addresses on Bitcoin and Ethereum transaction graph

Rank	Bitcoin	Ethereum
1	1HckjUpRGcrrRAtFaaCAUaGjsPx 9oYmLaZ Huobi	0x3f5ce5fbfe3e9af3971dd833d26ba9b 5c936f0be Binance
2	1NDyJtNTjmwk5xPNhjgAMu 4HDHigtobu1s Binance	0xdac17f958d2ee523a2206206994597 c13d831ec7 Bitfinex
3	1NxaBCFQwejSZbQfWcYNwgqML 5wWoE3rK4 LuckyB.it	0x70faa28a6b8d6829a4b1e649d26ec 9a2a39ba413 ShapeShift
4	1dice8EMZmqKvrGE4Qc9bUFf9 PX3xaYDp Satoshi Dice	0xf4a2eff88a408ff4c4550148151c33c 93442619e Plus Token
5	1FoWyxwPXuj4C6abqwhjDWdz6 D4PZgYRjA Binance	0xac08809df1048b82959d6251fbc 9538920bed1fa MSD Token
6	1G47mSr3oANXMafVrR8UC4pzV7 FEAzo3r9	0xfa52274dd61e1643d2205169732f 29114bc240b3 Kraken
7	1dice97ECuByXAvqXpaYzSa QuPVvrtmz6 Satoshi Dice	0xbcf935d206ca32929e1b887a07ed240f 0d8ccd22 Million Money
8	37Tm3Qz8Zw2VJrheUUhArDAoq58S6 YrS3g OKEx	0x689c56aef474df92d44a1b70850 f808488f9769c KuCoin
9	17kb7c9ndg7ioSuzMWEHWECdEVUeg NkcGc	0x86fa049857e0209aa7d9e616f7eb3b 3b78ecfdb0 EOS Token
10	3CD1QW6fjgTwKq3Pj97nty28 WZAVkziNom The Shadow Brokers	0x0d8775f648430679a709e98d 2b0cb6250d2887ef BAT Token

Table 14.6 Total number of distinct addresses at the tail/head of incoming/outcoming transactions to/from k most important addresses for the datasets given in Table 14.3

Ranking range	No. of addresses		Percentage of addresses	
	Bitcoin	Ethereum	Bitcoin	Ethereum
Incoming transactions				
	442,218	2,153,020	0.05	2.73
1–10	8,069,235	12,308,446	1.01	15.59
1–100	18,406,620	24,623,414	2.30	31.19
1–1000	56,795,882	38,110,110	7.10	48.27
1–10000	105,329,533	47,182,270	13.17	59.76
Outcoming transactions				
1	1,012,509	1,524,220	0.13	1.93
1–10	12,315,379	1,689,862	1.54	2.14
1–100	21,608,153	5,234,996	2.70	6.63
1–1000	43,235,102	15,698,561	5.40	19.88
1–10000	63,023,109	24,008,947	7.88	30.41

7   Discussion and Conclusions

In this chapter, we have covered the blockchain transaction graph analysis system we are currently developing, its architecture, and the results obtained from running various parallel graph algorithms on the massive Bitcoin and Ethereum blockchain datasets. Some graph algorithms such as pagerank and forest of tracing trees operate on the whole graph. This approach then introduces the problem of dynamically growing blockchain datasets not fitting the memory of a single node. Our parallel and distributed approach solves this (i) single node memory bottleneck problem as well as (ii) speeds up the computations. The work presented and the results obtained in this chapter actually demonstrate that we achieve both, (i) and (ii) on the implemented graph algorithms.

The computing infrastructure that is needed to run such analyses is available readily to everyone. A low-cost cluster can be started on a cloud provider like Amazon. The software libraries that we have used, i.e., MPI, for development are open source and free to use. Therefore, our system can be run even by small businesses with very low investments. In fact, a cluster can also be set up easily using local machines on a local area network in the premises of a company and MPI can run on this cluster.

Acknowledgments The research leading to the results presented in this chapter has received funding from the European Union's funded Project INFINITECH under grant agreement no: 856632.

References

1. Hornuf, L., Kück, T., & Schwienbacher, A. (2021). Initial coin offerings, information disclosure, and fraud. *Small Business Economics*, 1–19. https://doi.org/10.1007/s11187-021-00471-y
2. FATF. (2019). *Guidance for a risk-based approach, virtual assets and virtual asset service providers*. Paris: FATF.
3. Bing, C. (2021, June). Exclusive: U.S. to give ransomware hacks similar priority as terrorism.
4. Office of Public Affairs. (2021, June). Department of justice seizes 2.3 million in cryptocurrency paid to the ransomware extortionists darkside.
5. del Castillo, M. (2020, November). Bitcoin investigation giant to raise 100 million at 1 billion valuation.
6. The Graph. (2021). https://thegraph.com/
7. Coinmarketcap. (2021). https://coinmarketcap.com/
8. Ron, D., & Shamir, A. (2013). Quantitative analysis of the full bitcoin transaction graph. In *International Conference on Financial Cryptography and Data Security* (pp. 6–24). Springer.
9. Victor, F., & Lüders, B. K. (2019). Measuring Ethereum-based ERC20 token networks. In *Financial Cryptography*.
10. Somin, S., Gordon, G., Pentland, A., Shmueli, E., & Altshuler, Y. (2020). Erc20 transactions over Ethereum blockchain: Network analysis and predictions. Preprint, arXiv:2004.08201.
11. Chen, T., Zhang, Y., Li, Z., Luo, X., Wang, T., Cao, R., Xiao, X., & Zhang, X. (2019). TokenScope: Automatically detecting inconsistent behaviors of cryptocurrency tokens in Ethereum. In *Proceedings of the 2019 ACM SIGSAC Conference on Computer and Communications Security*.
12. Nerurkar, P., Patel, D., Busnel, Y., Ludinard, R., Kumari, S., & Khan, M. K. (2021). Dissecting bitcoin blockchain: Empirical analysis of bitcoin network (2009–2020). *Journal of Network and Computer Applications, 177*, 102940.

13. Kılıç, B., Özturan, C., & Sen, A. (2020). A cluster based system for analyzing Ethereum blockchain transaction data. In *2020 Second International Conference on Blockchain Computing and Applications (BCCA)* (pp. 59–65).
14. Guo, D., Dong, J., & Wang, K. (2019). Graph structure and statistical properties of Ethereum transaction relationships. *Information Sciences, 492*, 58–71.
15. BDVA. (2017, October). European big data value strategic research and innovation agenda.
16. Androulaki, E., et al. (2018). Hyperledger fabric: A distributed operating system for permissioned blockchains.
17. Buterin, V. (2020, June). Twitter post. https://twitter.com/VitalikButerin/status/1277961594958471168
18. Chakravarty, M. M. T., Coretti, S., Fitzi, M., Gazi, P., Kant, P., Kiayias, A., & Russell, A. (2020, May). Hydra: Fast isomorphic state channel.
19. Cusce, C. (2020, April). The avalanche platform — a tech primer.
20. Starcluster. (2013). http://star.mit.edu/cluster/
21. Message passing interface (MPI) documents (2015). https://www.mpi-forum.org/docs/
22. Karypis, G., & Kumar, V. (2009). MeTis: Unstructured Graph Partitioning and Sparse Matrix Ordering System, Version 4.0. http://www.cs.umn.edu/~metis
23. Khatri, Y. (2019, March). Singapore-based crypto exchange DragonEX has been hacked.
24. Wikipedia Contributors. (2021). 2020 twitter account hijacking — Wikipedia, the free encyclopedia. Accessed June 18, 2021.
25. Özturan, C., Şen, A., & Kılıç, B. (2021, April). Transaction Graph Dataset for the Ethereum Blockchain.
26. Paquet-Clouston, M., Haslhofer, B., & Dupont, B. (2018, April). Ransomware payments in the bitcoin ecosystem.
27. Masarah, P., Matteo, R., Bernhard, H., & Tomas, C. (2019, August). Spams meet cryptocurrencies: Sextortion in the bitcoin ecosystem.

Chapter 15
Cybersecurity and Fraud Detection in Financial Transactions

Massimiliano Aschi, Susanna Bonura, Nicola Masi, Domenico Messina, and Davide Profeta

1 Overview of Existing Financial Fraud Detection Systems and Their Limitations

Financial institutions deliver several kinds of critical services usually managing high-volume transactions, just to mention, among others, card payment transactions, online banking ones, and transactions enabled by PSD2 and generated by means of open banking APIs [1]. Each of these services is afflicted by specific frauds aiming at making illegal profits by unauthorized access to someone's funds. The opportunity for making conspicuous, easy earnings, often staying anonymous – therefore in total impunity – incentivizes the continuous development of novel and creative fraud schemas, verticalized on specific services, forcing fraud analysts in playing a never-ending "cat-and-mouse" game, in a labored attempt of detecting new kind of frauds before valuable assets get compromised or before the damage becomes significant.

The situation has been made with no doubt more complicated by the considerable scale of the problem and by its constant trend growth rate, driven by the exponential increment of payments for e-commerce transactions recorded in recent years [2] and by the European support policy to the digital transformation and to the "Digital Single Market" strategy [3]. Over the years, the academic world and the industry players of the cybersecurity market have seen in this context an opportunity: the former to create innovative solutions to the problem, while the latter to make

M. Aschi
POSTE ITALIANE, Rome, Italy
e-mail: massimiliano.aschi@posteitaliane.it

S. Bonura (✉) · N. Masi · D. Messina · D. Profeta
ENGINEERING, Rome, Italy
e-mail: susanna.bonura@eng.it; nicola.masi@eng.it; domenico.messina@eng.it; davide.profeta@eng.it

© The Author(s) 2022
J. Soldatos, D. Kyriazis (eds.), *Big Data and Artificial Intelligence in Digital Finance*,
https://doi.org/10.1007/978-3-030-94590-9_15

business by proposing technological solutions for supporting the analysts and for automating the detection processes. Rule-based expert systems (RBESSs) are a very simple implementation of artificial intelligence (AI), which leverages rules for encoding and therefore representing knowledge from a specific area in the form of an automated system [4].

RBESSs try to mimic the reasoning process of a human being expert of a subject matter when trying to solve a knowledge-intensive problem. This kind of systems consists of a set of decisional rules (*if-then-else*) which are interpreted and applied to specific features extracted from datasets. A wide set of problems can be managed by applying these very simple models, and lots of commercial products/services have been built leveraging on RBESSs for detecting frauds in banking, financial, and e-commerce suspicious transactions [5], by calculating a risk score based on users' behaviors such as repeated log-in attempts or "too-quick-for-being-human" operations, unusual foreign or domestic transactions, unusual operations considering user's transactions history (abnormal amounts of money managed), and unlikely execution day/time (e.g., weekend/3 am) [6].

Based on the risk score, the rules deliver a final decision on each analyzed transaction, therefore *blocking* it, *accepting* it, or putting it *on hold* for analyst's revision. The rules can be easily updated over time, or new rules can be inserted following specific needs to address new threats. Nevertheless, as the number of fraud detection rules increases and the more you combine rules for the detection of complex fraud cases, the more the rules could conflict with each other based on semantic inconsistencies. When this happens, the rule-based system performs inefficiently [7], for example, by automatically accepting in-fraud transactions (the "infamous" *false negatives*) or by blocking innocuous ones (*false positives*). Furthermore, the decisional process lead by anti-fraud analysts could be severely affected by an increasingly high and rapidly changing set of rules which could compromise their inference capabilities during investigations.

The standard practice in the cybersecurity industry has long been blocking potentially fraudulent traffic by adopting a set of rigid rules. A fraud detection rule-based engine aims at identifying just high-profile fraudulent patterns. This method is rather effective in mitigating fraud risks and in giving clients a sense of protection by discovering well-known fraud patterns. Nevertheless, rule-based fraud detection solutions demonstrated in the field that they can't keep pace with the increasingly sophisticated techniques adopted by fraudsters to compromise valuable assets: malicious actors easily reverse-engineer a preset of fixed thresholds, and fixed rules would not be of help in detecting emerging threats and would not adapt to previously unknown fraud schemes. These drawbacks should be taken in adequate consideration, especially if we consider the operational costs (OPEX) determined by erroneous evaluations of fraud detection engines (false positives and false negatives).

Example Rule 1: Block Transaction by IP Location

Input: u, user; t_u, user's transaction under analysis; $t_u[]$, user's transactions' collection
Output: t_s, transaction status is blocked, await or pass
1: **if** t_u.IPAddress.GeoLocate.Country **is not in** $t_u[]$.IPAddress.GeoLocate.Country **and** t_u.amount > €200 **then**
2:          t_s = await   // transaction is temporarily stopped by rule
3: **else** t_s = pass   // transaction is considered not dangerous

---

Example Rule 2: Block Frequent Low-Amount Transaction

Input: u, user; t_u, user's transaction under analysis; $t_u[]$, user's transactions' collection
Output: t_s, transaction status is blocked, await or pass
1: **if** t_u.amount < €5 **then**
2:    **for** i = 10 to 1 **do**
3:       **if** $t_u[i]$.amount < €5 **then** // check for existing recent similar transactions
4:          t_s = await   // transaction is temporarily stopped by rule
5:          **exitfor**
5:       **else** t_s = pass   // transaction is considered not dangerous
6: **endfor**

Another strong limitation of the rule-based detection engines is the potential lack of data to analyze: the more innovative is the fraud scheme, the fewer data you will find in analyzed transactions. This lack of data could simply mean that necessary information is not collected and stored or that information is present, but the necessary details are missing or that information cannot be correlated to other information. As fraud rates intensify, so their complexity: complementary methodologies for fraud detection need to be implemented.

It is well-known that fraud phenomenon is endemic and could never be totally eradicated, but only mitigated more or less effectively. In this framework, wide areas for improvement still currently exist: one need only partially addressed is banally to improve the fraud detection rate; decrease the number of false positives – analyzing them requires considerable resources; and at the same time reduce the number of false negatives which impacts negatively organizations first by the costs of the undetected frauds but also generate a misleading sense of security in users. If the system does not detect frauds, this doesn't mean there aren't. By improving the efficiency and effectiveness of a fraud detection system, it will allow us to implement mechanisms for automated or semi-automated decision, ensuring high-impact business results and a significant reduction of CAPEX and OPEX.

Over the years, several approaches have been developed and tested in order to improve the effectiveness of the rule-based detection methodologies, but current trends suggest that promising results could be obtained by adopting analytics at scale based on an agile data foundation and solid machine learning (ML) technologies.

2   On Artificial Intelligence-Based Fraud Detection

Artificial intelligence can address many of these limitations and more effectively identify risky transactions. Machine learning methods show better performance along with the growth of the dataset to which they are adapted, which means that the more fraudulent operations samples are trained, the better they recognize fraud. This principle does not apply to rule-based systems as they never evolve by learning. Furthermore, a data science team should be aware of the risks associated with rapid model scaling; if the model did not detect the fraud and marked it incorrectly, this will lead to false negatives in the future. By the machine learning approach, machines can take on the routine tasks and repetitive work of manual fraud analysis, while specialists can spend time making higher-level decisions.

From a business perspective, a more efficient fraud detection system based on machine learning would reduce costs through efficiencies generated by higher automation, reduced error rates, and better resource usage. In addition, the finance/insurance stakeholders could address new types of frauds, minimizing disruptions for legitimate customers and therefore increasing client trust and security.

Over the past decade, intense research on machine learning for credit card fraud detection has resulted in the development of *supervised* and *unsupervised* techniques [8, 9].

Supervised techniques are based on the set of past operations for which the label (also called *outcome* or *class*) of the transaction is known. In credit card fraud detection problems, the label is "trusted" (the transaction was made by the cardholder) or "fraudulent" (the transaction was made by a scammer). The label is usually assigned a posteriori, either following a customer complain or after thorough investigations on suspect transactions. Supervised techniques make use of past transactions labeled by training a fraud prediction model, which, for each new analyzed transaction, returns the likelihood that it is a fraudulent one.

Unsupervised outlier detection techniques do not make use of the transaction label and aim to characterize the data distribution of transactions. These techniques work on the assumption that outliers from the transaction distribution are fraudulent; they can therefore be used to detect types of fraud that have never seen before because their reasoning is not based on transactions observed in the past and labeled as fraudulent. It is worth noting that their use also extends to clustering and compression algorithms [10]. Clustering allows for the identification of separate data distributions for which different predictive models should be used, while compression reduces the dimensionality of the learning problem. Several works have adopted one or other techniques to address some specific issues of fraud detection, such as class imbalance and concept drift [11–13].

Both these approaches are needed to work together so that supervised techniques can learn from past fraudulent behavior, while unsupervised techniques aim at detecting new types of fraud. By choosing an optimal combination of supervised and unsupervised artificial intelligence techniques, it could be possible to detect previously unseen forms of suspicious behavior, quickly recognizing the more subtle patterns of fraud that have previously been observed across huge amounts of accounts. In literature, there are already paper works showing the combination of unsupervised and supervised learning as a solution for fraud detection in financial transactions [14–16].

But this is not enough: it is well known that the more effective machine learning techniques, the greater the required volume and variety of data. Fraud models that are trained using big data are more accurate than models that rely on a relatively thin dataset; to be able to process big data volumes so to properly train AI models while transaction data streams are detected to find fraud attempts, open-source solutions based on cloud computing and big data frameworks and technologies may come to aid.

In the next section, we present an innovative microservice-based system that leverages ML techniques and is designed and developed on top of the most cutting-edge open-source big data technologies and frameworks.

It can improve significantly the detection rate of fraud attempts while they are occurring by the real-time ML detection of the financial transactions and continuous batch ML retraining. It integrates two layers, stream and batch, to handle periodic ML retraining while real-time ML fraud detection occurs.

3 A Novel AI-Based Fraud Detection System

To overcome the aforementioned limits, this chapter describes the design, development, and deployment of a batch and stream integrated system to handle automated retraining while real-time ML prediction occurs and combine supervised and unsupervised AI models for real-time big data processing and fraud detection.

More specifically, it addresses the challenge of financial crime and fraud detection in the scope of the European Union-funded INFINITECH project, under Grant Agreement No. 856632.

To this end, Alida solution is adopted (https://home.alidalab.it/). Alida is an innovative data science and machine learning (abbreviated to DSML) platform for the rapid big data analytics (BDA) application prototyping and deployment.

DSML solutions are entering a phase of greater industrialization. Organizations are starting to understand that they need to add more agility and resilience to their ML pipelines and production models. DSML technologies can help improve many of the processes involved and increase operational efficiency. But to do that, they must be endowed with automation capabilities (automation enables better dissemination of best practices, reusability of ML artifacts, and enhanced productivity of data science teams), support for fast prototyping of AI and big data analytics applications, and operationalization of ML models to accelerate the passage from proof of concept (PoC) to production. With the Alida solution, the

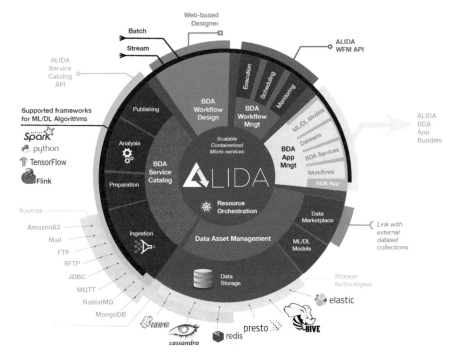

Fig. 15.1 Alida platform

R&D lab of engineering[1] aims to respond promptly to these needs in the field of data science and ML as well as the operationalization of data analytics.

Through Alida, users can design their stream/batch workflows by choosing the BDA services from the catalog, which big data set to process, run, and monitor the execution. The resulting BDA applications can be deployed and installed in another target infrastructure with the support of a package manager that simplifies the deployment within the target cluster. Alida is designed and developed on top of the most cutting-edge open-source big data technologies and frameworks. Being cloud-native, it can scale computing and storage resources, thanks to a pipeline orchestration engine that leverages the capabilities of Kubernetes (https://kubernetes.io/) for cloud resource and container management.

Alida (Fig. 15.1) provides a web-based graphical user interface (GUI) for both stream and batch BDA workflow design. It will thus be possible to design and directly run and monitor the execution or schedule the execution. In addition, the execution, scheduling, and monitoring functionalities are also available through the Alida APIs. Alida provides an extensible catalog of services (the building blocks of the BDA workflows) which covers all phases, from ingestion to preparation to

---

[1] https://www.eng.it/en/

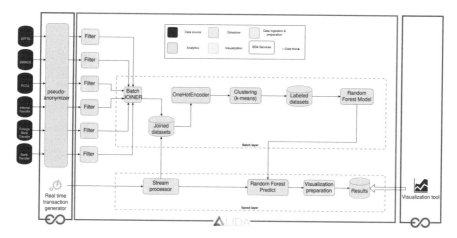

Fig. 15.2 AI-based fraud detection system architecture

analysis and data publishing. The catalog can be used both through a web interface and through specific APIs, which will allow the registration of new BDA services. In summary, Alida is a very useful and straightforward tool for a rapid BDA application prototyping: user designs his own stream/batch WF by choosing the BDA services from the catalog, chooses which big data set he/she wants to process, and runs and monitors the execution and the successful execution of a WF results in the creation of a BDA application which can be deployed and installed in another target infrastructure with the support of package managers that simplify deployment activities within the cluster. These applications will offer appropriate APIs to access the services offered by the BDA apps.

Thanks to Alida, it was possible to design, execute, and deploy a bundle of two workflows according to the architecture showed in Fig. 15.2.

The AI-based fraud detection system presents in the batch layer:

- A preprocessing step on transaction data where datasets on several kinds of transactions are properly filtered and joined to get only one unlabeled dataset
- Clustering of such unlabeled data to create labeled samples
- The random forest model training to periodically retrain it with new data (it is worth noting that the analyst's feedbacks also contribute to the training)
- Feeding the supervised model retraining, by means of fraud analyst's feedback

In the stream layer, the real-time fraud detection is handled by means of the random forest algorithm on the basis of new input transaction data, to analyze all the transactions and report the suspicious ones as a resulting streaming data ready to be properly visualized. The fraud attempts are investigated by the analyst who marks them as "suspicious" if he confirms the transaction was fraudulent and such feedback feeds the retrain within the batch layer. In this way, the fraud detection performance is improved over time.

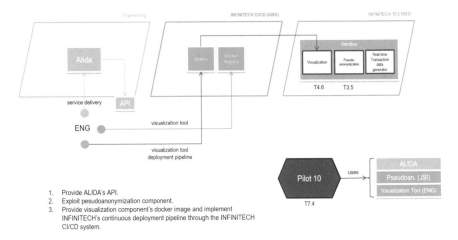

1. Provide ALIDA's API.
2. Exploit pesudoanonymization component.
3. Provide visualization component's docker image and implement
   INFINITECH's continuous deployment pipeline through the INFINITECH
   CI/CD system.

Fig. 15.3 AI-based fraud detection system deployment scheme

4 Real-Time Cybersecurity Analytics on Financial Transactions' Data Pilot in INFINITECH

In order to end-to-end test the overall system, the real-time transaction generator (Fig. 15.2) produces synthetic transactions, emulating the behavior of a traditional financial transaction software. The objective is to obtain a stream of data that feeds the fraud detection system whose effectiveness is demonstrated by the cybersecurity and fraud detection pilot of INFINITECH.

Moreover, a generated, realistic dataset was created that is consistent with the real data present in the data operations environment. The synthetic dataset is provided by Poste Italiane starting from randomly generated personal data with no correlation with real people.

It contains about one million records including fraudulent transactions (Bank Transfer SEPA Transactions) associated with ten thousand users and occurred in a 1-year period (from 1 January 2020 to 30 September 2020).

Nevertheless, the invocation of a pseudo-anonymization tool is necessary to make the execution of the abovementioned pilot more realistic and complying to data protection regulations; thus, transaction data will be pseudo-anonymized before leaving the banking system by using a tool available through the INFINITECH platform.

Finally, a visualization tool aims to address the advanced visualization requirements by offering to fraud analysts a variety of visualization formats, which span from simple static charts to interactive charts offering several layers of information and customization. It consists of a set of functionalities which support the execution of the visualization process. This set of functionalities includes the dataset selection, the dataset preview generation, the visualization-type selection, the visualization configuration, the visualization generation, and an interactive dashboard. The visualization tool is a project hosted in the presentation group of the official INFINITECH GitLab repository, it will be integrated with other components adopted by the pilot in a dedicated environment, and its development process is backed by a CI/CD pipeline (Fig. 15.3) as per the INFINITECH blueprint reference[2].

5 Conclusions

This chapter addresses one of the major challenges in the financial sector which is real-time cybersecurity analytics on financial transactions' data, presenting an innovative way to integrate supervised and unsupervised AI models by exploiting proper technological tools able to process large amounts of transaction data.

A previously stated, a supervised model is a model trained on a rich set of properly "tagged" transactions. This happens by ingesting massive amounts of tagged transaction details in order to learn patterns that best reflect legitimate behaviors. When developing a supervised model, the amount of clean, relevant training data is directly correlated with model accuracy. Unsupervised models are designed to spot anomalous behavior. In these cases, a form of self-learning is adopted to surface patterns in the data that are invisible to other forms of analytics.

A lambda architecture was designed where both real-time and batch analytics workflows are handled in an integrated fashion. Such architecture is designed to handle massive quantities of data by taking advantage of both batch and stream processing methods. This approach is designed to balance latency, throughput, and fault tolerance by using batch processing to provide comprehensive and accurate views of batch data while simultaneously using real-time stream processing to provide views of online data. In our case, data preprocessing and model training are handled in the batch layer, while the real-time fraud detection is handled on the basis of new input transaction data within the speed layer. The solution presented in this chapter aims at supporting in an innovative and effective way fraud analysts and automating the fraud detection processes.

Acknowledgments The research leading to the results presented in this chapter has received funding from the European Union's funded Project INFINITECH under Grant Agreement No. 856632.

---

[2] https://gitlab.infinitech-h2020.eu/presentation/veesualive.git

References

1. Commission, E. (2021, June). *Payment services (PSD 2) – Directive (EU) 2015/2366*. Tratto da An official website of the European Union: European Commission. https://ec.europa.eu/info/law/payment-services-psd-2-directive-eu-2015-2366_en
2. Eurostat. (2021, February). *E-commerce statistics*. Taken from Eurostat: statistics explained. https://ec.europa.eu/eurostat/statistics-explained/index.php?title=E-commerce_statistics
3. Commission, E. (2014–2019). *Shaping the digital single market*. Tratto da European Commission: Digital Single Market Strategy. https://ec.europa.eu/digital-single-market/en/shaping-digital-single-market
4. Durkin, J. (1998). *Expert systems: Design and development* (1st ed.). Prentice Hall PTR.
5. Ketar, P. S., Shankar, R., & & Banwet, K. D. (2014). Telecom KYC and mobile banking regulation: An exploratory study. *Journal of Banking Regulation*, 117-128.
6. Hand, D. J., & Blunt, G. (2009). Estimating the iceberg: How much fraud is there in the UK? *Journal of Financial Transformation*, 19–29.
7. Crina Grosan, A. A. (2011). Rule-based expert systems. In *Intelligent systems: A modern approach* (pp. 149–185). Springer.
8. Sethi, N., & Gera, A. (2014). A revived survey of various credit card fraud detection techniques. *International Journal of Computer Science and Mobile Computing, 3*(4), 780–791.
9. Shimpi, P. R., & Kadroli, V. (2015). Survey on credit card fraud detection techniques. *International Journal of Engineering and Computer Science, 4*(11), 15010–15015.
10. Chandola, V., Banerjee, A., & Kumar, V. (2009). Anomaly detection: A survey. *ACM Computing Surveys (CSUR), 41*(3), 15.
11. Krawczyk, B. (2016). Learning from imbalanced data: Open challenges and future directions. *Progress in Artificial Intelligence, 5*(4), 221–232.
12. Dal Pozzolo, A., Caelen, O., Johnson, R. A., & Bontempi, G. (2015). *Calibrating probability with undersampling for unbalanced classification*. In Symposium series on computational intelligence (pp. 159–166). IEEE.
13. Gama, J. A., Zliobaite, I., Bifet, A., Pechenizkiy, M., & Bouchachia, A. (2014). A survey on concept drift adaptation. *ACM Computing Surveys, 46*(4), 44.
14. Carcillo, F., Le Borgne, Y.-A., Caelen, O., & Bontempi, G. (2018b). Streaming active learning strategies for real-life credit card fraud detection: Assessment and visualization. *International Journal of Data Science and Analytics*, 1–16.
15. Yamanishi, K., & Takeuchi, J.-i. (2001). *Discovering outlier filtering rules from unlabeled data: Combining a supervised learner with an unsupervised learner*. In Proceedings of the seventh ACM SIGKDD international conference on Knowledge discovery and data mining, pp. 389–394. ACM.
16. Zhu, X. (2005). *Semi-supervised learning literature survey* (Technical report, Computer Sciences TR 1530). University of Wisconsin Madison.

Part IV
Applications of Big Data and AI in Insurance

Chapter 16
Risk Assessment for Personalized Health Insurance Products

Aristodemos Pnevmatikakis, Stathis Kanavos, Alexandros Perikleous, and Sofoklis Kyriazakos

1 Introduction

Personalization has always been a key factor in health insurance product provision. Traditionally, it involves a screening based on static information from the customer: their medical record and questionnaires they answer. But their medical history, enumerated by clinical data, can be scarce, and it certainly is just one determinant of health. The way people live their lives, enumerated by behavioral data, is the second determinant, and for risk assessment of chronic, age-related conditions of now seemingly healthy individuals, it is the most important, as indicated by several studies. A study on diabetes prevention [1] gives evidence to the importance of lifestyle for the outcomes in youths and adults. Another study [2] correlates health responsibility, physical activity, and stress management to obesity, a major risk factor for cardiovascular diseases, type 2 diabetes, and some forms of cancer. The 2017 Global Burden of Disease Study [3] considers behavioral, environmental, occupational, and metabolic risk factors.

Risk assessment has always been an integral part of the insurance industry [4]. Unlike risk assessment in medicine that is based on continuous estimation of risk factors, its insurance counterpart is usually static, done at the beginning of a contract with a client. Dynamic personalized products are only recently

A. Pnevmatikakis (✉) · S. Kanavos · A. Perikleous
Innovation Sprint Sprl, Brussels, Belgium
e-mail: apnevmatikakis@innovationsprint.eu; skanavos@innovationsprint.eu; aperikleous@innovationsprint.eu

S. Kyriazakos
Innovation Sprint Sprl, Brussels, Belgium

Business Development and Technology Department, Aarhus University, Herning, Denmark
e-mail: sofoklis@btech.au.dk; skyriazakos@innovationsprint.eu

J. Soldatos, D. Kyriazis (eds.), *Big Data and Artificial Intelligence in Digital Finance*,
https://doi.org/10.1007/978-3-030-94590-9_16

appearing as data-based digital risk assessment platforms. Such platforms start transforming insurance by disrupting the ways premiums are calculated [5] and are already being utilized in car insurance. In the scope of car insurance, continuous vehicle-based risk assessment [6] is already considered important for optimizing insurance premiums and providing personalized services to drivers. Specifically, driver behavior is analyzed [7] for usage-based insurance. Moreover, telematic driving profile classification [8] has facilitated pricing innovations in German car insurance.

Similarly, personalization of health insurance products needs to be based on continuous risk assessment of the individual, since lifestyle and behavior cannot be assessed at one instance in time; they involve people's habits and their continuous change. Health insurance products employing continuous assessment of customers' lifestyle and behavior are dynamically personalized.

Behavioral assessments, much like their clinical counterparts, rely on data. For behavior, the data collection needs to be continuous, facilitated by software tools for the collection of information capturing the important aspects of lifestyle and behavior. In the INFINITECH project [9], insurance experts define the data to be collected, and the Healthentia eClinical system [10] facilitates the collection. Specifically, Healthentia provider interfaces for data collection from medical and consumer devices, including IoT (Internet of Things) devices. Moreover, continuous risk assessment services are provided to health insurance professionals by training machine learning (ML) prediction models for the required health parameters. ML has been used in the insurance industry to analyze insurance claim data [11, 12]. Vehicle insurance coverage affects driving behavior and hence insurance claims [13]. These previous works employed ML to analyze data at the end of the insurance pathway, after the event. Instead, in this chapter, we follow the approach in [14]. We expand on the results presented therein, focusing on the continuous analysis of data at the customer side to personalize the health insurance product by modifying the insurance pathway.

Personalized dynamic product offerings benefit both the insurance companies and their customers, but the continuous assessment imposes a burden on the customers. Insurance companies gain competitive advantages with lower prices for low-risk customers. The customers have a direct financial benefit in the form of reduced premiums due to the lower risk of their healthy behavior. They also have an indirect benefit stemming from coaching about aspects of their lifestyle, both those that drive the risk assessment models toward positive decisions and those driving them toward negative decisions. The identification of these aspects is made possible by explainable AI techniques applied on the individual model decisions. The insurance companies need to balance the increased burden of the monitoring with the added financial and health benefits of using such a system.

The system for personalized health insurance products devised in the INFINITECH project is presented in Sect. 2 of this chapter. Then, its main components are detailed, covering the data collection (Sect. 3), the model training test bed (Sect. 4), and the provided ML services of risk assessment and lifestyle coaching (Sect. 5). Finally, the conclusions are drawn in Sect. 6.

2 INFINITECH Healthcare Insurance System

The healthcare insurance pilot of INFINITECH focuses on health insurance and risk analysis by developing two AI-powered services:

1. The risk assessment service allows the insurance company to adapt prices by classifying individuals according to their lifestyle.
2. The coach service advises individuals in their lifestyle choices, aiming at improving their health but also in persuading them to use the system correctly.

These two services rely on a model of health outlook trained on the collected data and used in the provision of the services.

An overview of pilot system for healthcare insurance is given in Fig. 16.1. It comprises two systems, the pilot test bed, built within the INFINITECH project, and the Healthentia eClinical platform, provided by Innovation Sprint. The data are collected by Healthentia, as detailed in Sect. 3. Toward this, the complete Healthentia eClinical platform is also presented in the same section. The pilot test bed facilitates secure and privacy-preserving model training as discussed in Sect. 4. The trained model is utilized for the risk assessment and the lifestyle coach ML services detailed in Sect. 5, and the results are finally visualized by the dashboards of the Healthentia portal web app.

3 Data Collection

Two types of data are collected to train the health outlook model: measurements and user reports. The measurements are values collected by sensors, which are automatically reported by these sensors to the data collection system, without the intervention of the user. They are objective data, since their quality only depends on the devices' measurement accuracy. They have to do with physical activity, the heart,

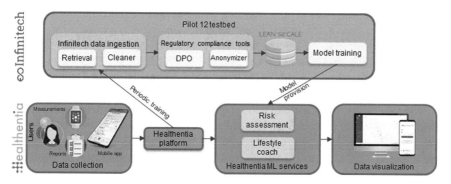

Fig. 16.1 INFINITECH healthcare insurance system comprising Healthentia and the respective test bed

and sleep. The physical activity measurements involve steps, distance, elevation, energy consumption, and time spent in three different zones of activity intensity (light, moderate, and intense). The heart measurements include the resting heart rate and the time spent in different zones of heart activity (fat burn, cardio, and peak). The sleep measurements include the time to bed and waking up time, so indirectly the sleep duration and the time spent in the different sleep stages (light, REM, and deep sleep).

The reports are self-assessments of the individuals; hence, they are subjective data. They cover common symptoms, nutrition, mood, and quality of life. The symptoms are systolic and diastolic blood pressure and body temperature (entered as numbers measured by the users), as well as cough, diarrhea, fatigue, headache, and pain (where the user provides a five-level self-assessment of severity from not at all up to very much). Regarding nutrition, the user enters the number of meals and whether they contain meat, as well as the consumption of liquids: water, coffee, tea, beverages, and spirits. Mood is a five-level self-assessment of the user's psychological condition, from very positive to neutral and down to very negative. Finally, quality of life [15] is reported on a weekly basis using the Positive Health questionnaire [16] and on a monthly basis using the EuroQol EQ-5D-5L questionnaire [17], which asks the user to assess their status in five degrees using five levels. The degrees are mobility, self-care, usual activities, pain/discomfort, and anxiety/depression, complemented with the overall numeric health self-assessment.

The data collection is facilitated by Healthentia [10]. The platform provides secure, persistent data storage and role-based, GDPR-compliant access. It collects the data from the mobile applications of all users, facilitating smart services, such as risk assessment, and providing both original and processed information to the mobile and portal applications for visualization. The high-level architecture of the platform is shown in Fig. 16.2. The service layer implements the necessary functionalities of the web portal and the mobile application. The study layer facilitates study management, organizing therein healthcare professionals, participants, and their data. They can be formal clinical studies or informal ones managed by pharmaceutical companies, hospitals, research centers, or, in this case, insurance companies.

The Healthentia core layer comprises of high-order functionalities on top of the data, like role-based control, participant management, participants' report management, and ML functionalities. The low-level operations on the data are hosted in the data management layer. Finally, the API layer provides the means to expose all the functionalities of the layers above to the outside world. Data exporting toward the pilot test bed and model importing from the test bed are facilitated by it.

The Healthentia mobile application (Fig. 16.3) enables data collection. Measurements are obtained from IoT devices, third-party mobile services, or a proprietary sensing service. User reports are obtained via answering questionnaires that either are regularly pushed to the users' phones or are accessed on demand by the users themselves. Both the measured and reported data are displayed to the users, together with any insights offered by the smart services of the platform.

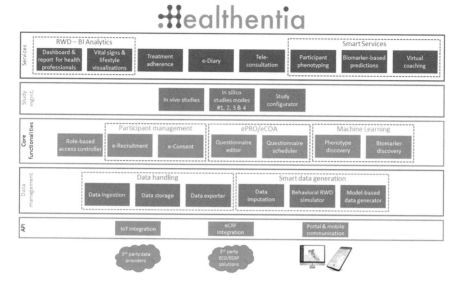

Fig. 16.2   Healthentia high-level architecture

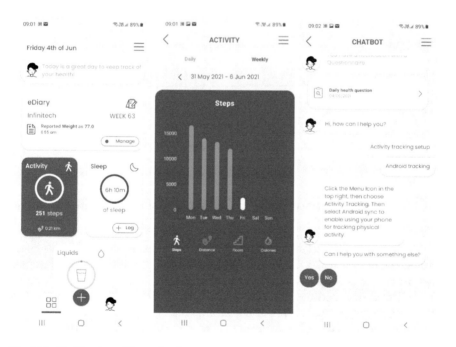

Fig. 16.3   Healthentia mobile application

Fig. 16.4 Healthentia portal application – viewing measurements and creating questionnaires

The Healthentia portal application (Fig. 16.4) targets the health insurance professionals. It provides an overview of the users of each insurance organization and details for each user. Both overview and details include analytics based on the collected data and the risk assessment insights. It also facilitates managing the organization, providing, for example, a questionnaire management system to determine the types of self-assessments and reports provided by the users.

4   INFINITECH Healthcare Insurance Pilot Test Bed

The INFINITECH healthcare insurance test bed facilitates model training by providing the necessary regulatory compliance tools and the hardware to run the model training scripts, whenever new models are to be trained. Its high-level architecture is shown in the upper part of Fig. 16.1. The test bed ingests data from Healthentia, processes it for model training, and offers the means to perform the model training. It then provides the models back to Healthentia for online usage in risk assessment.

The regulatory compliance tools provide the data in the form compliant for model training. The tools comprise the Data Protection Orchestrator (DPO) [18, 19], which among others regulates data ingestion, and the anonymizer, which are presented in detail in Chap. 20.

The INFINITECH data collection module [20] is responsible for ingesting the data from Healthentia utilizing the Healthentia API when so instructed by the DPO. It then provides data cleanup services, before handling the data to the INFINITECH anonymizer [21, 22]. The ingested data are already pseudo-anonymized, as only non-identifiable identifiers are used to designate the individuals providing the data, but the tool performs anonymization of the data itself. The anonymized data are stored in LeanXcale [23], the test bed's database. Different anonymized versions are to be stored, varying the effect of anonymization, aiming at determining its effect on the trained model quality. The model training is an offline process. Hence, the ML engineers responsible for model training will be instructing the DPO to orchestrate data ingestion at different anonymization levels. Models based on logistic regression [24], random forest [25], and (deep) neural networks [26] are trained to predict the self-reported health outlook variation on a weekly basis. Binary and tristate models have been trained using data collected in the data collection phase of healthcare insurance pilot, involving 29 individuals over periods of time spanning 3–15 months. The classification accuracy of the random forest models is best in this context, since the dataset is limited for neural networks of some depth. They are shown in Fig. 16.5.

Shapley additive explanations (SHAP) analysis [27] is employed to establish the impact of the different feature vector elements in the classifier decisions (either positive or negative). Average overall decisions, this gives the importance of the different attributes for the task at hand. Attributes of negligible impact on decisions can be removed, and the models can be retrained on a feature space of lower dimensions. Most importantly though, SHAP analysis is used in the virtual coaching service as discussed in Sect. 5.2.

These models are not the final ones. During the pilot validation phase that will start in September 2021, 150 individuals will be using the system for 12 months. Approximately two-thirds of these participants will be used to keep on retraining the models.

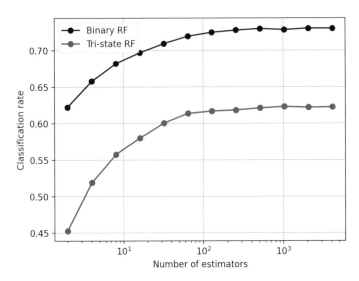

Fig. 16.5 Classification rate of random forest binary (improve, worsen) and tristate (improve, same, worsen) models for different number of estimators

5 ML Services

Two services are built using the models trained in Sect. 4. The model decisions accumulated across time per individual are used in risk assessment (see Sect. 5.1), while the SHAP analysis of the individual decisions is used in virtual coaching (see Sect. 5.2).

Since the actual data collected thus far are barely enough for training the models, a synthetic dataset is generated to evaluate the performance of both services. Five behavioral phenotypes are defined in the simulator of [14]. Two are extreme ones: at one end, the athletic phenotype that likes indoors and outdoors exercising and enjoys a good night's sleep and, at the other end, the gamer who is all about entertainment, mainly indoors, enjoys work, and is not too keen on sleeping on time. In between lies the balanced phenotype, with all behavioral traits being more or less of equal importance as they are allowed a small variance around the average. Two random variants of the balanced phenotype are also created, with the behavioral traits allowed quite some variance from the balanced state. The one random phenotype is associated with excellent health status, while the other one is associated with a typical health status. 200 individuals of each phenotype are simulated for a duration of 2 years.

5.1   Personalized Risk Assessment

Personalized risk assessment is based on the decisions of the models for each individual. The assessments are long term in the sense that they take into account all the model decisions over time intervals that are very long. In this study, we calculate the long-term averages of the different daily decisions with a memory length corresponding roughly to half a year. There are two such averaged outputs for models with binary decisions and three for those with tristate ones. In every case, the averages are run for the whole length of the synthetic dataset (two years), and for each day of decision, they sum to unity. At any day, the outlook is assessed as the sum of all the averaged positive outcomes from the beginning of the dataset up to the date of the assessment, minus the sum of the negative ones in the same time interval. In the tristate case, the difference is normalized by the sum of the constant ones. The resulting grade is multiplied by 100 and thus can be in the range of $[-100,100]$. Obviously, outlook grades larger than zero correspond to people whose well-being outlook has been mostly positive in the observation period, and outlook grades smaller than zero correspond to people whose well-being outlook has been mostly negative in the observation period.

The daily evolutions of the accumulated model decisions for the 2 years of observation for the first athletic, balanced, and gamer simulated person are shown in Fig 16.6. Clearly, the athletic person is doing great, and the balanced one is doing quite good. The gamer is not worsening but looks rather stagnant. The histograms of the outlook grades after 2 years are shown in Fig. 16.7 for each of the five behavioral phenotypes. It should be no surprise that the two extreme phenotypes are at opposite sides of the outlook spectrum, clearly separated by a range of values occupied by the balanced and random phenotypes. It is the actual activities done that determine the outlook grade, and in the balanced and random phenotypes, the selection of activities is quite different within each phenotype, so they exhibit quite a lot of spread in the outlook, as expected in real life.

5.2   Personalized Coaching

The SHAP analysis results for the individual decisions are shown in Fig. 16.8. Each row corresponds to a lifestyle attribute, and each dot in a specific row corresponds to the value of that element in one of the input daily vectors. The color of the dot indicates the element's value (from small values in blue to large values in red). The placement of the dot on the horizontal axis corresponds to the SHAP value. Values close to zero correspond to lifestyle attributes with negligible effect on the decision, while large positive or negative values correspond to lifestyle attributes with large effects. The vertical displacement indicates how many feature vectors fall into the particular range of SHAP values. Thus, thick dot cloud areas correspond to many input daily vectors in that range of SHAP values. Dots on the left correspond to

Fig. 16.6 Averaged outputs of the tristate random forest classifier with 2048 trees for the first athletic person (top) with an outlook grade after 2 years of 46.2, for the first balanced person (middle) with an outlook grade of 15.2, and for the first gamer (bottom) with an outlook grade of 2.9

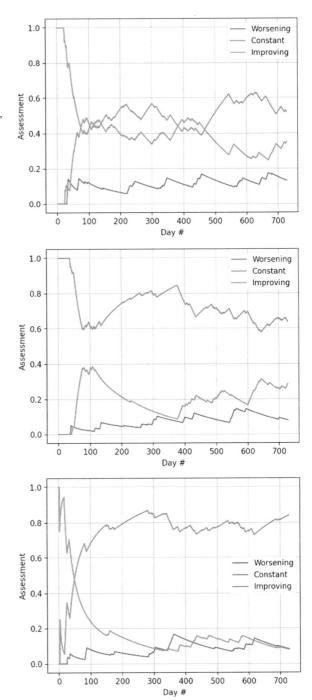

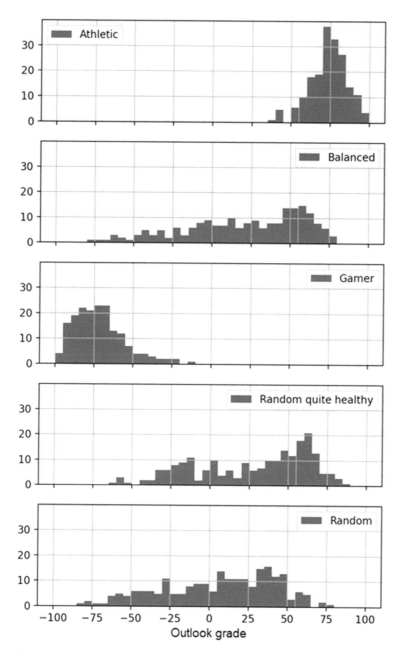

Fig. 16.7 Histograms of outlook grades for the five behavioral types. From athletic to gamer, the bulk of the grades move toward smaller values

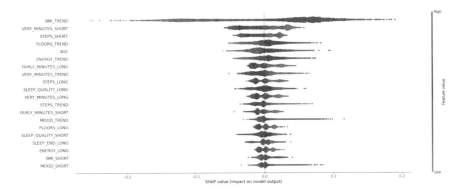

Fig. 16.8 SHAP analysis of the individual decisions, signaling the importance of small, medium, and large feature values in the final decision reached for this vector

attribute values that direct one toward a prediction that health is improving, while dots on the right suggest a worsening of health. For example, red dots of large values of the body mass index trend (increasing weight) are on the right indicating negative health outlook. Purple dots of moderate body mass index trend are around zero, indicating negligible effect on the decisions. Finally, blue dots indicating a trend to lose weight are on the left, indicating improved health outlook.

The individual SHAP coefficients per model decision are employed to establish per person importance of lifestyle attributes in positive or negative well-being outlook. The most influential lifestyle attributes for a positive outlook are collected over any short time interval, as are those with the largest positive SHAP coefficients. Similarly, the most influential negative attributes are obtained for the same interval. Then, the individual is coached about these positive and negative attributes. The personalized coach offers advice toward behaviors of the positive attributes and away from those of the negative attributes.

It is worth mentioning that the explainable AI technique for personalized coaching discussed here is only about the content of the advice. An actual virtual coach should also involve decisions on the timing, the modality, and the tone of the messages carrying the advice to the individuals.

6 Conclusions

The INFINITECH way of delivering personalized services for health insurance is discussed in this chapter. To that extent, the healthcare insurance pilot of the project is integrating Healthentia, an eClinical platform for collecting real-world data from individuals into a test bed for training classification models. The resulting models are used in providing risk assessment and personalized coaching services. The predictive capabilities of the models are acceptable, and their use in the services to analyze the simulated behavior of individuals is promising. The actual validation of the provided services is to be carried out in a pilot study with 150 individuals, which started in September 2021 and will last in 1 year.

Our future work on this usage-based healthcare insurance pilot will not be confined to validating the technical implementation of the pilot system, including its data collection and machine learning-based analytics parts. We will also explore how such usage-based systems can enable new business models and healthcare insurance service offerings. For instance, we will study possible pricing models and related incentives that could make usage-based insurance more attractive than conventional insurance products to the majority of consumers.

Acknowledgments This work has been carried out in the H2020 INFINITECH project, which has received funding from the European Union's Horizon 2020 Research and Innovation Programme under Grant Agreement No. 856632.

References

1. Grey, M. (2017). Lifestyle Determinants of Health: Isn't it all about genes and environment? *Nursing Outlook, 65*, 501–515. https://doi.org/10.1016/j.outlook.2017.04.011
2. Joseph-Shehu, E. M., Busisiwe, P. N., & Omolola, O. I. (2019). Health-promoting lifestyle behaviour: A determinant for noncommunicable diseases risk factors among employees in a Nigerian University. *Global Journal of Health Science, 11*, 1–15.
3. Stanaway, J. D., Afshin, A., Gakidou, E., Lim, E. S., Abate, D., Abate, K. H., Abbafati, C., Abbasi, N., Abbastabar, H., Abd-Allah, F., et al. (2018). Global, regional, and national comparative risk assessment of 84 behavioural, environmental and occupational, and metabolic risks or clusters of risks for 195 countries and territories, 1990–2017: A systematic analysis for the Global Burden of Disease Study 2017. *Global Health Metrics, 392*, 1923–1994.
4. Blackmore, P. (2016). *Easier approach to risk profiling*. Available online: https://www.insurancethoughtleadership.com/easier-approach-to-risk-profiling/. Accessed on 1/3/2021.
5. Blackmore, P. (2016). *Digital risk profiling transforms insurance*. Available online: https://www.insurancethoughtleadership.com/digital-risk-profiling-transforms-insurance/. Accessed on 1/3/2021.
6. Gage, T., Bishop, R., & Morris, J. (2015). The increasing importance of vehicle-based risk assessment for the vehicle insurance industry. *Minnesota Journal of Law, Science & Technology, 16*, 771.
7. Arumugam, S., & Bhargavi, R. (2019). A survey on driving behavior analysis in usage based insurance using big data. *Journal of Big Data, 6*, 1–21.
8. Weidner, W., Transchel, F. W. G., & Weidner, R. (2017). Telematic driving profile classification in car insurance pricing. *Annals of Actuarial Science, 11*, 213–236.
9. Infinitech H2020. (2021). *Infinitech—The flagship project for digital finance in Europe*. Available online: https://www.infinitech-h2020.eu/. Accessed on 7/6/2021.
10. Innovation Sprint. (2021). *Healthentia: Driving real world evidence in research & patient care*. Available online: https://innovationsprint.eu/healthentia. Accessed on 7/6/2021.
11. Bermúdez, L., Karlis, D., & Morillo, I. (2020). Modelling unobserved heterogeneity in claim counts using finite mixture models. *Risks, 8*, 10.
12. Burri, R. D., Burri, R., Bojja, R. R., & Buruga, S. (2019). Insurance claim analysis using machine learning algorithms. *International Journal of Innovative Technology and Exploring Engineering, 8*, 147–155.

13. Qazvini, M. (2019). On the validation of claims with excess zeros in liability insurance: A comparative study. *Risks, 7,* 71.
14. Pnevmatikakis, A., Kanavos, S., Matikas, G., Kostopoulou, K., Cesario, A., & Kyriazakos, S. (2021). Risk assessment for personalized health insurance based on real-world data. *Risks, 9*(3), 46. https://doi.org/10.3390/risks9030046
15. Revicki, D. A., Osoba, D., Fairclough, D., Barofsky, I., Berzon, R., Leidy, N. K., & Rothman, M. (2000). Recommendations on health-related quality of life research to support labeling and promotional claims in the United States. *Quality of Life Research, 9,* 887–900.
16. Huber, M., van Vliet, M., Giezenberg, M., Winkens, B., Heerkens, Y., Dagnelie, P. C., & Knottnerus, J. A. (2016). Towards a 'patient-centred' operationalisation of the new dynamic concept of health: a mixed methods study. *BMJ Open, 6,* e010091. https://doi.org/10.1136/bmjopen-2015-010091
17. Stolk, E., Ludwig, K., Rand, K., van Hout, B., & Ramos-Goñi, J. M. (2019). Overview, update, and lessons learned from the international EQ-5D-5L valuation work: Version 2 of the EQ-5D-5L valuation protocol. *Value in Health, 22,* 23–30.
18. Notario, N., Cicer, E., Crespo, A., Real, E. G., Catallo, I., & Vicini, S. (2017). *Orchestrating privacy enhancing technologies and services with BPM Tools. The WITDOM data protection orchestrator.* ARES'17, Reggio Calabria, Italy.
19. INFINITECH H2020 consortium. (2021). *D3.16 – Regulatory compliance tools – II*
20. INFINITECH H2020 consortium. (2020). *D5.13 – Datasets for algorithms training & evaluation – I*
21. Adkinson, O. L., Dago, C. P., Sestelo, M., & Pintos, C. B. (2021). A new approach for dynamic and risk-based data anonymization. In Á. Herrero, C. Cambra, D. Urda, J. Sedano, H. Quintián, & E. Corchado (Eds.), *13th International Conference on Computational Intelligence in Security for Information Systems (CISIS 2020)* (CISIS 2019. Advances in intelligent systems and computing) (Vol. 1267). Springer. https://doi.org/10.1007/978-3-030-57805-3_31
22. INFINITECH H2020 consortium. (2021). *D3.13 – Data governance framework and tools – II*
23. LeanXcale. (2021). *LeanXcale: The database for fast-growing companies.* Available online: http://leanxcale.com. Accessed on 4/6/2021.
24. Tolles, J., & Meurer, W. J. (2016). Logistic regression relating patient characteristics to outcomes. *JAMA., 316*(5), 533–534.
25. Breiman, L. (2001). Random forests. *Machine Learning, 45,* 5–32.
26. Schmidhuber, J. (2015). Deep learning in neural networks: An overview. *Neural Networks, 61,* 85–117.
27. Lundberg, S. M., Erion, G., Chen, H., DeGrave, A., Prutkin, J. M., Nair, B., Katz, R., Himmelfarb, J., Bansal, N., & Lee, S. I. (2020). From local explanations to global understanding with explainable AI for trees. *Nature Machine Intelligence, 2,* 56–67.

Chapter 17
Usage-Based Automotive Insurance

Ignacio Elicegui, Juan Carrasco, Carmen Perea Escribano, Jose Gato, Andrea Becerra, and Andreas Politis

1 Introduction

Connected objects are changing the world and the way organizations do business. Cutting-edge technologies in several disciplines, such as IoT, 5G, and AI, are paving the way to new business services. Thanks to these disparate technologies, advanced services, unimaginable previously, can be provided to the customers, bringing benefits to both companies and their customers. Present chapter introduces the INFINITECH Pilot 11, which exploits these concepts and technologies from the point of view of the car insurance companies and insured clients, the drivers. Although the idea of classifying a driver by reading technical data provided by their own vehicle is not really new, this pilot goes a couple of steps beyond: first by developing an IoT standards-based platform to homogenize and manage real-time data captured from vehicles while, at the same time, merging this with other relevant sources, like weather or traffic incidents, and, next, by using this aggregated information to evaluate cutting-edge ML/DL technologies and develop an accurate AI-powered model to infer different driving profiles. The final objective is to use

I. Elicegui (✉) · J. Carrasco
Research & Innovation, ATOS IT, Madrid, Spain
e-mail: ignacio.elicegui@atos.net; juan.carrascoa@atos.net

C. P. Escribano · J. Gato
Research & Innovation, ATOS Spain, Madrid, Spain
e-mail: carmen.perea@atos.net; jose.gato@atos.net

A. Becerra
CTAG – Centro Tecnológico de Automoción de Galicia, Pontevedra, Spain
e-mail: andrea.becerra@ctag.com

A. Politis
Dynamis Insurance Company, Athens, Greece
e-mail: a.politis@dynamis.gr

© The Author(s) 2022
J. Soldatos, D. Kyriazis (eds.), *Big Data and Artificial Intelligence in Digital Finance*,
https://doi.org/10.1007/978-3-030-94590-9_17

this driving profiling model to analyze the insured's driving behavior and tailor their offered services, while it helps insurance companies to better estimate the real-time associated risks.

In this line, this chapter presents a connected car as an IoT infrastructure itself and provides an overview of the usage-based automotive insurance solutions being developed within the INFINITECH project. This chapter is divided into six sections beyond this introduction: Section 2 "Insurance Premiums for Insured Clients" shows the current status in the automotive insurance sector. A general review of the different traditional methods to calculate automobile insurance premiums is also provided. Section 3 "Connected Vehicles' Infrastructures" provides an overview of the supported technologies by current vehicles to get connected and incoming scenarios to exploit these new connected vehicles (V2X infrastructures, autonomous vehicles, etc.). Section 4 "Data Gathering and Homogenization Process" describes the INFINITECH technologies used to collect the required data sources and the process undertaken to homogenize and standardize the data using the FIWARE NGSI standard for Context Information Management [1]. Section 5 "Driving Profile Model" explains the steps to develop the AI model to assist on route clustering. Section 6 "Customized Car Insurance Services Powered by ML" introduces the services that exploit this new created AI model and the new way to estimate the risks. Finally, Sect. 7, the concluding section of the chapter, gives a summary and critique of the findings. It also presents the found and envisioned challenges.

2   Insurance Premiums for Insured Clients

Motor Third Party Liability Insurance (MTPL) ensures that damage to third-party health and property caused by an accident for which the driver and/or the owner of the vehicle were responsible is covered. It was understood from the very beginning of the vehicle's circulation that this kind of insurance had to be compulsory due to the lack of solvency of first party who caused bodily injury or property damage following any event related to a car accident. For instance, in the United Kingdom, MTPL insurance has been compulsory since 1930 with the Road Traffic Act. In our days, MTPL insurance is compulsory in all countries of European Union and in most of the countries worldwide with the limits of the cover varying from region to region. Apart from the compulsory MTPL insurance, own damage (OD) is an additional cover in a car insurance policy. OD helps you stay covered against damage caused to your vehicle due to accidents like fire, theft, etc. In case of an accident, an own damage cover compensates you for expense to repair or replace parts of your car damaged in the accident.

In order for the insurance companies to be in a position to pay the claims and make profit out of their business, they need to assess the risk involved in every insurance policy. Insurance actuaries help insurance companies in risk assessment. Then they use this analysis to design and price insurance policies. The higher the risk for a certain group, the more likely it is for insurance companies to pay out a claim, meaning these groups will be charged higher for their motor insurance. Risk assessment involves measuring the probability that something will happen to cause a loss. In few regions in the world, compulsory MTPL insurance premiums are determined by the government; in most of the world, insurance premiums are determined by insurance companies, depending on many factors that are believed to affect the expected cost of future claims. These factors are:

- The *driving record of the driver or the vehicle*: the better the record, the lower the premium. If the insured had accidents or serious traffic violations, it is likely to pay more; accordingly, a new driver with no driving record pays more.
- The *driver* and especially *the gender* (male drivers, especially younger ones, are on average regarded to drive more aggressively), *the age* (teenage drivers are involved in more accidents because of no driving experience so they are charged more, and senior-experienced drivers are charged less. However, after a certain age, the premiums rise again due to slower reflexes and reaction times), the *marital status* (married drivers average fewer accidents than single drivers), and the *profession* (certain professions are proven to result in more accidents especially if travelling with a car is involved).
- The *vehicle*, focusing on specific aspects of it such as the cost of the car, likelihood of theft, the cost of repairs, its engine size, and the overall safety record of the vehicle. Vehicles with high-quality safety equipment may qualify for premium discounts. Insurance companies also take into consideration the damage a vehicle can inflict on other vehicles.
- The *location*, meaning the address of the owner or the usual place of circulation; areas with high crime rates (vandalism, thefts) generally lead to higher premiums as well as areas with higher rates of accidents. So, urban drivers pay more for an insurance policy than those in smaller towns or rural areas. Other area-related pricing factors are cost and frequency of litigation, medical care and car repair costs, prevalence of auto insurance fraud, and weather trends.
- And finally, the type and amount of motor insurance, the limits on basic motor insurance, the amount of deductibles if any, and the types and amounts of policy options.

This is the traditional way of pricing based on population-level statistics available to the insurance companies prior to the initial insurance policy. The contemporary approach in motor insurance is to consider the present patterns of driving behavior through *usage-based insurance* (UBI) schemes. There are three types of UBI:

- The simplest form of UBI bases the insurance premiums on the distance driven *by simply using the odometer reading* of the vehicle. In 1986, Cents Per Mile Now introduced in the insurance market classified odometer-mile rates. The insureds buy prepaid miles of insurance protection according to their needs. Insurance automatically ends when the odometer limit is reached. The insureds must keep track of miles on their own to know when to buy more. In the event of a traffic stop or an accident, the officer can easily verify that the insurance is valid by comparing the figure on the insurance card and the odometer. Critics of this UBI scheme point out that there is a potential odometer tampering to cheat the system, but newer electronic odometers are difficult to tamper with, and definitely it is more expensive to do so because of the equipment needed, so apart from risky, its uneconomical in terms of motor insurance.

- Another instance of UBI is based on distance driven by aggregating miles (or minutes of use) *from installed GPS-based devices* reporting the results via cellphones or RF technology. In 1998, Progressive insurance in Texas started a pilot program using this approach; although the pilot discontinued in 2000, many other products followed this approach, referred to as telematic insurance or black box insurance technology used for stolen vehicle and fleet tracking. Since 2010, GPS-based systems and telematic systems have become mainstream in motor insurance, and two new forms of UBI appeared: Pay As You Drive (PAYD) and Pay How You Drive (PHYD) insurance policies. In PAYD policies, insurance premiums are calculated from the distance driven, whereas in PHYD, the calculation is similar to PAYD, but also brings in additional sensors like accelerometer to monitor driving behavior. Since 2012, smartphone auto insurance policies are another type of GPS-based systems utilizing smartphones as a GPS sensor. Although this system lacks in reliability, it is used due to its availability as it only requires a smartphone that most of the insureds use and no other special equipment.
- The last type of UBI is based on *data collected directly from the vehicle* with a device connected to a *vehicle's On-Board Diagnostic* (OBD II) *Port*, transmitting speed, time of the day, and number of miles the vehicle is driven. Vehicles that are driven less often, in less risky ways, and at less risky times of the day can receive discounts and vice versa. This means drivers have a stronger incentive to adopt safer practices which may result in fewer accidents and safer roads for everyone. In more recent approaches, OBDII-based systems may record braking force, speed, and proximity to other cars, drunken drivers, or drivers using cellphones.

3   Connected Vehicles' Infrastructures

In the recent years, the functionalities and services around connected and autonomous vehicles have grown in a vast manner, by exploiting the technical datasets these vehicles can provide. Some examples of these new services are:

- Infotainment services such as HD video transmission, personal assistant, etc.
- HD maps for autonomous navigation
- Cooperative Intelligent Transport Systems (C-ITS) services [2] such as hazard warning for the drivers
- Autonomous maneuvers such as overtaking, motorway lane merge, etc.
- Algorithms such as pedestrian detection, parking slot mark detection, etc.

To implement these new features, these vehicles (and infrastructures) are demanding more and more bandwidth, based on the criticality of the latency and/or the potential large amount of data able to be transmitted. This fosters the development of new communication protocols and connection paradigms to support them, while new potential services grow around.

In order to gather information from the vehicles, their own protocols and systems have evolved based on the needs and the capabilities of their embedded devices and central processing units. These have been enhanced by IoT technologies and telecommunication systems that allow nowadays communication between different components and actors not only within the vehicle but also with other vehicles, infrastructure, and clouds.

Considering the new connected cars as a complete IoT infrastructure that feeds the Pilot 11 services, it relies on the following standards:

- *OBD* (on-board diagnostics) refers to the standard for exposing different vehicle parameters that can be accessed and checked, in order to detect and to prevent malfunctions. The way and the variety of parameters that can be accessed through this connector have evolved since the 1980s, when this technology was introduced. *OBD-I* interfaces came up at first instance and intend to encourage manufacturers to control the emissions in a more efficient way. *OBD-II* provides more standardized information and system check protocols, creating enriched failure logs and supporting wireless access, via Wi-Fi or Bluetooth. Nowadays, the standards used for monitoring and control of the engine and other devices of the vehicles are OBD-2 (USA), EOBD (Europe), and JOBD (Japan).
- *CAN* (Controller Area Network), born in the 2000s and evolved to support the *CAN BUS* protocol that broadcasts relevant parameter status and information to all the connected nodes within the vehicle. With a data rate up to 1 Mbps, it supports message priority, latency guarantee, multicast, error detection and signalization, and automatic retransmission of error frames. CAN is normally used in soft real-time systems, such as engines, power trains, chassis, battery management systems, etc. *CAN FD* (flexible data rate) fulfills the need of increasing the data transmission rate (up to 5 Mbps) and to support larger frames in order to take advantage of all the capabilities of the latest automotive ECUs (electronic control units).
- *MOST* (Media Oriented Systems Transports) is another protocol based on bus data transfer oriented to the interconnection of multimedia components in vehicles. It was created in 1997, and the main difference with other bus standards for automotive is that it is based on optic fiber, which allows higher data rates (up to 150 Mbps). MOST is normally used in media-related applications and control in automotive.
- *FlexRay* is considered more advanced than CAN regarding costs and features. The most significant are a high data rate (10 Mbps), redundancy, security, and fault tolerance. And while CAN nodes must be configured for a specific baud rate, FlexRay allows the combination of deterministic data that arrives in a predictable time frame, with dynamic (event-driven) data. FlexRay is normally used in hard real-time systems, such as powertrain or chassis.

- *Automotive Ethernet* provides a different option. It allows multipoint connection, and it is oriented to provide a secure data transfer handling large amounts of data. The most relevant advantages for the automotive sector are higher bandwidth and low latency. Traditional Ethernet is too noisy and very sensitive to interference. Automotive Ethernet was raised to overcome these two issues.

3.1   Vehicle to Everything: V2X Communication Technologies

Taking into account the evolution of the autonomous and connected vehicle functionalities, a large part of the data accessed and gathered using the technologies presented above can be uploaded into an upper infrastructure (Cloud, Road Side Units, etc.) or shared with other vehicles with the aim of improving road safety and traffic efficiency. In turn, information from cloud infrastructures and from other vehicles can be received and exploited by a connected car. These scenarios are known as vehicle-to-infrastructure (V2I) or vehicle-to-vehicle (V2V) connections, both grouped into V2X acronyms, and will allow the data gathering process required to feed this pilot.

The V2X communication technologies can be divided into short- and long-range communications. Based on the requirements of the use case and the scenario deployed, one or another would be selected.

3.1.1   Dedicated Short-Range Communications: 802.11p

802.11p, also known in Europe as ITS-G5 [3], is the standard that supports direct communication for V2V and V2I. This specification defines the architecture, the protocol at network and transport level, the security layer, and the frequency allocation, as well as the message format, size, attributes, and headers. The messages defined by the standard are Decentralized Environmental Notification Messages (DENM), Cooperative Awareness Messages (CAM), Basic Safety Message (BSM), Signal Phase and Timing Message (SPAT), In Vehicle Information Message (IVI), and Service Request Message (SRM).

The ETSI standard also defines the quality of service, regarding the transmission and reception requirements. In this way, the protocol is very robust, and it allows it to transmit very low volume of data with very low latency (order of millisecond) which makes it very appropriate for the execution of autonomous vehicle use cases or road hazard warnings.

On the other hand, ITS-G5 communications are defined for short-range communications, directly between vehicles or between vehicles and the near infrastructure as RSUs (Road Side Units) which can cover an area of 400 m radio for interurban environments and 150 m in urban areas. In this way, the more covered is the area, the wider infrastructure required.

3.1.2   Long Range: Cellular Vehicle to Everything (C-V2X)

In recent years, cellular networks for V2X communications have gained importance as an alternative or complement to 802.11p/Wi-Fi. They are supported by many organizations such as the 5GAA (5G Automotive Association) which counts with dozens of global members, including principal carmakers. This radio technology also operates on the 5.9 GHz ITS spectrum and promises very low latency in the transmission reaching the 4 milliseconds or even less, based on the scenario and the network capabilities, as well as overcoming (when possible) known issues or barriers that 802.11p faces.

C-V2X latest release is designed to support and take advantage of 5G network capabilities, regarding speed and data rate transmission.

With regard to the semantic content of the messages, the standard does not include a specification, but it is proposed to use the ITS-G5 one (SPAT, CAM, DENM, etc.) to be sent over the C-V2X data transport layer.

3.1.3   CTAG's Hybrid Modular Communication Unit

In order to communicate with the infrastructure and to publish the data gathered from the vehicle to the Cloud and the Road Side Units, as well as to other vehicles, the cars involved in this pilot need to install an On-Board Unit (OBU). This device is designed, manufactured, and installed by CTAG, and it's known as its Hybrid Modular Communication Unit (HMCU) (Fig. 17.1). This unit implements the link between the vehicle's internal network and the external cloud infrastructures, by supporting the main technologies and interfaces mentioned above: cellular channel (3G/4G/LTE/LTE-V2X/5G) and 802.11p (G5) channel, CAN channel, Automotive Ethernet, Wi-Fi (802.11n), and Ethernet 100/1000, allowing the possibility to modular different configuration for cellular, 802.11p, and D-GPS, according to specific needs.

In this way, the HMCU allows the vehicle's technical data compilation and its update on the corresponding remote infrastructure in real time, supporting novel

Fig. 17.1 CTAG's HMCU that captures vehicle's data and connects to the remote infrastructure

infotainment and telematic services development, plus advanced driving (AD) and IoT services deployment for the different communication technologies.

For this specific pilot, the datasets from inside the vehicle are gathered by using the CAN channel and published to the CTAG Cloud by using the 4G/LTE channel.

To perform the proposed scenario and evaluate the components involved, CTAG has provided 20 vehicles equipped with its HMCU which will be reporting data during pilot's duration, driving free mostly all along CTAG location and surroundings in Spain.

3.1.4 Context Information

Complementing the typical car data source, there are additional information sets which may have a relevant impact on the driving behavior. These available sources are mapped as context information, and the way the driver reacts within this context helps to define their corresponding driving profile. In this sense, the inclusion of these information sources in the driving profiling analysis and the AI models, plus its corresponding impact on the outcomes, represents one of the novelties included in this modeling. INFINITECH developed adaptors to capture this context information related to the driving area monitored in this pilot:

- *Weather information*, including temperature, humidity, visibility, precipitations, wind speed, etc., captured from the Spanish State Meteorological Agency (AEMET) [4] and referenced to the area of Vigo [5].
- *Traffic alerts and events*, which report traffic accidents, weather issues (fog, ice, heavy rain, etc.), or roadworks, providing location and duration. These are captured from the Spain's Directorate General of Traffic (DGT) [6] and cover the whole Spanish region of Galicia [7].
- From the OpenStreetMap's [8] databases, the pilot imports *information from roads*, such as their location, shape, limits, lanes, maximum allowed speeds, etc. The IDs of these roads will be key to relate and correlate data from vehicles, traffic alerts, weather, and roads themselves.

4 Data Gathering and Homogenization Process

To manage the required information sources and support the pilot's driving profiling AI model development, INFINITECH implements an open software infrastructure, known as the smart fleet framework, that integrates all these heterogeneous assets into standard and homogeneous datasets, so it can be served to the AI framework in a coherent and interrelated fashion.

The gathering and homogenization process is represented in Fig. 17.2 and is based on the FIWARE NGSI standard for Context Information Management [9], fostered by the European Commission and supported by the ETSI. This FIWARE

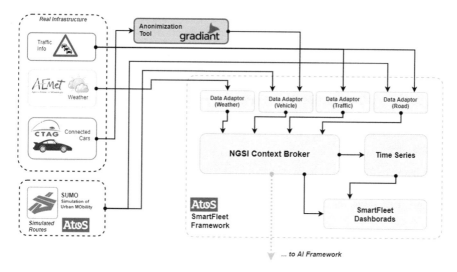

Fig. 17.2 Pilot 11 data sources and data gathering process

approach provides both a common API to update and distribute information and common data models to map and reference each involved data source. According to this, we can consider two sets of components within the smart fleet framework that leads the data collection and standardization:

- The *data adaptors*, also identified within the FIWARE ecosystem as IoT agents [10]. These are flexible components, specifically configured or developed for each data source, considering its original communication protocol (LoRa, MQTT, etc.) and the data format (text, JSON, UltraLight, etc.). They implement the NGSI adaptation layer, and its flexibility aims at integrating new data providers, such as vehicle manufactures or connected car infrastructures.
- The *context management* and *storage* enablers, with the Context Broker as its core. This layer covers data management, data protection, and data processing functionalities from the INFINITECH RA, linking with INFINITECH enablers. It stores and distributes the collected information, once standardized, supporting, among other options, the NGSI-LD protocol for context data and time series retrieval. In addition, they implement SQL and PostgreSQL native protocols to move large numbers of records with elaborated database requests from external layers.

Within this smart fleet framework, collected information is homogenized and normalized according to predefined FIWARE data models, based on its NGSI protocol. This way, smart fleet intends to be aligned with the current Connecting Europe Facility (CEF) [11] program, fostering IoT standardization and integration. The models currently used and enhanced in collaboration with the smart data model [12] initiative are:

- *Vehicle* [13] model to map all (real and simulated) data captured from connected cars.
- *WeatherObserved* [14] to capture data from AEMET's weather stations (area of Vigo's city).
- *Alert* [15] that represents an alert or event generated in a specific location. It is used to get data from DGT's reported traffic events (area of Vigo's city).
- *Road* [16]and *RoadSegment* [17] that map information from roads and lanes (captured from OpenStreetMap [18]) where both simulated and real connected cars are driving.

Linked to the context information and storage components, and based on Grafana [19] tools, the smart fleet framework implements a set of customized dashboards designed to show the data gathering process. An example of these dashboards is shown in Fig. 17.3, representing the last datasets reported by the connected cars, their location, the corresponding speed for last reporting vehicles, and the table with a subset of the captured raw data.

5 Driving Profile Model

Within the context of the INFINITECH project, the objective of Pilot 11 is to improve the vehicle insurance tariffications by using AI technologies, allowing customers to pay according to their driving behavior and routines. To support this, the concept of *driving profiles* will be used and developed. By gathering the data captured on each trip from a vehicle (speed, braking force, indicators' use, etc.), as well as the context information (weather data, traffic alerts, time of the day, road information, etc.), it is possible to group drivers into several clusters, as grounds of further profiles. Each of these profiles can later be analyzed and checked for consistency among different routes, providing critical info about the actual use of each insured vehicle.

Figure 17.4 attempts at depicting the dataflow followed throughout the pilot's AI framework to achieve the proper clustering of the different driving profiles. This section will aim at explaining the different parts of the full algorithm, as well as giving insight at how some of the phases could be improved.

First, datasets are served by the smart fleet framework. This information is aggregated and injected directly into the ATOS-powered EASIER.AI platform [20], the pilot's AI proper framework, by using a specific connector developed on top of the smart fleet APIs. This will greatly help with the *data gathering process*: to retrieve vehicles' data and context information from the smart fleet by exploiting its APIs in a seamless and simple way. This data will be properly encapsulated into JSON format files to be later retrieved by the service.

Once the data has been properly loaded, it undergoes several *data engineering stages*. First action is to ensure that all the data is complete and does not present wrong or empty values. Then, an analysis phase is entered, where the most valuable

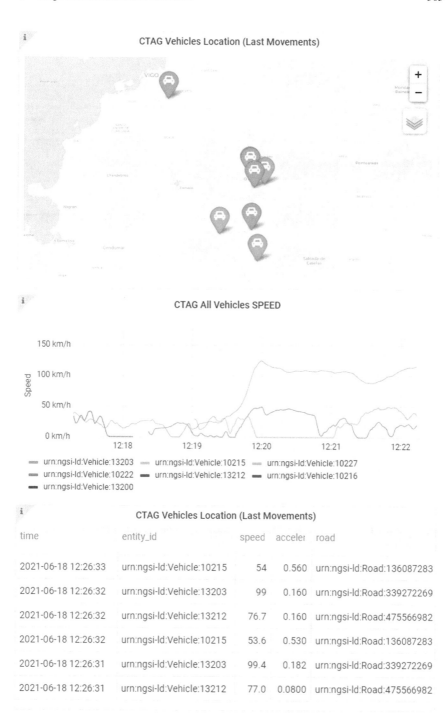

Fig. 17.3 Smart fleet dashboard showing latest reported data from vehicles

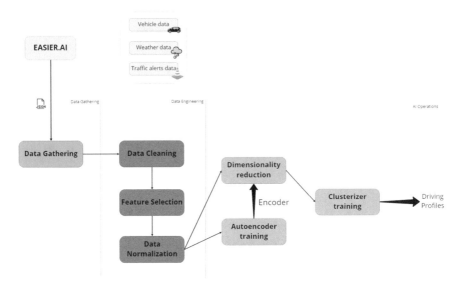

Fig. 17.4 Data clustering process

features for the matter at hand are selected and extracted. Currently, the used features are:

- Average trip velocity
- Trip distance
- Trip duration
- Mean CO2 emissions
- Mean fuel consumption
- Mean acceleration force

Last on the data engineering phase, the data is normalized to ensure that all value ranges are the same. This proves especially useful when dealing with features as different as distances and velocities, whose value ranges could bias the model in a wrong way.

When the data is already fit for inputting into the model, we will proceed *to pass it through an encoder block* in order to reduce its dimensionality. This action will improve the clustering algorithm's performance that will be trained as the final step. In order to achieve this dimensionality reduction, an AI autoencoder component, exploring linear autoencoders [21], will be trained. This type of unsupervised artificial neural network is illustrated in Fig. 17.5. It takes the same data as input and output, resulting in a model that recreates the injected values. This by itself does not prove very useful in this context, but by taking several of its layers, we can find an encoder block that outputs the data with a different number of dimensions.

Once the autoencoder is trained, the encoding layers are taken, and all the data are passed through it. These data, now scrambled and with fewer features, are sent to our

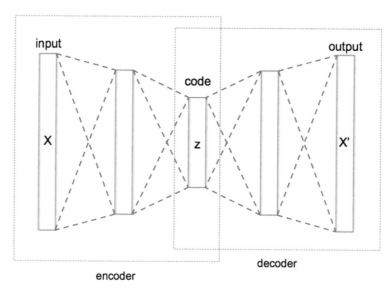

Fig. 17.5 Schematic structure of an autoencoder with three fully connected hidden layers [22]

K-means clustering algorithm [23], which will separate each driver's information into different clusters, according to the parameters of each route. Once the model and algorithm have been trained, the different clusters can be characterized and assigned to a driving profile, so they can be served in order to classify new drivers and routes.

The pilot will study different customizations of the mentioned sifting stages, encoding and clustering algorithms that produce the AI driving profiling model, testing this with tagged routes (real routes and simulated routes) to refine its accuracy and support the final Pay How You Drive and fraud detection services.

6 Customized Car Insurance Services Powered by ML

As it was mentioned in the first section of this chapter, the problem in motor insurance and in every insurance sector is accurate risk assessment, and the solution to this problem is usage-based insurance (UBI) products and services. Pay As You Drive and Pay How You Drive insurance products exist in the insurance market, but their level of integration is significantly low. One of the services developed within the pilot is a Pay How You Drive insurance product with higher level of accuracy than the existing ones, by exploiting, apart from the numerous data provided by the On-Board Unit connected to car's OBD-II, context information including weather conditions and traffic incidences. The innovation this pilot brings in the insurance business will make the services developed quite desirable.

As in all the UBI products, insurance premiums are not static and predefined based on historical data, but based on the actual usage and as consequence on the actual risk. Insurance companies will have the opportunity to provide to their clients personalized insurance products based on connected vehicles. Having all these data gathered as mentioned in previous sections and using artificial intelligence (AI) techniques and machine learning (ML) algorithms, there will be created different driving profiles from conservative and considerate drivers to aggressive, very aggressive, or extremely aggressive drivers. The more the driving profiles created by training the algorithms with real and simulated data, the more accurate will be the classification, and the more accurate will be the reflection of this classification in the premiums.

In the following lines, there will be presented some of the numerous scenarios to emphasize how important in creating driving profiles and classifying drivers in these profiles is the combination of data collected from the vehicles and data gathered with context information:

- Starting from some simple scenarios like driving above speed limit and making a turn or changing lanes without using blinkers, where depending on the deviation of the speed limit or the percentage of times using and not using blinkers the drivers can be classified in the different profiles.
- Another scenario may be the frequent change of lanes in a straight segment of the road, classifying drivers according to how many times they do so in a specific distance driven.
- Then, we have scenarios combining weather conditions with the driving behavior monitored, for example, driving with relatively high speed in severe weather conditions and driving with lights turned off after sunset or in low visibility conditions.
- Finally, using the data collected from vehicle's OBU and traffic incidences, we can built a scenario like driving without reducing speed when approaching area with a car accident or heavy traffic.

Having available all these data gathered and collected, it is easily understood that these are only few of the scenarios that can be exploited to generate different driving profiles.

Another problem to deal with in motor insurance is the fraud detection in a car incident, such as an accident or theft. So far, the means insurance companies have to detect fraud in a claim are confined to:

- Those related to police investigation (police reports, eyewitnesses, traffic control cameras, court procedure)
- Those related to experts' opinion, for example, experts or even investigators may find inconsistencies by examining the cars involved in an accident or the drivers involved
- Those related to social media (there are cases where a relationship between the drivers involved in a car accident is revealed or even a video with the actual accident and moments before and after the accident is uploaded)

Obviously, the data gathered and collected within the pilot can be used to provide to the insurance companies one more service, the fraud detection service. Here are some use cases that insurance companies can detect fraud with the process of all these data:

- The vehicle reported to be involved in the accident is different than the real one.
- The driver reported to be involved in the accident is different than the real one or different than the usual driver of the vehicle.
- To report an accident in a different location than where it occurred.
- To detect speed and driving reactions before the moment of the accident and generally all the circumstances of the accident.
- To detect the driving behavior in a period before the accident.
- Last but not least, to detect fake car thefts.

Some of these can even be detected with a simple GPS tracker, certainly not all of them, and some may need to be considered in forthcoming legislations, but this is the challenge of this service, and the result will be beneficial for the insurance companies, the insureds, and the society in general, by assuming responsibilities correctly, paying claims according to the real responsibility, and avoiding fraud.

7   Conclusions

Calculating the risk of the insured object is one of the most delicate tasks when determining the premium for car insurance products. Traditional methods use common statistical data, providing general and sometimes biased results. The connected car scenario provides the possibility of getting more precise data about the vehicle, the driver, and its context, which offer new ways to improve driving safety and risk estimations.

With this idea, personalized usage-based insurance products rely on real-world data collection from IoT devices and frameworks to tailor novel services adapted to actual risks and driving behaviors. The presented pilot within INFINITECH ecosystem explores new ML/DL approaches to exploit datasets from vehicles enriched with new and relevant context information and so and develop AI models that enhance the risk assessment for insurance services. This specialized data gathering process and the applied AI technologies support the business innovation that allows insurance companies to evolve the way they offer premiums and prices, according to their insured clients' real profile instead of using general statistics. In this line, the pilot implements a complete infrastructure to develop AI models and services for car insurance providers so the proposed business models, Pay How You Drive and fraud detection, can be explored. Real datasets captured from CTAG vehicles plus the synthetic and simulated data reported by our simulation environments are enough to get a first version of the driving profiling AI model, but for expanding this solution and obtaining evolved and more accurate outcomes, there are some challenges that must be considered, mainly related to three issues:

- Datasets and data sources: the wider the set of relevant data captured, the better the model will be. This requires involving as many connected cars as possible so the whole system proposed here relies on automotive and IoT standard, trying to make the integration easier for new car manufactures and IoT providers. Extra efforts should be done also to identify new sources that impact on driving profiling and remove those datasets from the AI modeling that have less relevance.
- Accuracy of the driving profiling, key factor that should be improved by adapting ML/DL techniques, reinforcing training, and evaluating the impact of anonymization.
- Evaluation and adaptation of the services offered to both insured clients and insurance companies, by involving related stakeholders on the new versions' development.

Acknowledgments This work has been carried out in the H2020 INFINITECH project, which has received funding from the European Union's Horizon 2020 Research and Innovation Program under Grant Agreement No. 856632.

References

1. ETSI, G. (2019). Context Information Management (CIM); NGSI-LD API. Tech. Rep.[Online]. Available: https://www.etsi.org/deliver/etsi_gs/CIM/001_099/009/01.01.01_60/gs_CIM009v010101p.pdf
2. About C-ITS. (2021). Retrieved 23 June 2021, from https://www.car-2-car.org/about-c-its/
3. ETSI, T. (2013, July). Intelligent Transport Systems (ITS); Access layer specification for Intelligent Transport Systems operating in the 5 GHz frequency band. EN 302 663 V1. 2.1.
4. State Meteorological Agency – AEMET – Spanish Government http://www.aemet.es/en/portada
5. https://goo.gl/maps/JhxAxXNHYX1axCTN6
6. Dirección General de Tráfico https://www.dgt.es/es/
7. https://goo.gl/maps/n1zpRY75bnukDMhQ8
8. OpenStreetMap https://www.openstreetmap.org/
9. ETSI, G. (2019). Context Information Management (CIM); NGSI-LD API. Tech. Rep.[Online]. Available: https://www.etsi.org/deliver/etsi_gs/CIM/001_099/009/01.01.01_60/gs_CIM009v010101p.pdf
10. IoT Agents – Academy https://fiware-academy.readthedocs.io/en/latest/iot-agents/idas/index.html
11. CEF Digital Home https://ec.europa.eu/cefdigital/wiki/display/CEFDIGITAL/CEF+Digital+Home
12. Smart Data Models – FIWARE https://www.fiware.org/developers/smart-data-models /
13. smart-data-models/dataModel.Transportation Vehicle https://github.com/smart-data-models/dataModel.Transportation/blob/master/Vehicle/doc/spec.md
14. smart-data-models/dataModel.Transportation Weather https://smart-data-models.github.io/dataModel.Weather/WeatherObserved/doc/spec.md
15. smart-data-models/dataModel.Transportation Alert https://smart-data-models.github.io/dataModel.Alert/Alert/doc/spec.md
16. smart-data-models/dataModel.Transportation Road https://github.com/smart-data-models/dataModel.Transportation/blob/master/Road/doc/spec.md

17. smart-data-models/dataModel.Transportation Road Segment https://github.com/smart-data-models/dataModel.Transportation/blob/master/RoadSegment/doc/spec.md
18. OpenStreetMap. (n.d.). http://www.openstreetmap.org/.
19. Grafana® Features. (n.d.). Retrieved from https://grafana.com/grafana/
20. We Do Easier AI. (n.d.). Retrieved from https://webapp.easier-ai.eu/
21. Plaut, E. (2018). *From principal subspaces to principal components with linear autoencoders.* ArXiv, abs/1804.10253.
22. By Chervinskii – Own work, CC BY-SA 4.0, https://commons.wikimedia.org/w/index.php?curid=45555552
23. https://sites.google.com/site/dataclusteringalgorithms/k-means-clustering-algorithm

Chapter 18
Alternative Data for Configurable and Personalized Commercial Insurance Products

Carlos Albo

1 Introduction: The State of Practice Regarding the Use of Data in the Insurance Sector

The personalization of insurance products is a peremptory necessity for insurance companies. This is because, until now, the scarce availability of information on the part of clients has made such personalization impossible. Furthermore, there is little or no automation in data entry, which leads to a significant amount of data errors, low quality, repeated data, and other data issues.

Despite these problems, insurance companies base all their activity on data management, so these types of processes are key in these companies. Therefore, these types of entities invest important resources in carrying out part of these tasks semi-manually and with obsolete data or data sent by the client, based on the experience of certain highly qualified personnel. However, if they really want to compete with an advantage in today's world, it is necessary to automate and improve these processes.

In the following paragraphs, we analyze the state of the art of different technological areas, along with related innovations.

C. Albo (✉)
WENALYZE SL, Valencia, Spain
e-mail: carlos@wenalyze.com

© The Author(s) 2022
J. Soldatos, D. Kyriazis (eds.), *Big Data and Artificial Intelligence in Digital Finance*,
https://doi.org/10.1007/978-3-030-94590-9_18

1.1 Predictive Underwriting and Automation of Underwriting Processes

Strangely enough, the finance and insurance industries still gather and complete information based on manual or quasi-manual procedures. This leads to great inefficiencies, reduced economies of scale, and poorer customer service. As an example, there are cases of insurance pricing for SMEs in the USA where the average time from product application to authorization and issuance takes on average between 30 min and 1 h.

In recent years, robotic information gathering techniques, automated ingestors, and machine learning hold the promise to offer significant advantages in this area. The use of such techniques has many advantages, including:

1. The automatic subscription system can be put into operation more quickly if it deals with only a subset of all possible instances.
2. The error rate that the automatic underwriting system will have. In easy cases, it will be lower than the error rate in non-automated cases.
3. Deployment issues (e.g., integration with other systems, obtaining user feedback and acceptance) can be addressed at an early stage of development.

1.2 Customization of Financial Product Offerings and Recommendation According to the SME's Context

Product recommendation systems determine the likelihood that a particular product will be of interest to a new or existing customer. The rationale behind the recommendations is to enable targeted marketing actions addressed only to those business customers who might be interested, thus avoiding the traditional bombardment of untargeted advertising.

Recommendation systems are basically of three types:

- Characteristic-based: In this case, the system learns to select potential customers by studying the differentiating characteristics of existing customers.
- Grouping-based: Customers are divided into groups based on their characteristics using unsupervised learning algorithms, and the system recommends products oriented to each group.
- Based on collaborative filtering: Such systems recommend products that are of interest to customers with similar profiles, based exclusively on the preferences expressed by users.

Collaborative filtering systems are the most widely used. However, these systems have been developed for cases where the number of products to be recommended is very large, for example, the case of Netflix movies. This is not typically the case with insurance products, where the number of customers is much larger than the number of products.

There are not many documented applications of these techniques to the insurance sector. For example, Qazi [1] proposes the use of Bayesian networks, while Rokach [2] uses simple collaborative filtering to obtain apparently good practical results. Mitra [3] uses hybrid methods combining feature-based recommendation with preference matching, and Gupta [4] uses another mixed approach, which however includes clustering algorithms.

1.3   Business Continuity Risk Assessment

A correct assessment of the business continuity risk assumed when offering a financial or insurance product is a foundation for doing business in the financial and insurance sector. Hence, all financial institutions have systems for estimating this risk.

This is another area where the use of machine learning can offer significant advantages. However, although the use of these techniques is common in credit scoring activities, their use in insurance remains quite limited. As an example, in [5], Guelman compares traditional linear models with decision tree algorithms with good results.

Up to date, financial and insurance companies leverage machine learning models based on internal and traditionally financial data. Hence, the use of alternative data sources for the improvement of machine learning models in insurance applications is completely new. Despite the implementation of personalization features in predictive underwriting and process automation, open data and data from alternative sources are not currently used, and their innovation potential remains underexploited.

2   Open Data in Commercial Insurance Services

Nowadays, the most common business data sources available include information about their financial statements, property data, criminality levels, catastrophic risks, news, opinions from clients, information from the managers, number of locations, etc. While some of these sources are useful to update our current data, other sources can be useful to automate processes and increase the risk assessment accuracy.

	Insurer's Data (2018)		Open Data (2020)
Name	Weanalyze	✕	Wenalyze
Address	Av. de Secundino Zuazo, 111, 28055 Madrid, Spain	✕	Calle Indústria, 22, 46022 Valencia, Spain
Employees	3	✕	7
Activity	Consulting Services	✕	FinTech Start-up

Fig. 18.1 Information inaccuracy example

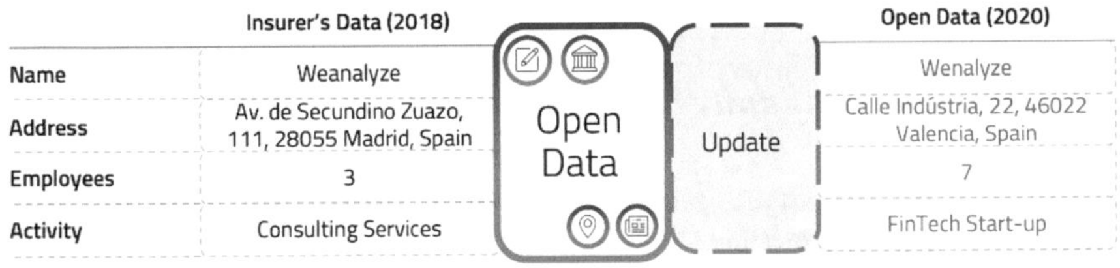

	Insurer's Data (2018)			Open Data (2020)
Name	Weanalyze			Wenalyze
Address	Av. de Secundino Zuazo, 111, 28055 Madrid, Spain	Open Data	Update	Calle Indústria, 22, 46022 Valencia, Spain
Employees	3			7
Activity	Consulting Services			FinTech Start-up

Fig. 18.2 Linking open data to insurer's database

2.1 Improving Current Processes

Data quality is the first aspect that must be improved. It is important to know the reality of the customer or potential customer to improve the results of daily tasks. Understanding the customers' context is important prior to deploying sophisticated processes. The problem is rather common: According to our projects, an average of 45% of insurer's commercial customers has inaccuracies in their business activities or/and their addresses, as illustrated in Fig. 18.1.

Figure 18.1 illustrates the data sent to the insurer when buying a business insurance policy right. However, this business information changes over time. Insurers tend to renew policy rights without updating the business data as required. Therefore, open data are likely to comprise inaccurate or obsolete information. To alleviate this issue, there is a need for linking open data to the actual data sources, so as to ensure that open data are properly updated and remain consistent with the actual data (Fig. 18.2).

2.2 Process Automation

Once the data are adjusted to the customers' reality, process automation becomes feasible. By establishing a connection with our systems and open data sources, process automation can be applied to underwriting and renewals to reduce times and increase efficiency. Figure 18.3 illustrates the change from a traditional underwriting process for business property insurance to automated underwriting. The traditional underwriting process for a business property insurance requires the potential

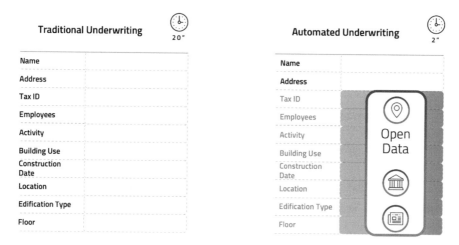

Fig. 18.3 Traditional vs. automated underwriting

customer to collect an extensive list of data points. In the case of Fig. 18.3, our system is connected to a property register and can perform the data collection instead of the customer. This can greatly simplify and accelerate the underwriting process. Specifically, a two-question underwriting process can be offered to the customer.

2.3   Ensuring Data Quality

Data quality is a major concern as insurers base risks and premiums on the data they receive. The information that can be found online about a business is typically shared by themselves, their clients, or public institutions' business databases. Hence, the quality of the data must be assessed differently depending on the source and type of data.

Businesses share information on their website, on their social media profiles, and on their Google Business Card. Data quality can be audited by checking when it was published and whether it matches the different data sources. For example, if the business address instances that are found on Google Maps, the business website, and their social media profiles are the same, it can be safely assumed that this data point is correct and updated.

When the data is found on public registers or opinions, it often comes with the date on which the information was published. This enables the use of the information to assess whether the insurer's data is more recent or not.

Another problem is illustrated in Fig. 18.4, which presents the rating of two different businesses. Business 1 has a higher rating than Business 2. However, it only received seven opinions, and all of them were published in only 1 day. Therefore, the number of opinions and dates must be considered, to avoid potential doubts that

	Business 1	**Business 2**
Opinions Number	7	3,048
Rating	4.9/5	4.6/5
First Opinion Date	December 1, 2020	January 13, 2005
Last Opinion Date	December 2, 2020	December 22, 2020

Fig. 18.4 Ratings for two different businesses

can lead to a customer's dissatisfaction and complaints. In this case, opinions for Business 1 will be considered, as it seems that either it is either a new business or clients do not like to share opinions about it. On the other hand, Business 2 is strong and popular and has a good rating.

In the future, commercial insurance processes will become more automated. The automation of underwriting and renewal processes will provide significant benefits and increased profitability, especially in the case of small-medium enterprises (SMEs) that usually have small premiums. SMEs represent a market segment where competition among insurers is high. Hence, insurance companies that will manage to offer SMEs lower prices will be more competitive in this market.

Automation is however more challenging for bigger risks and other segments of the insurance market. Insurers that handle big premiums can invest on human resources to perform the underwriting process with the human in the loop. Most importantly, they can do so without any substantial increase in the premium. Also, automated systems are sometime prone to errors or inefficiencies that cannot be tolerated for certain types of risks. In this context, machine learning techniques will be used on conjunction with human supervision. Human supervision is not only useful to prevent these mistakes but also to find new ways to apply open data.

Overall, increasing data quality is a key prerequisite for leveraging open data in insurance. With quality data at hand, one can consider implementing and deploying more complex processes, including the development of more automated and efficient data pipelines for insurance companies.

3   Open Data for Personalization and Risk Assessment in Insurance

To enable users to leverage open data for personalization and risk assessment in insurance applications, a new big data platform has been developed. The main functional elements of the platform are outlined in following paragraphs.

3.1   User Interface Application

One of the main modules of the platform is its user interface, which enables users to manage their information. The interface is developed with the IONIC v3 framework using languages such as Angular 6 with the TypeScript layer, HTML5, and CCS3 with Flexbox. This setup enables the delivery of the user interface in Web format while at the same time facilitating its operation on mobile platforms (i.e., Android, iOS) as well. This client environment is hosted in the Amazon EC2 service (Amazon Elastic Compute Cloud), using a Nginx server, being able to increase/scale the power of the server depending on the load of the clients.

3.2   Information Upload

The platform enables insurance and financial entities to upload their company's information in the system. The information can be uploaded either via an appropriate API (application programming interface) or using a panel on the platform in a CSV format. The latter option is important, as many SME companies are not familiar with open API interfaces. Once uploaded, the information is structured and automatically converted to a JSON (Javascript Object Notation) format. Accordingly, the information is stored in a specific "bucket" for that company in Amazon S3, which includes separate JSON files for each client of the company that uses the platform.

3.3   Target Identification: Matching Targets with Sources

As soon as the information is uploaded to Amazon S3, triggers are automatically activated. Triggers are a series of algorithms that analyze the information uploaded by the client and identify the different possible sources from which, according to the uploaded data, the user can be identified on the Internet. The targets are identified and uploaded to a non-relational database using the AWS (Amazon Web Service) DynamoDB module.

Once the target is loaded, a set of matching robots are activated. These robots go through all the possible sources identified in the loaded data and associated the client information with the possible Internet sources that comprise the same pieces of information about the company. The marching is performed with good accuracy, yet there is never absolute 100% success in the matching. Once matched, the relevant information is stored to identify the customer in the source.

3.4 Information Gathering, Collection, and Processing: ML Algorithms

The collection and processing of information from open sources are carried out using cognitive algorithms technology, namely, WENALYZE Cognitive Algorithms (WCA). These algorithms process in real time only the information that belongs to the users or targets without storing any data as it is found in the source of origin. Hence, only the result of the analysis is retained or stored.

Once the matching of the target with the different Internet sources has been completed, the platform unifies all the information about the company, leveraging all the different sources in a single node per user. Analytical algorithms are applied to the stored information toward associating the clients with the information obtained from external sources. This association is based on a hash map search algorithm, which forms an associative matrix. The latter relates catalogued words from a dictionary with those of the previously stored client information. This relationship generates a series of results that are stored to facilitate processes like fraud detection, risk assessment, and product recommendation.

For the automation of the underwriting processes, classification algorithms are configured and trained, notably algorithms applying binary classification such as SVM (support vector machines).

For the product recommendation system, a hybrid algorithm between feature-based filtering and collaborative filtering has been designed to maximize the use of all available information.

For risk assessment, regression algorithms are used to predict a numerical value corresponding to the risk of the transaction. This regressor is based on the characteristics obtained from all the data sources mentioned above.

4 Alternative and Automated Insurance Risk Selection and Insurance Product Recommendation for SMEs

The platform is installed, deployed, and operated in one of the data centers and sandboxes provided by the H2020 INFINITECH project. The platform is destined to operate as a software as a service (SaaS) tool that will be used by different customers based on a fair additional customization and parameterization effort for each new use case. It also offers a visual part that will allow the creation of a demonstrator and an API for connectivity.

The platform automates the processes of data extraction from different sources (i.e., from insurance companies, from banks, and from public data records or social networks) while boosting the ingestion and the preprocessing of this data. Moreover, it facilitates the storage and processing of distributed data, along with the execution of analytical techniques over large amounts of data toward improving business processes.

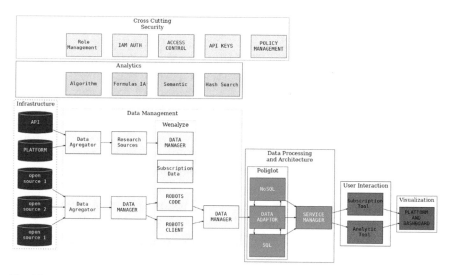

Fig. 18.5 Elements of the personalized insurance product platform in line with the INFINITECH-RA

The data pipelines of the platform are illustrated in Fig. 18.5. The platform comprises of the following main modules:

- A cloud infrastructure, which provides scalability, distribution, and fault tolerance
- Data extractors and data ingestion modules for the extraction and loading of data from insurance companies and banks
- Data extractors from external sources, for the extraction and loading of data from external sources
- Multi-language and multicountry data processing modules, which enable adaptive processing and extraction of different structured and unstructured data sources toward extraction of relevant characteristics
- Storage modules that facilitate the creation of a distributed and secure data lake
- Data processing and analysis modules that perform anonymization and transformation functions, including functions that identify the relationship between different sources and extract relevant characteristics
- Microservices, which enable the scalable implementation of the business processes of the platform

Moreover, the platform provides a dashboard including different views for the end customer and for internal use by the company. Also, integration with API and through online browser access is supported, to ensure connectivity with client financial institutions.

5   Conclusions

This chapter addresses one of the fundamental problems of the insurance industry: the use of data beyond the traditional data sources, toward broadening the spectrum of data collection in a scalable way and with open-source processing. The implementation and use of the results of this project in the INFINITECH project represent a quantum leap in improving the quality and variety of data. The presented solution improves the efficiency of insurance processes and provides significant efficiency benefits to the insurance industry. In this direction, advanced technologies in the areas of robotic processing, cloud computing, and machine learning are combined, integrated, and used.

Leveraging machine learning techniques, the presented solution can support the following use cases: (i) automation of (banking) financial product underwriting, (ii) risk estimation and product recommendation, and (iii) prediction of the company's business continuity.

Acknowledgments The research leading to the results presented in this chapter has received funding from the European Union's funded Project INFINITECH under Grant Agreement No. 856632.

References

1. Qazi, M., et al. (2017). An insurance recommendation system using Bayesian networks. In *Proceedings of the eleventh ACM conference on recommender systems* (pp. 274–278).
2. Rokach, L., et al. (2013). Recommending insurance riders. In *Proceedings of the 28th annual ACM symposium on applied computing* (pp. 253–260).
3. Sanghamitra Mitra, B. P., & Chaudhari, N. (2014, October). Leveraging hybrid recommendation system in insurance domain. *International Journal of Engineering and Computer Science, 3*.
4. Gupta, & Jain, A. (2013, October). Life insurance recommender system based on association rule mining and dual clustering method for solving cold-start problem. *International Journal of Advanced Research in Computer Science and Software Engineering, 3*.
5. Guelman, L. (2012). Gradient boosting trees for auto insurance loss cost modeling and prediction. *Expert Systems with Applications, 39*(3), 3659–3667.

Part V
Technologies for Regulatory Compliance in the Finance Sector

Chapter 19
Large Scale Data Anonymisation for GDPR Compliance

Ines Ortega-Fernandez, Sara El Kortbi Martinez, and Lilian Adkinson Orellana

1 Introduction

Data privacy refers to the rights of individuals over their personal information. That is, it is concerned with the data collection, its purpose, and how is it handled. In recent years, data privacy has become a major issue due to the growth of data generation, as well as due to the interest of third parties, such as business or researchers, in collecting and exploiting that information. It is important to differentiate it from the concept of data security, whose main objective is to protect personal information from being accessed by unauthorised third parties or attackers. Data security is, however, a prerequisite for data privacy [1].

One way to ensure data privacy is through anonymisation, a process that enables being non-identifiable within a set of individuals. It is a Privacy Enhancing Technique (PET) that results from transforming personal data to irreversibly prevent identification, and it comprises a set of techniques to manipulate the information to make data subjects (i.e., the persons to which the data refers) less identifiable. The robustness of each anonymisation technique can be analysed in terms of different criteria, such as if is it possible to identify a single person, link different records regarding the same individual, or the quantity of information that can be inferred regarding the data subject [2]. As a result, once the data is properly anonymised it cannot be linked back to the individual, and therefore it is not considered personal data anymore, according to the General Data Protection Regulation (GDPR): *"The principles of data protection should therefore not apply to anonymous information, namely information which does not relate to an identified or identifiable natural person"*.

I. Ortega-Fernandez (✉) · S. El Kortbi M. · L. Adkinson O.
Fundacion Centro Tecnoloxico de Telecomunicacions de Galicia (GRADIANT), Vigo, Spain
e-mail: iortega@gradiant.org; selkortbi@gradiant.org; ladkinson@gradiant.org

© The Author(s) 2022
J. Soldatos, D. Kyriazis (eds.), *Big Data and Artificial Intelligence in Digital Finance*,
https://doi.org/10.1007/978-3-030-94590-9_19

It is important to highlight the differences between anonymisation and other PETs, such as pseudonymisation and encryption, since these terms are often misunderstood: pseudonymisation protects an individual by replacing its identifiers with a pseudonym, making the individual less (but still) identifiable. It is a reversible process, so it does not remove the re-identification risk, since the mapping between the real identifiers and pseudonyms still exists. Article 25 of the GDPR highlights the role of pseudonymisation *"as a technical and organisational measure to help enforce data minimisation principles and compliance with Data Protection by Design and by Default obligations"* [3].

On the other hand, encryption is a security measure to provide confidentiality in a communication channel or with the data at rest (data saved in a physical stable storage), to avoid disclosure of information to unauthorised parties. The goal of encryption is not related to making the data subject less identifiable: the original data is always available to any entity that has access to the encryption key (or can break the encryption protocol, recovering the original information), and therefore the possibility to identify the subject remains [2]. In addition, a key management system must be in place to protect and manage the encryption keys, which introduces complexity in the system.

The main advantage of anonymisation techniques over other PETs such as encryption is that it does not involve key management, and depending on the technique it requires less computational resources. However, data anonymisation is an irreversible process and only provides privacy, meaning that other security properties (e.g., confidentiality or integrity) must be implemented through other means. In addition, adequate anonymisation allows data processing without the need to be compliant with data privacy regulations, reducing the organisational cost of using and exploiting the data.

However, in order to ensure that data is properly anonymised multiple factors need to be taken into account: anonymisation of data to lower the re-identification risk to a specific extent is not always possible without also losing the utility of the data. Moreover, the anonymisation process requires a deep analysis of the original dataset to find the best anonymisation procedure that fit our needs. The anonymisation process that was suitable for a particular dataset might not work for a second dataset, since the nature and scope of the data will differ, as well as the later use of the anonymised data. In addition, it is also necessary to consider that additional datasets might be available in the future, which could be used for making cross-referencing with the anonymous data, affecting the overall re-identification risk.

Despite how useful the data anonymisation can be from a compliance perspective, it can be a daunting task, especially if we try to minimise the risks and ensure that the data is being properly anonymised. This chapter explores how anonymisation can be used as a Regulatory Compliance Tool, addressing common issues to introduce data anonymisation in a Big Data context. The complexity of a data anonymisation procedure is explored in Sect. 2, and its relationship with GDPR compliance. Finally, the specific challenges of data anonymisation in a Big Data context are analysed in Sect. 3, highlighting the differences with a Small Data environment.

2 Anonymisation as Regulatory Compliance Tool

With regards to the GPDR, encryption and pseudonymisation are considered security measures that need to be implemented to allow the processing of personal data. On the other hand, anonymisation makes the individuals within a particular dataset non-identifiable, and therefore anonymised data is not considered personal data anymore.

However, ensuring that the data anonymisation is being correctly applied is a challenging task. The use of data anonymisation techniques implies, in most cases, a certain loss of the utility of the data, as it relies typically on modifying the data values in order to make them less unique. While privacy aims to avoid sensitive data disclosure and the possibility of making certain deductions from a given dataset, the utility describes the analytical value of the data. In other words, the utility seeks to find correlations in a real-world scenario from a certain dataset, and the goal of privacy is to hide those correlations [4]. When applying anonymisation operations, data privacy improves, but there is a risk of reducing the analytical value of the dataset. For this reason, it is necessary to find a suitable trade-off between privacy and utility, as illustrated in Fig. 19.1, where some utility is sacrificed to reach an acceptable level of privacy.

D'Acquisto et al. distinguish two anonymisation approximations to find this trade-off point [5]: the Utility-first and the Privacy-first anonymisation. In the

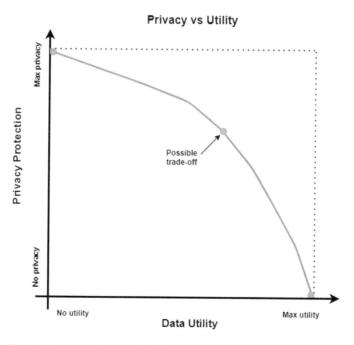

Fig. 19.1 Trade-off between privacy and utility during an anonymisation process

former, an anonymisation method with a heuristic parameter and utility preservation properties is run, and the risk of disclosure is measured afterwards. On the contrary, in the latter, an upper bound on the re-identification disclosure risk and/or the attribute disclosure risk is set.

Independently on which anonymisation approach is followed, it is essential to ensure that the individuals are not identifiable after the process. To verify this property, it is necessary to analyse the data and classify it into direct identifiers (data attributes that unequivocally identify a data subject, such as the name, telephone number or ID card number), quasi-identifiers (attributes which by themselves do not reveal an identity, but they can suppose a privacy risk when they are combined with others; e.g., postal code or birth date) and sensitive data (attributes that should be preserved as they present a certain value for later analysis, such as a medical condition or salary).

It has been proven that removing the direct identifiers from a dataset is not enough to preserve privacy: an attacker might have access to additional data sources or background knowledge that could lead to the re-identification of the anonymised individuals [6]. Montjoye et al. [7] demonstrated that 95% of the individuals of an anonymised dataset of fifteen months of mobility data (containing records from around 1.5M individuals) could be identified when taking into consideration only four spatial-temporal points. This probes how simple approaches to anonymisation are not enough and more complex solutions are needed.

The GDPR sets a high standard for data to be considered truly anonymous since it implies that data protection regulations do not apply anymore: in Recital 26, the GDPR states that the organisation should not only consider if the individual is re-identifiable, but also consider: *all the means reasonably likely to be used, such as singling out, either by the controller or by another person to identify the natural person directly or indirectly. To ascertain whether means are reasonably likely to be used to identify the natural person, account should be taken of all objective factors, such as the costs of and the amount of time required for identification, taking into consideration the available technology at the time of the processing and technological developments* [3].

The sentence above can be better understood with a real example: in October 2018, a Danish service was fined 1.2 million kroner (around €160.000) for not deleting users' data once it wasn't needed anymore for the company's business activity [8]. The company argued that they could keep the data since it was properly anonymised and it could not be considered personal data anymore: they were anonymising the dataset by deleting the names from the database, but other data such as telephone numbers were neither removed nor masked.

The assumption of the company about how anonymised data cannot be considered personal data was correct. However, they failed to analyse the GDPR requirements for data to be considered anonymous. At Opinion 05/2014 on Anonymisation Techniques, the Article 29 Working party clearly states that *"removing directly identifying elements in itself is not enough to ensure that identification of the data subject is no longer possible. It will often be necessary to take additional measures to prevent identification, once again depending on the context and purposes of the processing for which the anonymised data are intended"* [2].

According to Article 29 Working Party, an effective anonymisation solution should satisfy the following criteria, preventing all parties from:

- Singling out an individual in a dataset (i.e., isolating some records that could point out the identity of the person).
- Linking two records within a dataset that belong to the same person, even though her identity remains unknown.
- Inferring any information in such dataset by deducing the value of an attribute from another set of attributes.

Therefore, re-identification does not only mean retrieving a person's name but also being able to "single out an individual in a dataset". Coming back to our previous example, now the issues of the company's anonymisation solution are clear: by keeping direct identifiers such as telephone numbers in the dataset, they did not satisfy the singling nor the linkability criterion: first, telephone numbers can be considered direct identifiers since each person will have a different telephone number, singling out all individuals in the dataset. Second, a telephone number can be easily linked to a natural person by using other datasets and publicly available information. Therefore, the data could not be considered anonymous, and they were not fully compliant with GDPR.

This example demonstrates the complexity of data anonymisation to be GDPR compliant: basic anonymisation techniques are insufficient to guarantee the privacy of the data subjects; the risk that the anonymised data retains must be analysed to ensure a proper level of privacy protection, and how the anonymisation solution is able to meet the requirements stated above.

Therefore, the anonymisation process must be adapted on a case-by-case basis, ideally adopting a risk-based approach. An analysis of the re-identification risk must be performed to assess if the anonymisation solution meets the criteria, or further measures are needed [2]. Moreover, this analysis must be performed continuously since it is subject to change, and new datasets might be published that allow cross-referencing anonymous data: for instance, in Sweden taxpayers' information is publicly available [9], while in other countries is not. This circumstance might change in the future, making previous assumptions about data linkability erroneous. In addition, a dataset containing information about both Swedish and Spanish citizens cannot be anonymised following the same procedure, since the re-identification risks might be different [10].

As a result, the analysis of the re-identification risk is a useful procedure that will allow us to identify the residual risk of the anonymised data, as well as easing the GDPR compliance process. The re-identification risk of the data can be assessed using multiple techniques, mainly focused on analysing the uniqueness of the data subjects (the first criteria for a good anonymisation solution) but that do not take into account other properties such as linkability or inference. To address this issue, Adkinson et al. [11] propose a dynamic risk-based approach and a set of privacy metrics that take into account both the re-identification risk and the presence of an adversary with certain background knowledge trying to disclose information from a dataset.

To summarise, anonymisation can be considered as a non-trivial process that needs to be performed cautiously to become GDPR compliant. It is essential to understand that basic anonymisation procedures are not usually sufficient for most of the real-world applications and datasets, and that it is necessary not just to anonymise the data but also to estimate the remaining privacy risks. Therefore, it is essential to perform a continuous analysis of the re-identification risk of the anonymised dataset, taking into account the uniqueness, linkability, and inference properties of the anonymisation solution applied. Furthermore, the anonymised datasets need to guarantee an adequate balance between data privacy and utility, preserving the analytical value of the data, which requires analysing different approximations to decide the adequate trade-off point between data privacy and utility for the future data processing purpose.

3   Anonymisation at Large Scale: Challenges and Solutions

The rapid increment of the amount of data that business and researchers handle nowadays is a consequence of the technological advances in fields such as Cloud Computing, the Internet of Things (IoT) or Machine Learning, the increase of computational power and the lower cost of data storage. This has led to the concept of Big Data: large volumes of data, whose complexity and velocity hinders the use of conventional technologies [12]. This definition remarks three properties of Big Data, also known as 3V's: volume, velocity, and variety. However, later studies improved this definition by adding the properties of veracity, variability, visualisation, and value [13]. The term veracity refers to the reliability or the quality of the data, that is, the truthfulness of the data. Variability alludes to the non-homogeneity of the data. Finally, the value property describes the remaining utility of the data after its processing. The opposite to Big Data is known as Small Data, where the amount of data and its complexity is way lower, and therefore, easier to process.

As introduced earlier, ensuring data privacy while preserving some utility is a hard challenge. However, the increment of data volume, as well as the complexity, variety, and velocity typical of a Big Data scenario introduce even more complications. Firstly, due to the large volume of the data, computational efficiency may be a critical issue when selecting an anonymisation technique or privacy model [5]. Furthermore, evaluating the identification risks in a Big Data context, as well as measuring the utility and information loss, is also computationally complex. Secondly, aside from the computational issues derived from working with a huge amount of data, other problems arise in Big Data scenarios: the variety of the data also plays an important role, since most of the current algorithms that can be applied to preserve privacy, such as clustering, are designed for homogeneous data. These algorithms work well in Small Data scenarios, however, in Big Data, the information is usually heterogeneous. Therefore, traditional data privacy techniques are outdated, and there is a lack of scalable and efficient privacy algorithms.

Classical anonymisation techniques such as k-anonymity [6], l-diversity [14] or t-closeness [15] are not completely adequate to ensure Big Data privacy, as in most of the cases the data to be anonymised can be unstructured or live streaming. However, other well-known techniques such as differential privacy [16] can be more easily adapted to a Big Data context [17]. This method introduces an intermediary between the data processor and the database, which acts as a privacy guard. The data processor does not get direct access to the full version of the data. Instead, the privacy guard evaluates the privacy risk (according to different factors such as the sensitivity of the query to be executed and/or the size of the dataset) and introduces some distortion within the partial information retrieved from the database, which will be proportional to the current privacy risk. Differential privacy benefits from larger datasets: if the dataset is large then less noise will be needed to protect privacy [17]. However, one drawback of this technique is that the amount of noise to be introduced for sensitive queries (with a high privacy risk) is large, and retaining the utility of the data may be challenging or impossible.

Regarding privacy models, the main difference with anonymisation techniques is that they do not specify the set of specific transformations that need to be performed on the data, but the conditions that a dataset must satisfy to maintain the disclosure risk under control [18]. According to Soria-Comas et al. [19], a privacy model needs to satisfy three properties to be usable in a Big Data environment: composability, computability, and linkability. A privacy model is composable if its privacy guarantees are preserved for a dataset resultant from merging several datasets, for each of which the privacy guarantees of the model holds. Computability refers to the cost of computation of the anonymisation. Finally, linkability is the ability to link records relating to an individual. In their work, they evaluate k-anonymity and differential privacy in terms of those properties.

While k-anonymity is not composable (the combination of two k-anonymous datasets does not guarantee the privacy preservation of k-anonymity) differential privacy is strongly composable (combining two differentially private datasets increases the risk of disclosure, but the differential privacy is still preserved). In terms of linkability, in k-anonymity it is possible to at least link the groups of k-anonymous records of the individuals. In differential privacy, datasets are not linkable if noise addition is used. Finally, the computability property cannot be compared, as the performance of the privacy models depends on the method used to anonymise the data.

Therefore, there is not an absolute solution to protect data privacy in a Big Data context since each method has its advantages and flaws, and the selection of the privacy model should be analysed on a case-by-case basis. The existing techniques, practices, and methodologies for data privacy protection are ineffective in Big Data if they are not used in an integrated manner [20]. In any case, any solution dealing with sensitive data should analyse the privacy risks to address the possible data privacy challenges correctly.

Another relevant issue associated with Big Data privacy is the complexity of data anonymisation in real-time, as in many cases the data has to be processed immediately as it arrives at the system. This concept is known as stream processing, and it occurs in many Big Data scenarios where the data is usually generated by many sources at a high speed. Unlike batch processing, where the data is collected into batches and then fed into the system, in streaming processing the data is dynamic since it has a temporal dimension. Therefore, there is a maximum acceptable delay between the in-flowing data and the processed output [21].

Data processing in streaming adds certain difficulties when it comes to making a prior analysis of the data to select the best anonymisation strategy. Since the data arrives in portions to the system, the information is always incomplete, and performing a correct privacy risk assessment and utility evaluation is not an easy task. Moreover, traditional k-anonymity schemes are designed for static datasets and therefore are not suitable for streaming contexts. Furthermore, these techniques assume that each person appears in the dataset only once, assumption that cannot be made within a streaming context [21]. These challenges become especially hard in the context of Big Data streams.

Some solutions have been proposed to solve the issues that arise when anonymising in such scenarios. Sakpere et al. [22] made a review of the state of the art of the existing methods to anonymise Big Data streams, which will be briefly explored hereunder.

Li et al. developed in 2007 a perturbative method to achieve streaming data privacy based on adding random noise to the incoming data [23]. However, this method has certain flaws, as it can only handle numerical data, complicating the analysis of the anonymised dataset due to a large amount of artificial noise.

Other proposed methods are based on tree structures, such as SKY (Stream K-anonYmity) [24], SWAF (Sliding Window Anonymisation Framework) [25] and KIDS (K-anonymIsation Data Stream) [26]. The main disadvantages of these methods are the time and space complexity for data streams, the risk of re-identification if the data hierarchy used for the generalisation process is discovered by an attacker, and its use for the anonymisation of numerical values, which becomes considerably complicated due to the difficulty of finding a suitable hierarchy for the specialisation tree [27].

Clustering (grouping of similar data points) algorithms can also be useful for data anonymisation in a streaming context. Some examples are CASTLE (Continuously Anonymising STreaming Data via adaptive cLustEring) [21], B-CASTLE [28], which is an improvement of CASTLE, FAANST (Fast Anonymising Algorithm for Numerical STreaming data) [29] and FADS (Fast clustering-based k-Anonymisation approach for Data Streams) [30]. FADS is the best choice for anonymisation in streaming, due to its low time and space complexity. However, since it handles tuples sequentially, it is not suitable for Big Data streams. Mohammadian et al. introduced a new method, FAST (Fast Anonymization of Big Data Streams) based on FADS, which uses parallelism to increase the effectiveness of FADS and make it applicable in big data streams [31].

Last but not least, is important to take into account that the growth of Big Data in recent years has facilitated cross-referencing information from different databases, increasing the risk of re-identification. Databases that contain information that even though at first might seem insensitive, can be matched with other publicly available datasets to re-identify users [32]. For example, it has been shown that only a 5-digit zip code, birth date and gender can identify 80% of the population in the United States [33]. The inherent properties of Big Data, such as the high volume and its complexity, aggravate even more this problem.

In brief, the characteristics of Big Data make it especially difficult to preserve privacy according to the GDPR. Most of the traditional anonymisation algorithms are ineffective and/or inefficient in such scenarios, and further research in the field of Big Data privacy is necessary. Moreover, the velocity of the streaming data (usually also present in Big Data scenarios) introduces additional complications. Even though some anonymisation algorithms have been developed to address this issue, there are still unsolved challenges.

4 Conclusions

Through the course of this chapter, we introduced some of the existing data privacy protection approaches, highlighting the main differences between them. Different real-world examples have been provided to emphasise the importance of the correct application of the GDPR, as well as the consequences of a wrong anonymisation approach, which are critical for both individuals and companies. Anonymisation is a process that requires a deep analysis and constant monitoring of the re-identification risks to be performed correctly. In addition, we presented the privacy-utility problem and the metrics that can be used to monitor the re-identification risk of a dataset.

Furthermore, an introduction to Big Data and streaming processing and its anonymisation difficulties were presented. Different methods for data anonymisation in Big Data scenarios have been reviewed, highlighting their strengths and their drawbacks, as well as their applicability in different contexts.

Acknowledgment This work is funded by the EU H2020 Programme INFINITECH (grant agreement No. 856632).

References

1. Data privacy vs. data security [definitions and comparisons] (2021, January). https://dataprivacymanager.net/security-vs-privacy/
2. Article 29 Data Protection Working Party. (2014, April). Opinion 05/2014 on Anonymisation Techniques. In *Working Party Opinions* (pp. 1–37).
3. European Commission. (2016). Regulation (EU) 2016/679 of the European Parliament and of the Council of 27 April 2016 on the protection of natural persons with regard to the processing

of personal data and on the free movement of such data, and repealing Directive 95/46/EC (General Da. https://eur-lex.europa.eu/eli/reg/2016/679/oj

4. Srivastava, S., Namboodiri, V. P., & Prabhakar, T. (2020, February). Achieving privacy-utility trade-off in existing software systems. *Journal of Physics: Conference Series, 1454*, 012004.

5. D'Acquisto, G., Domingo-Ferrer, J., Kikiras, P., Torra, V., de Montjoye, Y.-A., & Bourka, A. (2015, December). Privacy by design in big data: An overview of privacy enhancing technologies in the era of big data analytics. https://op.europa.eu/en/publication-detail/-/publication/20492499-ce2e-11e5-a4b5-01aa75ed71a1/language-en

6. Sweeney, L. (2002, October). K-anonymity: A model for protecting privacy. *International Journal of Uncertainty, Fuzziness and Knowledge-Based Systems, 10*, 557–570.

7. de Montjoye, Y.-A., Hidalgo, C. A., Verleysen, M., & Blondel, V. D. (2013). Unique in the Crowd: The privacy bounds of human mobility. *Scientific Reports, 3*(1), 1376.

8. Data anonymization and GDPR compliance: The case of Taxa 4x35 - GDPR.eu (2020). https://gdpr.eu/data-anonymization-taxa-4x35/.

9. Sweden - Information on Tax Identification Numbers. https://www.oecd.org/tax/automatic-exchange/crs-implementation-and-assistance/tax-identification-numbers/Sweden-TIN.pdf

10. European Data Protection Supervisor and Agencia Espanola Protection Datos. (2021). 10 Misunderstandings related to anonymisation. https://edps.europa.eu/system/files/2021-04/21-04-27_aepd-edps_anonymisation_en_5.pdf

11. Adkinson Orellana, L., Dago Casas, P., Sestelo, M., & Pintos Castro, B. (2021). A new approach for dynamic and risk-based data anonymization. In Á. Herrero, C. Cambra, D. Urda, J. Sedano, H. Quintián, & E. Corchado (Eds.), *13th International Conference on Computational Intelligence in Security for Information Systems (CISIS 2020)* (pp. 327–336). Cham: Springer International Publishing.

12. Laney, D. (2001, February). 3D data management: Controlling data volume, velocity, and variety. Tech. rep., META Group.

13. Tsai, C.-W., Lai, C.-F., Chao, H.-C., & Vasilakos, A. V. (2015). Big data analytics: A survey. *Journal of Big Data, 2*(1), 21.

14. Machanavajjhala, A., Gehrke, J., Kifer, D., & Venkitasubramaniam, M. (2006, April). L-diversity: Privacy beyond k-anonymity. In *22nd International Conference on Data Engineering (ICDE'06)* (pp. 24–24).

15. Li, N., Li, T., & Venkatasubramanian, S. (2007, April). t-closeness: Privacy beyond k-anonymity and l-diversity, in *2007 IEEE 23rd International Conference on Data Engineering* (pp. 106–115).

16. Dwork, C., & Roth, A. (2013). The algorithmic foundations of differential privacy. *Foundations and Trends in Theoretical Computer Science, 9*(3–4), 211–487.

17. Shrivastva, K. M. P., Rizvi, M., & Singh, S. (2014). Big data privacy based on differential privacy a hope for big data. In *2014 International Conference on Computational Intelligence and Communication Networks* (pp. 776–781).

18. Pawar, A., Ahirrao, S., & Churi, P. P. (2018). Anonymization techniques for protecting privacy: A survey, in *2018 IEEE Punecon* (pp. 1–6).

19. Soria-Comas, J., & Domingo-Ferrer, J. (2016, March). Big data privacy: Challenges to privacy principles and models. *Data Science and Engineering, 1*, 21–28.

20. Moura, J., & Serrão, C. (2015). Security and privacy issues of big data. In *Handbook of research on trends and future directions in big data and web intelligence* (pp. 20–52). IGI Global.

21. Cao, J., Carminati, B., Ferrari, E., & Tan, K.-L. (2011, July). Castle: Continuously anonymizing data streams. *IEEE Transactions on Dependable and Secure Computing, 8*, 337–352.

22. Sakpere, A. B., & Kayem, A. V. (2014). A state-of-the-art review of data stream anonymization schemes. In *Information security in diverse computing environments*. IGI Global.

23. Li, F., Sun, J., Papadimitriou, S., Mihaila, G. A., & Stanoi, I. (2007). Hiding in the crowd: Privacy preservation on evolving streams through correlation tracking. In *2007 IEEE 23rd International Conference on Data Engineering* (pp. 686–695).

24. Li, J., Ooi, B. C., & Wang, W. (2008, April). Anonymizing streaming data for privacy protection. In *2008 IEEE 24th International Conference on Data Engineering* (pp. 1367–1369).
25. Wang, W., Li, J., Ai, C., & Li, Y. (2007, November). Privacy protection on sliding window of data streams. In *2007 International Conference on Collaborative Computing: Networking, Applications and Worksharing (CollaborateCom 2007)* (pp. 213–221).
26. Zhang, J., Yang, J., Zhang, J., & Yuan, Y. (2010, May). Kids:k-anonymization data stream base on sliding window. In *2010 2nd International Conference on Future Computer and Communication* (Vol. 2, pp. V2-311–V2-316).
27. Zakerzadeh, H., & Osborn, S. L. (2013, October). Delay-sensitive approaches for anonymizing numerical streaming data. *International Journal of Information Security, 12*, 423–437.
28. Wang, P., Lu, J., Zhao, L., & Yang, J. (2010). B-castle: An efficient publishing algorithm for k-anonymizing data streams. In *2010 Second WRI Global Congress on Intelligent Systems* (Vol. 2, pp. 132–136).
29. Zakerzadeh, H., & Osborn, S. L. (2010). Faanst: Fast anonymizing algorithm for numerical streaming data. In *Proceedings of the 5th International Workshop on Data Privacy Management, and 3rd International Conference on Autonomous Spontaneous Security, DPM'10/SETOP'10* (pp. 36–50). Berlin: Springer.
30. Guo, K., & Zhang, Q. (2013, July). Fast clustering-based anonymization approaches with time constraints for data streams. *Knowledge-Based Systems, 46*, 95–108.
31. Mohammadian, E., Noferesti, M., & Jalili, R. (2014). Fast: Fast anonymization of big data streams. In *Proceedings of the 2014 International Conference on Big Data Science and Computing, BigDataScience '14*. New York, NY: Association for Computing Machinery.
32. Narayanan, A., & Shmatikov, V. (2008, May). Robust de-anonymization of large sparse datasets. In *2008 IEEE Symposium on Security and Privacy (SP 2008)* (pp. 111–125).
33. Lu, R., Zhu, H., Liu, X., Liu, J. K., & Shao, J. (2014, July). Toward efficient and privacy-preserving computing in big data era. *IEEE Network, 28*, 46–50.

Chapter 20
Overview of Applicable Regulations in Digital Finance and Supporting Technologies

Ilesh Dattani Assentian and Nuria Ituarte Aranda

1 Applicable Regulations in Digital Finance

Regulators and financial supervisory authorities have increasingly begun to worry about the increased risks posed by the rise of digital technologies/fintech. They can be characterized as risks to consumers and investors, financial services firms and financial stability. Specific actions in response to these risks have resulted in the development of international standards, implementation of very detailed and prescriptive national regulations and guidance and increased priority from supervisory authorities.

These actions cover a broad spread of areas, including technology risk, cybersecurity and operational resilience more generally, data privacy, consumer protection, firms' governance and risk governance and amendments to anti-money laundering requirements [1].

The regulatory response has taken a number of forms [2]:

- *Regulatory perimeter*: Some developments, such as the mainstream use of cryptocurrencies, outsourced cloud computing services and the introduction of non-financial services firms as providers of products and services such as lending to SMEs and retail payment systems, bring into question where the boundary of the regulatory perimeter should be. As the network widens, an increasing number of firms will find themselves subject to regulatory requirements.

I. D. Assentian
Assentian Europe Limited, Dublin, Ireland
e-mail: ilesh.dattani@assentian.com

N. Ituarte Aranda (✉)
ATOS Spain, Madrid, Spain
e-mail: nuria.ituarte@atos.net

© The Author(s) 2022
J. Soldatos, D. Kyriazis (eds.), *Big Data and Artificial Intelligence in Digital Finance*,
https://doi.org/10.1007/978-3-030-94590-9_20

- *Retail conduct*: For consumer protection, regulators are using a mixture of (a) transparency and disclosure, (b) prohibiting or limiting the sale of some products and services to retail customers and (c) adapting codes of conduct to consider fintech developments.
- *Data and artificial intelligence*: Data protection legislation, including the EU General Data Protection Regulation (GDPR), already covers most issues arising from the use of digital technologies. However, rapid change resulting from the use of artificial intelligence (AI) and distributed ledger technologies means further regulation on the collection, management and use of data may be needed. The increased data gathering activities of broad ranges (of financial and non-financial data), and the sharing of such data between a growing pool of organizations, have intensified the debate on which further regulatory control may be required.
- *Governance*: Regulators are seeking to ensure that boards and senior managers have complete awareness and understanding of the digital technologies and applications used, and of the risks resulting from them, aiming to increase the focus on risk management and accountability.
- *Cybersecurity*: The focus from regulators is on enforcing the implementation of existing international standards.
- *Open banking*: Regulation has managed the creation of a market for new fintechs in open banking by establishing the principles and protocols on which data can be shared between different parties, usually through an application programming interface (API).

2   Main Digital Finance Regulations

Banks, fintechs and other financial organizations must consider and comply with several regulations, including general purpose regulations and regulations specific to the finance sector. The following paragraphs provide an overview of popular regulations, including their impact.

2.1   The General Data Protection Regulation (GDPR)

GDPR is a regulatory framework aimed at providing the means by which citizens can have control over their personal data. Organizations must make sure that:

- Personal data is gathered legally with the appropriate consent and declarations.
- Collected data is not misused or exploited for purposes other than for which it was collected.
- Rights of the data owner(s) are respected in line with the controls as set out in the regulation.

The regulation relates specifically from the perspective of the H2020 INFINITECH project to the processing of 'personal data', meaning:

any information relating to an identified or identifiable natural person ('data subject'); an identifiable natural person is one who can be identified, directly or indirectly, in particular by reference to an identifier such as a name, an identification number, location data, an online identifier or to one or more factors specific to the physical, physiological, genetic, mental, economic, cultural or social identity of that natural person. [3]

Processing within this context means 'any form of automated processing of personal data consisting of the use of personal data to evaluate certain personal aspects relating to a natural person, in particular to analyse or predict aspects concerning that natural person's performance at work, economic situation, health, personal preferences, interests, reliability, behaviour, location or movement' [4].

Data collection is directly impacted by GDPR. Fintechs need to have the procedures, policies and mechanisms in place to demonstrate that their technologies and services are compliant with the GDPR. This means keeping integrity and making sure they have valid and proper consent for the customer's data they hold, share, use and process.

Compliance breaches can result in financial penalties ranging between 2% and 4% of global turnover depending on the severity and impact of the breech. The use of technologies like blockchain requires considerable oversight to make sure that the way it is used and deployed still facilitates a core principle of GDPR, being the 'right to be forgotten'.

Other important aspects that should be considered is the need to understand data flows within systems and applications as these days most financial applications do not sit in silos – client data passes to and from multiple systems.

Finally, pseudonymization rules are critical to make sure that data access is only ever on the basis of the 'need-to-know' principles [5].

3 The Market in Financial Instruments Directive II (MIFID II)

MIFID II is focused on all financial instruments and covers services including advice, brokerage, dealing, storage and financial analysis. The directive tries to harmonize oversight and governance across the industry within all EU member states.

It has introduced more stringent reporting requirements and tests to ensure transparency and to crack down on the use of 'dark pools' (mechanisms by which trading takes place without identities of individuals or organizations involved being revealed). The amount of trading that can be done using 'dark pools' is as a result now limited to 8% of the overall trading in any 12-month period.

Algorithms used for automated trading under the directive now need to be registered with the applicable regulator(s), rigorously tested, and circuit breakers must be in place.

MIFID II also seeks to increase transparency around the cost of services. In so doing, limitations are set up on the payments made to investment firms or advisors by third parties for services that they have or are providing to clients in return. Charges for different services can no longer be aggregated into a single fee. More detailed reporting is required from brokers on the trades they carry out. Storage of all communications is recommended, thereby providing a clear audit trail of all activities [6].

MIFID II impacts digital finance technology development and deployment in the following way:

- *Data storage, aggregation and analytical requirements*: All information related to trades must be retained. A data retention and archiving strategy is required to support the likely volumes of data and information.
- *Integration between disparate applications*: Integration of applications with trading platforms so that key data can flow through becomes a key requirement resulting from MIFID II. API-based integration is seen as the most efficient approach in most cases.
- *Enhanced and transparent client portal*: In order to provide the appropriate protection to investors, it is a requirement to maintain comprehensive client classification and client data inventories.
- *Mobile device management (MDM) strategy*: This relates to the need to maintain a record of all telephone calls and electronic communications. The MDM strategy should ensure that the appropriate technology is in place to facilitate this and also restrict the use of mediums of communication such as social media where communications are encrypted, thereby making compliance difficult [7].

Alongside data retention, security and integrity of data are also critical and could pose a challenge. Appropriate mechanisms to support access control including the use of multifactor authentication should be in place.

Monitoring and audit trails of data throughout the data's operational lifecycle are also critical to keep integrity. Regular audits should be put in place to make sure all controls, policies and procedures are being followed.

3.1 Payment Services Directive 2 (PSD2)

PSD2 [8] seeks to strengthen the regulations governing online payments within the EU. At its core is the aim of developing a more integrated and seamless payment approach across all the member states of the EU.

A key requirement of the directive is the use of strong customer authentication (SCA). The objective is to improve the levels of security around payments and to reduce fraud.

It updates PSD1 (adopted in 2007) and, as in SD1:

- Opens the payment market to new entrants which until now had been limited to those institutions with a banking license.
- Transparency over services offered and the resulting fees that will be incurred has been improved. This also covers both maximum execution times and exchange rates. Development of the Single Euro Payments Area (SEPA) to ease the execution of payments has been accelerated as a result of the directive.

PSD2 has a strong impact on the whole financial services ecosystem and the entire infrastructure behind payments. Furthermore, it has impact on all businesses that are making use of payment data to better serve their customers.

Security requirements are introduced for both the initiation and processing of all electronic payments alongside the continued need to protect data belonging to customers (including specific financial data). Third-party providers also fall under the remit of PSD2, including all providers who have the right to access and/or aggregate accounts and provide payment services. In essence, PSD2 is designed to give all customers access to their own data and to support greater levels of innovation and resulting competition by encouraging the incumbent banks to engage in secure customer data exchange with third parties. It should open the market for organizations in other verticals to access data with their customers' prior consent as long as the appropriate controls are in place.

3.2 PSD II Security Requirements/Guidelines

PSD2 maintains the need for strong customer authentication (SCA), secured communication, risk management and transaction risk analysis (TRA).

In the endeavour of increasing protection for the consumer, the directive makes a requirement for banks to implement multifactor authentication for all transactions performed by any channel.

This requirement means making use of two of the following three features:

- Knowledge: information that the customer should only know such as a password, pin code, or a personal ID number
- Possession: an artefact that only the customer has such as a smart card or a mobile handset
- Inherence: the relation of an attribute to its subject, e.g. a fingerprint

The elements used need to be mutually independent so that a breach of one cannot inherently mean a breach of any of the others.

Payment service providers (PSPs), as a result of the directive, will need to establish a framework implementing the required mitigation measures. In addition, controls to effectively manage all operational and security risks with respect to services they provide should be implemented. These controls formally come under the remit of the directive. These measures should include, at least:

- The maintenance of an inventory of all business functions, roles and processes, thereby making it possible to map functions, roles and processes and to understand their interdependencies.
- Maintain an inventory of information assets, infrastructure and the interconnection with other systems (both internal and external) so that all assets critical to the core business functions are effectively managed minimizing disruption.
- Regularly monitor threats and vulnerabilities to assets critical to the effective delivery of business functions and processes.

3.3   4th Anti-money Laundering (AMLD4)

AMLD4 [9] is designed to ensure that accountability of companies is increased with respect to any connections that can be attributable to money laundering and/or terrorist financing. Failure to comply can bring both sanctions and reputational damage. Whilst the directive is targeted essentially at banks and other financial institutions, all corporations need to have the appropriate measures and controls in place to maintain compliance.

AMLD4 catches more business types than AMLD3. Gambling institutions; real estate letting companies, individuals or companies making single transactions over €10,000; virtual exchange currency platforms; and foreign and domestic politically exposed persons (PEPs) come under the scope of the directive.

The directive requires companies to be more aware of all risk factors when assessing business activities and functions. Some of the key risks include the following: (1) Check before engaging in business dealings that owners or people of significant control (PSC) are not PEPs (politically exposed persons). (2) Make more rigorous checks when dealing with business sectors where there is excessive cash. Such industries are known to be key targets of money launderers. (3) Make more rigorous checks when there is a connection with or when dealing with high-risk sectors. This covers sectors like construction, pharmaceuticals, arms, extractive industries and public procurement. (4) Check reputable credible media sources for any allegations of 'criminality of terrorism'. (5) Investigate into any potential risk of dealings with an organization or an individual who may have a record of frozen assets. (6) Business ownership and control should be clear – if there are any suspicions, complexities or lack of transparency in the structure of ownership, then a thorough investigation should be instigated. Proposed business partners should be a 'legal person'. (7) Doubts over identity of a beneficial owner should be further investigated. (8) Income and other sources of wealth for any potential partners should be clearly traceable back to their origin. (9) Increased scrutiny and investigation should be undertaken when dealing with companies who have, or are suspected of having, ties to countries that appear on the sanctions list.

Moreover, according to AMLD4, companies working in high-risk countries should be placed under increased scrutiny, and they should undergo a more rigorous set of checks. The Financial Action Task Force (FATF) releases a list three times a year which details the qualifying countries. Disclosure: Jurisdictions are increasingly encouraged to disclose organizations who continue to break the rules and/or break regulatory frameworks.

Also, AMLD4 imposes a need for identification, closer scrutiny and increased monitoring of people who are beneficial owners of companies. This can be determined by their shareholding in the business or if they are a PSC. With this in mind, a national register linked to other national registers should be kept so that anyone who needs it can see and access the required information. Compliance professionals as a result need to be able to decide what risk a company that they are working with poses before they go on to investigate their beneficial owners. National registers not being up to date is no excuse for non-compliance.

3.4 Basel IV

Basel IV focuses on operational risk management within the context of capital savings. Whilst it is not a direct stipulation, there is a recommendation that supervisory authorities should still maintain a strong focus on making operational risk improvements through strengthening operational resilience. This entails making sure that critical functions within banks are able to continue to operate through any periods of recovery or stress. This incorporates effective management of cyber risks and requires business continuity plans that are regularly reviewed and tested.

3.5 EU Legislative Proposal on Markets in Crypto-Assets (MiCA)

MiCA is being developed to streamline distributed ledger technology (DLT) and virtual asset regulation in the European Union (EU) whilst still keeping strong protection for both users and investors. MiCA aims to supply a single harmonized, consistent framework for the issuance and trading of crypto tokens [10].

The proposal has four broad objectives:

- Legal certainty for crypto-assets not covered by existing EU legislation.
- Uniform rules for crypto-asset service providers and issuers.
- Replace existing national frameworks applicable to crypto-assets not covered by existing EU legislation.
- Establish rules for 'stablecoins', including when these are e-money.

3.6 EU Draft Digital Operational Resilience Act (DORA)

DORA aims to set up clear foundation for EU regulators and supervisors to expand their focus to make sure that financial institutions can maintain resilient operations through a severe operational disruption.

The main reforms include [11]:

- Bringing 'critical ICT (Information and Communication Technology) third-party providers' (CTPPs), including cloud service providers (CSPs), within the regulatory perimeter. The European Supervisory Authorities would have mechanisms to request information, conduct off-site and on-site inspections, issue recommendations and requests and impose fines in certain circumstances.
- EU-wide standards for digital operational resilience testing.
- Creating consistency in ICT risk management rules across financial services sectors, based on existing guidelines.
- Consistency in ICT incident classification and reporting.
- Future potential for a single EU hub for major ICT-related incident reporting by financial institutions.

3.7 EU Draft AI Act

Under the AI Act, a number of AI practices will be forbidden. AI which makes use of subliminal components or exploits human vulnerabilities to distort a person's behaviour in a harmful way will be forbidden. In addition, social scoring systems and real-time remote biometric identification systems are not allowed.

The AI act includes a list of high-risk AI systems, including those systems that could result in adverse outcome to health and safety of persons, and stand-alone AI systems which may pose a threat to the fundamental rights of persons. Before a high-risk AI system is put on the market or deployed to service, it must undergo a conformity assessment procedure. Once completed, the system will have to be labelled with a CE marking. Certain high-risk AI systems should be also registered in an EU database maintained by the Commission.

ICT systems can only be placed on the EU market or put into service if they comply with strict mandatory requirements, in particular [12]:

- Establish a risk management system.
- Maintain clear quality criteria for the datasets used for training and testing of those systems.
- Design of high-risk AI systems should enable them to automatically log events whilst the system is operating.
- Operation of the high-risk AI systems must be sufficiently transparent to allow the users to interpret the system's output and use it appropriately.
- Systems must come with instructions.
- High-risk AI systems must be designed to enable humans to oversee them effectively, including understanding the capacities and limitations of the AI.
- Oversight features allowing users to override the decisions of the AI or to interrupt the system by means of a 'stop' button.
- Design and develop in such a way that they achieve, in the light of their intended purpose, an appropriate level of accuracy, robustness and cybersecurity, perform consistently in those respects throughout their lifecycle and are resilient, in particular against errors, inconsistency of the operating environment and malicious actors.

4 Supporting Technologies for Regulations in Digital Finance

Focusing on financial and insurance sectors, the INFINITECH project contributes supplying technologies that help to fulfil the regulations exposed in the previous section. Given the number of pilots that are being conducted, a large number of technologies that provide solutions for this sector are being developed by INFINITECH partners.

Among the technologies provided in INFINITECH, this chapter exposes those that help fulfilling the regulations in digital finance. The list of INFINTIECH technologies is provided in Table 1, which is based on INFINITECH's work on regulatory compliance tools [13]:

Taking regulations of digital finance exposed in the first section of this chapter and the mentioned technologies that help to fulfil the regulations, a mapping between regulations and technologies has been performed. Table 2 shows the technologies that can be deployed to address different regulations applicable to digital finance.

Table 20.1 INFINITECH technologies for regulatory support

Name tool/platform	Relevance and applications for regulatory compliance
Data Protection Orchestrator (DPO) (Atos (https://atos.net) is the provider of this technology)	It allows embedding of automating tools for assessing security and privacy by design and by default in business flows, these being heterogeneous and complex. It orchestrates various privacy and security management functions (such as access control, encryption and anonymization)
Digital User Onboarding System (DUOS) (Atos (https://atos.net) is the provider of this technology)	This solution allows management of virtual identities in a mobile device. It makes use of electronic ID (eID) or passport for remote user registration
	DUOS uses eIDs issued by European national authorities according to the EU eID schemas: eID cards and passports
	In order to integrate DUOS, it is necessary to adapt and customize it for a user's context of need (e.g., bank application) that requires user authentication
	This technology could be used in digital finance to implement 'anonymous' user onboarding. The user can be securely identified by eID or e-Passport without revealing any detail about his/her identity to a third party
Botakis Chatbot Development Network (Crowd Policy (https://crowdpolicy.com) is the provider of this technology)	A tool for rapid development of chatbot applications, which can be used for the development of chatbot's features in digital finance
	Botakis Chatbot Platform is expected to be enhanced in this way:
	Built-in dialogs that utilize and are integrated with existing natural language processing (NLP) frameworks (open or proprietary) provided by partners or every interested party
	Powerful dialog system with dialogs that are isolated and composable
	Built-in prompts for simple actions like yes/no, strings, numbers or enumerations [14]
Crowdpolicy Open (innovation) banking solution (Crowd Policy (https://crowdpolicy.com) is the provider of this technology)	As part of the available chatbot functionality, it will be possible to include GDPR consent and manage requests from people exercising the right to be informed, the right of access, the right to rectification, the right to erasure, the right to restrict processing, the right to data portability and the right to object
	'Crowdpolicy Open (innovation) banking platform is a set of predefined and customizable banking web services and data models integrated with our own API Manager that supports access control, monitoring and authentication. This solution puts the bank (or any monetary financial institution) in control of the third-party partner relation' [14]
	Crowdpolicy Open (innovation) banking platform mainly covers the requirements for open banking APIs as part of the PSD2 directive, which has several modules that also are API-based

Provider	Description
Anonymization tool (Gradiant (https://gradiant.org) is the provider of this technology)	'The anonymization tool is based on a risk-based approach that modifies data in order to preserve privacy. The tool includes different anonymization algorithms and it will determine automatically which of them (generalization, randomization, deletion, etc.) should be applied in order to preserve the maximum level of privacy for the data. It also includes a set of privacy and utility metrics that allow to measure the risk that remains after anonymizing the dataset, and the impact of the anonymization process on the quality of the data.
	The component requires two inputs: the data that has to be anonymized and a configuration file that defines the structure of the data, its location and the privacy requirements' [14]
	The anonymization tool is intended to be used in two modes, batch or streaming. In the case of using it in batch mode, the output of the component (anonymized data) is stored in a database. The location of the database has to be known beforehand (through the configuration file that is taken as an input). If the streaming mode is used, the output will be the queue of the service
Blockchain-enabled Consent Management System (Ubitech Limited (https://ubitech.eu), IBM (https://il.ibm.com) and Innov-Acts Limited (https://innov-acts.com) are the providers of this technology)	The blockchain-enabled Consent Management System offers a decentralized and robust consent management mechanism that enables the sharing of the customer's consent to exchange and utilize their customer data across different banking institutions. The solution enables the financial institutions to effectively manage and share their customer's consents in a transparent and unambiguous manner. It is capable of storing the consents and their complete update history with complete consents' versioning in a secure and trusted manner. The integrity of customer data processing consents and their immutable versioning control are protected by the blockchain infrastructure [15]
	To achieve this, the solution exploits the key characteristics of blockchain technology to overcome the underlying challenges of trust improvement, capitalizing on its decentralized nature and immutability due to the impossibility of ledger falsification. The usage of blockchain enables extensibility, scalability, confidentiality, flexibility and resilience to attacks or misuse, guaranteeing the integrity, transparency and trustworthiness of the underlying data [15]

Table 20.2 INFINITECH regulatory support tools

Regulation	Regulatory compliance need	Technology applied
GDPR	Consent management: set of tools that allow the data subject what data they permit to share	Botakis Chatbot Development Network
		Privacy dashboards
		CMS (Content Management System) for storing digitized documents
		Blockchain-enabled Consent Management System
	Anonymized data: dataset in which it has removed personally identifiable information, in an irreversible way	Anonymization tool
		ICARUS
	Pseudonymized data: dataset in which the fields within a data record are replaced by one or more artificial identifiers or pseudonyms; this makes the data less identifiable whilst remaining suitable for data analysis and data processing	Pseudonymization tool
MIFID II	Recording and auditing system: allows the review and evaluation of computer systems and processes that are running, as well as their use, security and privacy whilst processing information. It provides security and adequate decision-making	Ad hoc logging implementations
PSD II	Online payment services meeting the regulations	CrowdPolicy Open banking solution
		SIEM
AMLD4	Inclusion on local databases of PEPs	Ad hoc solutions for each country
General	Authentication: the process of identifying a user or process, proving that they are valid or genuine	Specific solutions
		CrowdPolicy Open solution
		DUOS for Digital User Onboarding
	Authorization: the process of specifying access rights to use resources	Specific solutions
		CrowdPolicy Open solution
		Identity and access management
	Privacy and security services orchestration: execution of processes involved in creating a service that provides compliance with regulations	Data Protection Orchestrator

5 Regulatory Compliance Tool and Data Protection Orchestrator (DPO)

A regulatory compliance tool is a set of software components that provides added functionalities needed to comply with the regulations applicable to a particular use case [13].

A common situation is the implementation of regulatory compliance mechanisms on existent tools and solutions: when the need for regulatory compliance is identified on existing tools, implementation of additional security, privacy and data protection mechanisms is challenging and usually requires a complete redesign of the solution. INFINITECH proposes a general regulatory compliance tool based on the Data Protection Orchestrator (DPO) [16] introduced previously on this chapter. The DPO has been designed in WITDOM Project [17] and will be demonstrated in INFINITECH. The DPO can be customized and applied to multiple use cases, providing the needed protection, security and privacy mechanisms for regulatory compliance.

The main goal of the DPO is to coordinate the invocation of components that implement privacy, security or data protection techniques, as well as other external services, in order to provide a suitable privacy, security and data protection level (specified by a secure service provider compliant to regulations).

The Data Protection Orchestrator uses the Business Process Model and Notation 2 (BPMN2) format. The business process guides and establishes all the steps in the adequate order that have to happen to ensure the security, privacy and data protection required for a particular use case.

The use of the Data Protection Orchestrator provides the following benefits:

- Helps secure service developers and protection component's providers to ease the provision of the process for protection configurations
- Allows the combination of individual privacy, security or data protection components creating complex protection processes
- Supplies the needed business logic to ensure that regulations are fulfilled
- BPMN diagrams can be visually exposed, providing a clear view of the protection process

6 Conclusions

The development of digital finance applications and financial technology must comply with the rich set of regulations that are applicable to the finance sector. Regulatory compliance is mandatory for launching novel data-intensive applications in the market. As such, it is also a key prerequisite for the development and deployment of big data and AI applications, such as the ones presented in earlier chapter in the book. To facilitate regulatory compliance, the INFINITECH project has developed a rich set of tools, which help financial organizations to address the mandates of the various regulations. These tools aim at lowering the effort and costs of regulatory compliance for financial institutions and fintechs.

Acknowledgments The research leading to the results presented in this chapter has received funding from the European Union's funded Project INFINITECH under Grant Agreement No. 856632.

References

1. Regulation and Supervision of Fintech, KPMG, March 2019. https://assets.kpmg/content/dam/kpmg/xx/pdf/2019/03/regulation-and-supervision-of-fintech.pdf
2. Regulation 2030: What Lies Ahead, KPMG, April 2018. https://assets.kpmg/content/dam/kpmg/xx/pdf/2018/03/regulation-2030.pdf
3. GDPR – IT Governance, https://www.itgovernance.co.uk/data-privacy/gdpr-overview/gdpr-faq/gdpr-scope. Accessed Nov 2020.
4. GDPR scope – IT Governance, https://www.itgovernance.co.uk/data-privacy/gdpr-overview/gdpr-faq/gdpr-scope. Accessed Nov 2020.
5. Patel, M. (2017, November). Top Five Impacts of GDPR on Financial Services. *Fintech Times.*
6. Kenton, W, & Mansa, J., (2020, July). MiFID II – Laws & Regulations, https://www.investopedia.com/terms/m/mifid-ii.asp. Accessed Nov 2020.
7. Johnson, P. (2018, February). WhiteSource, "MiFID II Reforms and Their Impact on Technology and Security", https://resources.whitesourcesoftware.com/blog-whitesource/mifid-ii-reforms-and-their-impact-on-technology-and-security. Accessed Nov 2020.
8. Final Report: Guidelines on the security measures for operational and security risks of payment services under Directive (EU) 2015/2366 (PSD2), European Banking Authority, Dec 2017.
9. Mark Halstead, "What is the Fourth AML Directive (AML4D)-RedFlagAlert", https://www.redflagalert.com/articles/data/what-is-the-fourth-aml-directive-aml4d
10. MiCA: A Guide to the EU's Proposed Markets in Crypto-Assets Regulation https://www.sygna.io/blog/what-is-mica-markets-in-crypto-assets-eu-regulation-guide/. Accessed June 2021.
11. The EU's Digital Operational Resilience Act for financial services New rules, broader scope of application, Deloitte, Authors: Quentin Mosseray, Scott Martin, Simon Brennan, Sarah Black, Rick Cudworth. https://www2.deloitte.com/lu/en/pages/financial-services/articles/the-eus-digital-operational-resilience-act-for-financial-services-new-rules.html. Accessed June 2021.
12. What do you need to know about the AI Act? Magdalena Kogut-Czarkowska, 28/04/2021 https://www.timelex.eu/en/blog/what-do-you-need-know-about-ai-act. Accessed June 2021.
13. INFINITECH consortium, "INFINITECH D3.15 – Regulatory Compliance Tools – I", 2020
14. INFINITECH consortium, "INFINITECH D2.5 – Specifications of INFINITECH Technologies – I", 2020
15. INFINITECH consortium, "INFINITECH D4.7 – Permissioned Blockchain for Finance and Insurance – I", 2020
16. INFINITECH consortium, "INFINITECH D3.17 – Regulatory Compliance Tools – III", 2021
17. H2020 WITDOM (empoWering prIvacy and securiTy in non-trusteD envirOnMents) project, Contract Number: 644371, https://cordis.europa.eu/project/id/644371

Index

Printed in the United States
by Baker & Taylor Publisher Services